Operator Theory
Advances and Applications
Vol. 91

Elliptic Functional Differential Equations and Applications

Alexander L. Skubachevskii

Birkhäuser Verlag
Basel · Boston · Berlin

Author's address:

Alexander L. Skubachevskii
Moscow State Aviation Institute
Volokolamskoe shosse 4
Moscow 125 871
Russia

1991 Mathematics Subject Classification 39A05, 35J99

A CIP catalogue record for this book is available from the Library of Congress, Washington D.C., USA

Deutsche Bibliothek Cataloging-in-Publication Data
Skubachevskij, Aleksandr L.:
Elliptic functional differential equations and applications /
Alexander L. Skubachevskii. - Basel ; Boston ; Berlin :
Birkhäuser, 1997
(Operator theory ; Vol. 91)
ISBN-13:978-3-0348-9877-5 e-ISBN-13:978-3-0348-9033-5
DOI: 10.1007/978-3-0348-9033-5

NE: GT

Softcover reprint of the hardcover 1st edition 1997

Printed on acid-free paper produced from chlorine-free pulp. TCF ∞
Cover design: Heinz Hiltbrunner, Basel

ISBN-13:978-3-0348-9877-5

To my mother Sof'ya M. Skubachevskaya
and to the memory of my father Leonid S. Skubachevskiĭ

Contents

Acknowledgments

In preparing this book, I am indebted to many people. I would like to express my hearty thanks to Professors M. S. Agranovich, A. V. Bitsadze, V. A. Il′in, A. G. Kamenskiĭ, G. A. Kamenskiĭ, A. N. Kozhevnikov, M. A. Krasnosel′skiĭ, S. G. Krein, A. D. Myshkis, G. G. Onanov, and V. G. Veretennikov for their constant interest in my work. I am deeply indebted to Professors A. K. Gushchin, V. P. Mikhailov, V. A. Kondrat′ev, E. M. Landis, and O. A. Oleĭnik for the discussions of results at the seminars of the Steklov Institute of Mathematics and Moscow State University. I am very thankful to Professors I. C. Gohberg and S. Verduyn Lunel. Their advice has helped me to make many improvements to this book.

I would like to express my gratitude to Professor J. Kato and the Kawai Foundation for the Promotion of Mathematical Science for their financial support and hospitality during my visit to Japan in 1992. I am very thankful to Mr. V. V. Pronin for his financial and moral support.

I am grateful to the editorial staff of Birkhäuser for their highly qualified assistance in the preparation of the manuscript.

The research described in this book was also made possible in part by Grant JH6100 from the International Science Foundation and Russian Government, and by Grant 94-2187 from INTAS.

Notation

$\mathbb{R}$	real numbers
$\mathbb{C}$	complex numbers
$\mathbb{R}^n$	n-dimensional real space
$\mathbb{C}^n$	n-dimensional complex space
$[a, b]$	closed interval $\{x \in \mathbb{R} : a \leq x \leq b\}$
(a, b)	open interval $\{x \in \mathbb{R} : a < x < b\}$
$\overline{Q}$	closure of Q
$\oplus$	orthogonal sum
$\square$	end of a proof

Other notation introduced in the text is listed in the List of Symbols.

Introduction

I. This book is devoted to the theory of boundary value problems for elliptic functional differential equations. This new field of differential equations has grown out of the theory of functional differential equations and modern partial differential equations theory.

Boundary value problems for elliptic functional differential equations have some astonishing properties. For example, unlike elliptic differential equations, the smoothness of the generalized solutions can be violated in a bounded domain and is preserved only in some subdomains. A symbol of a self-adjoint semibounded functional differential operator can change its sign. This theory has important applications to elasticity theory, control theory, and diffusion processes. Elliptic functional differential equations are closely associated with differential equations with nonlocal boundary conditions, which arise in plasma theory.

In the one-dimensional case, functional differential equations describe processes depending on the history of a system. Some results for such equations were obtained more than 200 years ago. The new classes of functional differential equations arising in mechanics and biology were studied by V. Volterra [1, 2], and a general theory of functional differential equations was put forward by A. D. Myshkis [1], R. Bellman and K. Cooke [1], J. Hale [1], and others. Much research in this field is connected with applications to control systems with delay (see R. Bellman and J. M. Danskin [1], N. N. Krasovskiĭ [1], Yu. S. Osipov [1]).

Elliptic functional differential equations containing transformations of arguments were studied by A. B. Antonevich [1], D. Przeworska-Rolewicz [1], and V. S. Rabinovich [1]. These authors assume that the transformations of arguments map a domain onto itself and generate a finite group. Therefore, their results are similar to well-known results for elliptic differential equations. The situation changes if the equation has these shifts in the highest derivatives, and the shifts map the points of the boundary into the domain. The influence of such shifts on the solvability and smoothness of generalized solutions was studied only in one dimension in the papers of G. A. Kamenskiĭ and A. D. Myshkis [1] and A. G. Kamenskiĭ [1].

The theory of elliptic differential-difference equations was constructed by A. L. Skubachevskiĭ [1–3, 5, 8–10]. He considered necessary and sufficient conditions for ellipticity, solvability, spectrum and smoothness of generalized solutions.

These problems are connected with nonlocal boundary value problems for elliptic differential equations. We note that ordinary differential equations with nonlocal boundary conditions were studied by M. Picone [1, 2], Ya. D. Tamarkin [1, 2], W. Feller [1, 2], A. M. Krall [1, 2], and others. T. Carleman [1] considered the problem of finding a holomorphic function in a domain Ω satisfying the following nonlocal boundary condition: the value of the unknown function at a point t of the boundary $\partial\Omega$ is connected with the value at the point $\alpha(t)$, where $\alpha(\alpha(t)) = t$ and $\alpha(\partial\Omega) = \partial\Omega$. This problem is closely associated with further investigations of elliptic boundary value problems with shifts which map the boundary onto itself, abstract elliptic problems, and singular integral equations with shifts. The appropriate references are contained in F. Browder [1], J. L. Lions, and E. Magenes [1] and Yu. I. Karlovich, V. G. Kravchenko, and G. S. Litvinchuk [1].

A new nonlocal boundary value problem for elliptic differential equation, which arises in plasma theory, was formulated by A. V. Bitsadze and A. A. Samarskiĭ [1]:

$$\begin{aligned} Aw &= -\sum_{i,j=1}^{n} a_{ij}(x) w_{x_i x_j}(x) + \sum_{i=1}^{n} a_i(x) w_{x_i}(x) + a_0(x) w(x) \\ &= f_0(x) \qquad (x \in Q), \end{aligned} \tag{0.1}$$

$$\left.\begin{aligned} w(x)|_{\Gamma_1} &= w(\omega(x))|_{\Gamma_1} + f_1(x) \qquad (x \in \Gamma_1), \\ w(x)|_{\Gamma_2} &= f_2(x) \qquad (x \in \Gamma_2). \end{aligned}\right\} \tag{0.2}$$

Here $\sum_{i,j=1}^{n} a_{ij}(x)\xi_i\xi_j > 0$ $(0 \neq \xi \in \mathbb{R}^n,\ x \in \overline{Q})$, $Q \subset \mathbb{R}^n$ is a bounded domain with boundary ∂Q; $\Gamma_1 \subset \partial Q$ is an $(n-1)$-dimensional manifold open in the topology of ∂Q, $\Gamma_2 = \partial Q \setminus \Gamma_1$; $\omega(x)$ is an infinitely differentiable nondegenerate transformation mapping some neighborhood Ω_1 of the manifold Γ_1 onto the set $\omega(\Omega_1)$ in such a way that $\omega(\Gamma_1) \subset Q$.

In this paper A. V. Bitsadze, A. A. Samarskiĭ studied the following two problems:

$$-\Delta w(x) = f_0(x) \qquad (x \in Q = (0,2) \times (0,1)), \tag{0.3}$$

$$\left.\begin{aligned} &w(x_1, 0) = w(x_1, 1) = 0 && (0 \le x_1 \le 2), \\ &w(0, x_2) = \gamma_1 w(1, x_2),\ w(2, x_2) = \gamma_2 w(1, x_2) && (0 \le x_2 \le 1) \end{aligned}\right\} \tag{0.4}$$

for $\gamma_1 = 0$, $\gamma_2 = 1$, and

$$-\Delta w(x) = f_0(x) \qquad (x \in Q = (0,2) \times (0,1)), \tag{0.5}$$

$$\left.\begin{aligned} &w(x_1, 0) = w(x_1, 1) = 0 && (0 \le x_1 \le 2), \\ &w(x_1, x_2) = w(x_1 + 1, x_2) && (0 \le x_1 \le 1,\ 0 \le x_2 \le 1), \end{aligned}\right\} \tag{0.6}$$

where Δ is the Laplace operator, $x = (x_1, x_2)$.

It was an open problem to study the solvability of elliptic equations with nonlocal boundary conditions (see A. M. Krall [1], A. A. Samarskiĭ [1]). Various versions and generalizations of elliptic problems with nonlocal boundary conditions of the type (0.2) were studied by A. V. Bitsadze [1, 2], B. P. Paneyakh [1], Ya. A. Roitberg and Z. G. Sheftel′ [1, 2], S. D. Eĭdel′man and N. V. Zhitarashu [1], and others. Most of the publications are devoted to the case $\Gamma_2 = \emptyset$, $\omega(\partial Q) \cap \partial Q = \emptyset$. In the others, the authors assume that the set $\omega(\Gamma_1)$ satisfies some rigid geometrical conditions near the boundary ∂Q. For example, $\overline{\omega(\Gamma_1)} \cap \overline{\Gamma_1} = \emptyset$. Only recently have several developments in the theory of partial differential equations and functional differential equations made possible further progress in the study of elliptic problems with nonlocal boundary conditions (see A. L. Skubachevskiĭ [1, 4, 6–9, 11, 12, 14–16]).

II. In order to illustrate a connection between boundary value problems for elliptic differential-difference equations and elliptic differential equations with nonlocal boundary conditions, we consider the following example:

$$-\Delta R_Q u = f_0(x) \qquad (x \in Q), \tag{0.7}$$

$$u(x) = 0 \qquad (x \in \partial Q). \tag{0.8}$$

Here

$$Ru(x) = u(x_1, x_2) + \gamma_1 u(x_1 + 1, x_2) + \gamma_2 u(x_1 - 1, x_2), \tag{0.9}$$

$Q = (0,2) \times (0,1)$, $\gamma_1, \gamma_2 \in \mathbb{R}$, $f_0 \in L_2(Q)$. Since the difference operator R is nonlocal, we must put the boundary conditions not only on the boundary ∂Q but also in some neighborhood of ∂Q. Therefore, we introduce a bounded operator $R_Q = P_Q R I_Q : L_2(Q) \to L_2(Q)$, where I_Q is the extension operator of functions from $L_2(Q)$ by zero in $\mathbb{R}^2 \setminus Q$, P_Q is the restriction operator of functions from $L_2(\mathbb{R}^2)$ to Q.

Let $W^k(Q)$ be the Sobolev space of order k, and let $\mathring{W}^1(Q) = \{u \in W^1(Q) : u|_{\partial Q} = 0\}$, where $u|_{\partial Q}$ is a trace of u (see Appendix B). We suppose that $u \in \mathring{W}^1(Q)$. Let $w = R_Q u$. Then $w \in W^1(Q)$, and

$$w|_{x_2=0} = w|_{x_2=1} = 0,$$

$$w|_{x_1=0} = \gamma_1 u|_{x_1=1}, \quad w|_{x_1=1} = u|_{x_1=1}, \quad w|_{x_1=2} = \gamma_2 u|_{x_1=1}.$$

Hence, $w(x)$ satisfies the nonlocal conditions (0.4). In other words, $R_Q(\mathring{W}^1(Q)) \subset W^1_\gamma(Q)$, where $W^1_\gamma(Q) = \{w \in W^1(Q) : w \text{ satisfies the conditions (0.4)}\}$. We note that the operator $R_Q : L_2(Q) \to L_2(Q)$ has a bounded inverse if and only if $\gamma_1\gamma_2 \neq 1$. In Example 8.4, we shall prove that if $\gamma_1\gamma_2 \neq 1$, R_Q maps $\mathring{W}^1(Q)$ onto $W^1_\gamma(Q)$ continuously and in a one-to-one manner. This statement is a particular case of a theorem on isomorphism for a nondegenerate difference operator (see Theorem 8.1). We define the unbounded operators $\mathcal{A}_R, \mathcal{A}_\gamma : L_2(Q) \to L_2(Q)$ acting in the space of distributions $\mathcal{D}'(Q)$ (see Appendix B) by the formulas

$$\mathcal{A}_R u = -\Delta R_Q u \qquad (u \in \mathcal{D}(\mathcal{A}_R) = \{u \in \mathring{W}^1(Q) : \mathcal{A}_R u \in L_2(Q)\}),$$

$$\mathcal{A}_\gamma u = -\Delta w \qquad (w \in \mathcal{D}(\mathcal{A}_\gamma) = \{w \in W^1_\gamma(Q) : \mathcal{A}_\gamma w \in L_2(Q)\}).$$

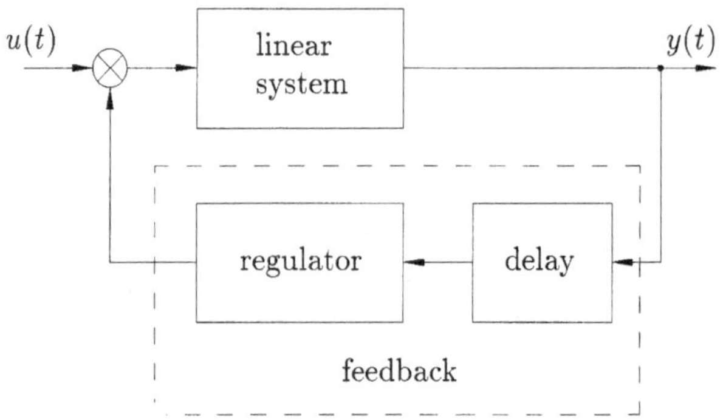

Fig. 0.1

If $\gamma_1\gamma_2 \neq 1$, then $\mathcal{A}_R = \mathcal{A}_\gamma R_Q$. Thus, the problems (0.3), (0.4) and (0.7), (0.8) are equivalent for $\gamma_1\gamma_2 \neq 1$. Therefore, we can apply some results concerning the problem (0.7), (0.8) to the study of the problem (0.3), (0.4) (see Examples 13.3, 13.4). Conversely, some statements on the solvability of the problem (0.3), (0.4) can be used in the investigation of the problem (0.7), (0.8) (see Example 23.1).

Now let $\gamma_1 = \gamma_2 = 1$. In this case the operator $R_Q\colon L_2(Q) \to L_2(Q)$ has an infinite-dimensional null space. Suppose that $u \in \mathring{W}^1(Q)$. Then $w = R_Q u \in W^1(Q)$, and

$$\begin{aligned} w|_{x_2=0} &= w|_{x_2=1} = 0, \\ w(x_1,x_2) &= u(x_1,x_2) + u(x_1+1,x_2) \qquad (0 \le x_1 \le 1,\ 0 \le x_2 \le 1), \\ w(x_1,x_2) &= u(x_1,x_2) + u(x_1-1,x_2) \qquad (1 \le x_1 \le 2,\ 0 \le x_2 \le 1). \end{aligned}$$

Changing the variables $x_1' = x_1 - 1$, $x_2' = x_2$ in the last relation, it is easy to see that $w = R_Q u$ satisfies the conditions (0.6). Hence, we can apply the results concerning the problem with degeneration (0.7), (0.8) to the study of the problem (0.5), (0.6).

Therefore, the elliptic equations with nonlocal conditions give us a very interesting and important application of the theory of boundary value problems for elliptic differential-difference equations. On the other hand, these two problems represent different fields of the theory of general boundary value problems for elliptic functional differential equations.

III. We now consider the applications of this theory.

It is well known that feedback in a control system can lead to signal delay (see Fig. 0.1).

Controllability of delay-differential systems was studied by F. M. Kirillova and S. V. Churakova [1], L. Weiss [1], R. Gabasov and S. V. Churakova [1], R. Gabasov and F. M. Kirillova [1], and others. A large number of papers have been written on the optimal control of functional differential equations of retarded type to target sets in $\mathbb{R}^n$ (for a bibliography, see L. S. Pontryagin, V. G. Boltyanskiĭ, R. V. Gamkrelidze, and E. F. Mishchenko [1], A. Halanay [1], R. Gabasov and F. M. Kirillova [1], G. L. Kharatishvili and T. A. Tadumadze [1]). References concerning the optimal periodic control of retarded functional differential equations can be found in F. Colonius [1]. Optimal control of functional differential equations of retarded type and neutral type to a target set in a function space was investigated by N. N. Krasovskiĭ [1] and G. A. Kent [1], H. T. Banks and G. A. Kent [1], and H. T. Banks and M. Q. Jacobs [1]. In the case of neutral functional differential equations it was assumed that the dominant terms with delay are sufficiently small. Their methods are quite different from those employed here. Our approach allows the study of the control problem for a linear functional differential equation of neutral type with arbitrary dominant terms. If a linear control system contains a dominant term with delay, it can be described as follows:

$$y'(t) + ay'(t-\tau) + by(t) + cy(t-\tau) = u(t) \qquad (0 < t), \tag{0.10}$$

where a, b and c are real constants, $u(t)$ is a control function, $y(t)$ is a state of system, $\tau > 0$ is a constant.

A previous history of the system is defined by the initial condition

$$y(t) = \varphi(t) \qquad (t \in [-\tau, 0]), \tag{0.11}$$

where $\varphi(t)$ is a given function.

We shall study the problem of how to reduce the system (0.10), (0.11) to equilibrium. Let us find a control function $u(t)$ $(0 < t < T)$ such that

$$y(t) = 0 \qquad (t \in [T-\tau, T]). \tag{0.12}$$

If we set $u(t) = 0$ $(t > T)$, then the solution of the problem (0.10), (0.12), $y(t) \equiv 0$ $(t > T)$. We also assume that the energy

$$\int_0^T |u(t)|^2\, dt \to \min. \tag{0.13}$$

Without loss of generality, we assume that $\tau = 1$. Then we obtain the problem of minimizing functional

$$J(y) = \int_0^T \{y'(t) + ay'(t-1) + by(t) + cy(t-1)\}^2\, dt \to \min \tag{0.14}$$

with boundary conditions (0.11), (0.12). A function $y(t)$ yields a minimum of the variational problem (0.14), (0.11), (0.12) if and only if it satisfies the differential-difference equation

$$\begin{aligned} -[(1+a^2)y'(t) + ay'(t-1) + ay'(t+1)]' + (ab-c)(y'(t-1) - y'(t+1)) \\ +(b^2+c^2)y(t) + bc(y(t-1) + y(t+1)) = 0 \qquad (t \in (0, T-1)) \end{aligned} \tag{0.15}$$

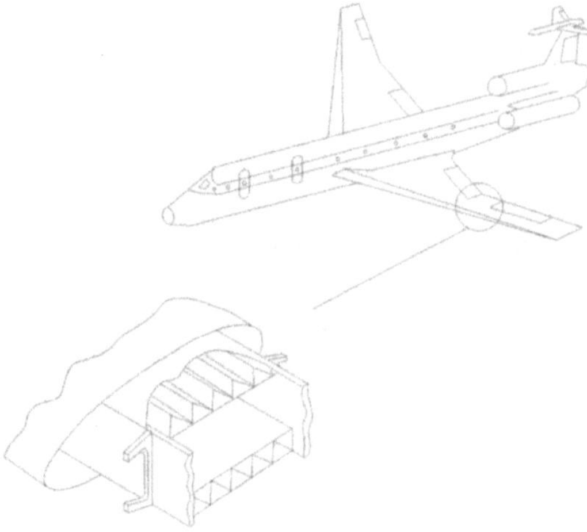

Fig. 0.2

with boundary conditions (0.11), (0.12). The differential-difference equation (0.15) contains both delayed and advanced arguments. In Section 5, we consider the variational problem (0.14), (0.11), (0.12) and the boundary value problem (0.15), (0.11), (0.12). Some results in this section can be extended to multidimensional control systems with delay (see A. Baumstein and A. L. Skubachevskiĭ [1]).

We note that variational problems with delay also arise in relativistic electrodynamics (see R. P. Feynman and J. A. Wheeler [1], L. S. Schulman [1]). Since an electromagnetic field has a finite propagation velocity, a delay depends on an unknown function. A corresponding boundary value problem for functional differential equation contains both delayed and advanced arguments.

Modern aircraft technology is based on constructions containing sandwich shells and plates. Fig. 0.2 shows a wing which has a panel with a goffered filler. We can consider this panel as an elastic system consisting of two parallel plates connected by two regular systems of ribs. It is natural to reduce this discrete-continuous model, "spreading out" both systems of ribs in the space between the plates. As a result, we arrive at a three-layered plate with a "two-phase" model of a filler uniformly distributed in the space between the plates.

Using the Lagrange principle, we can reduce this elastic model to a variational problem for a quadratic functional

$$E(u_x, u_y, u, (u_{\pm\tau})_x, (u_{\pm\tau})_y, u_{\pm\tau}) \to \min, \tag{0.16}$$

where $E(\,\cdot\,)$ is the total potential energy of the three-layer plate, a vector-valued function $u(x,y) = (u^1(x,y), \dots, u^4(x,y))$ corresponds to the elastic displacements of the ribs, $u_{\pm\tau} = (u^1(x, y \pm \tau), \dots, u^4(x, y \pm \tau))$, 2τ is the distance between two ribs at the surface of a plate (see Section 14). The displacements of the filler

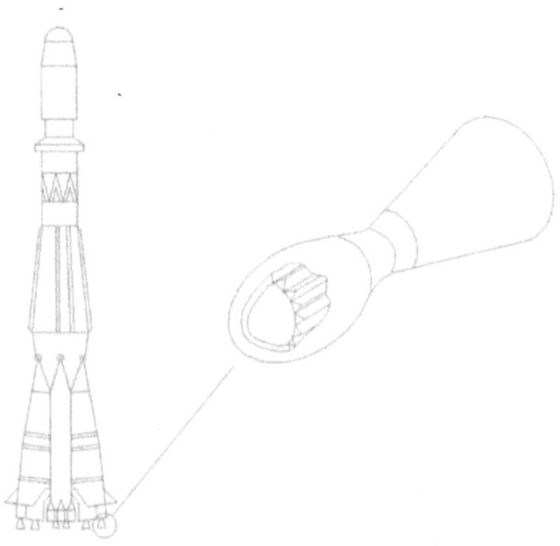

Fig. 0.3

are connected with the displacements of plates at different points. Therefore, the kinematic connection of the filler with the supporting layers leads to the appearance of shifts of argument. The variational problem (0.16) can be reduced to a boundary value problem for a system of four differential-difference equations of the type (0.7), (0.8) (see (15.5)). The elastic model and the corresponding boundary value problem for a strongly elliptic system of differential-difference equations were studied by G. G. Onanov and A. L. Skubachevskiĭ [1]. We consider these results in Chapter III.

Figure 0.3 shows a rocket engine, whose cooling system has the form of a sandwich shell. An elastic model of this system can be reduced to a boundary value problem for a system of four differential-difference equations.

In [1, 2], W. Feller completely characterized the analytic structure of one-dimensional diffusion processes. He gave an explicit representation of the infinitesimal generator $\mathcal{A}$ of a one-dimensional diffusion process and determined all possible boundary conditions which describe the domain $\mathcal{D}(\mathcal{A})$. An analogous problem for multidimensional diffusion processes in a bounded domain $Q \subset \mathbb{R}^n$ was studied by A. D. Ventsel′ [1]. He obtained a general form of boundary conditions for an infinitesimal generator of a Feller semigroup. We note that a Feller semigroup is a strongly continuous, non-negative and contractive semigroup. In general, these conditions contain the values of a function and its derivatives up to the second order and an integral over $\overline{Q}$ (see (24.5)). This problem arises in biophysics (see W. Feller [2]). An integral corresponds to diffusion in a cell in which a particle arriving at the membrane can later jump to a point $x \in \overline{Q}$ (see Fig. 0.4).

The values of a function and its derivatives correspond to the absorption, reflection, viscosity, and diffusion along the boundary. The problem of constructing Feller semigroups was studied only in the so-called "transversal" case (see K. Sato and T. Ueno [1], J. M. Bony, P. Courrege, and P. Priouret [1], C. Cancelier [1],

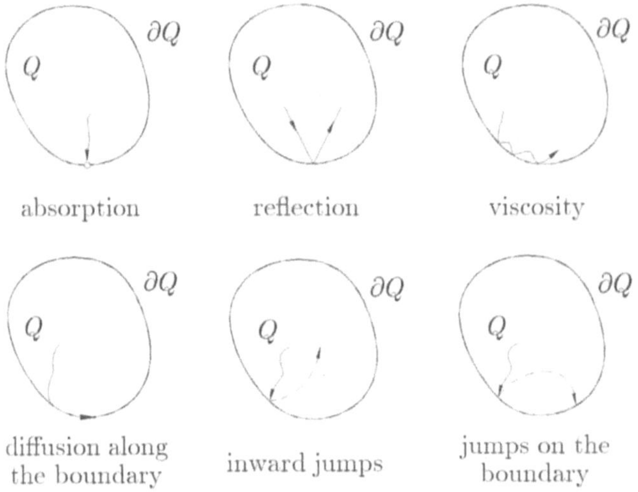

Fig. 0.4

S. Watanabe [1], K. Taira [2], Y. Ishikawa [1]). Analytically, in the transversal case the boundary conditions contain derivatives of a function. Therefore, a nonlocal perturbation has lower order with respect to dominant terms. *In the non-transversal case the problem of construction of Feller semigroups is unsolved.*

In this case, by virtue of the Hille–Yosida theorem, the problem of the existence of a Feller semigroup can be reduced to the following nonlocal elliptic problem

$$Aw(x) + \lambda w(x) = f_0(x) \qquad (x \in Q), \tag{0.17}$$

$$\gamma(x)w(x) + \int_{\overline{Q}} [w(x) - w(y)]\, m(x, dy) = 0 \qquad (x \in \partial Q). \tag{0.18}$$

Here A is the elliptic operator of the form (0.1), $a_0(x) \geq 0$ $(x \in \overline{Q})$, $\lambda > 0$, $\gamma(x) \geq 0$ $(x \in \partial Q)$, $m(x, \cdot)$ is a non-negative Borel measure. In particular, if $\gamma(x) = 1$, $m(x, \overline{Q}) = 0$ $(x \in \Gamma_2)$, $\gamma(x) = 0$, $m(x, \omega(x)) = 1$ and $m(x, \overline{Q} \backslash \omega(x)) = 0$ $(x \in \Gamma_1)$, then we obtain the homogeneous nonlocal conditions (0.2). Thus, the theory of multidimensional diffusion processes is closely connected with the theory of elliptic functional differential equations. In [13], A. L. Skubachevskiĭ applied the theorems on the solvability of elliptic functional differential equations to the problem of existence of Feller semigroups in both transversal and non-transversal cases. We consider these results in Sections 24, 25.

IV. We now discuss the main goals and methods of this book. The book deals with the boundary value problems for elliptic differential-difference equations, elliptic problems with nonlocal boundary conditions, and their applications.

Our approach is based on the properties of elliptic operators and difference operators.

We consider three types of differential-difference operator: 1) strongly elliptic operators (Chapters II, III), 2) symmetric semi-bounded operators with degeneration (Chapter IV), and 3) elliptic operators (Chapter V).

In cases 1) and 2), we study the following basic questions: properties of corresponding sesquilinear forms, solvability and spectrum, and smoothness of generalized solutions. The Friedrichs extension of sectorial operators (see Appendix A) and the theory of difference operators provide a most convenient tool for the study of these problems.

In case 3), we reduce a boundary value problem for elliptic differential-difference equation to an elliptic differential equation with nonlocal conditions. Therefore, we study a priori estimates, solvability and spectrum for elliptic differential equations with nonlocal boundary conditions. Combining these results and properties of difference operators, we consider solvability, spectrum and smoothness of generalized solutions of boundary value problems for elliptic differential-difference equations.

The most interesting properties of elliptic differential-difference equations are connected with smoothness of generalized solutions. Unlike elliptic differential equations, the smoothness of generalized solutions of elliptic differential-difference equations can be violated in a bounded domain $Q \subset \mathbb{R}^n$, even for infinitely differentiable right-hand sides. In general, the smoothness of solutions is conserved only in certain subdomains. The presentation of these new results is one of the main purposes of this book.

V. The book consists of five chapters.

Chapter I is devoted to boundary value problems for differential-difference equations and to differential equations with nonlocal boundary conditions in one dimension. In this chapter we have attempted to demonstrate some methods in the simplest case.

Section 1 is devoted to solvability and spectrum of ordinary differential equations with nonlocal boundary conditions. Section 2 deals with the properties of difference operators. In this section we study the connection between ordinary differential equations with nonlocal boundary conditions and boundary value problems for differential-difference equations. In the case when a differential-difference operator is not sectorial, we apply the results of Sections 1 and 2 to the study of Fredholm solvability and smoothness of generalized solutions of differential-difference equations (see Sections 3, 4). In Section 3, we investigate solvability and smoothness of generalized solutions. *Unlike differential equations, the smoothness of solutions of differential-difference equations can be violated in the interval* $(0, d)$. In Section 4, we obtain the necessary and sufficient conditions providing the existence of smooth solutions on the closed interval $[0, d]$. Applications of the results of Sections 2, 3 and 4 to control systems with delay are studied in Section 5. Section 6 is devoted to the boundary value problem for differential-difference equation with degeneration. The following diagram demonstrates the generalizations to a multidimensional case:

Section 1 → Sections 21, 22; Section 2 → Section 8; Section 3 → Sections 9–11, 15, 16, 23; Section 6 → Sections 18–20.

In Chapter II, we consider the strongly elliptic differential-difference equation

$$\mathcal{A}_R u(x) = \sum_{|\alpha|,|\beta| \le m} \mathcal{D}^\alpha R_{\alpha\beta Q} \mathcal{D}^\beta u(x) = f_0(x) \qquad (x \in Q) \tag{0.19}$$

with boundary conditions

$$\mathcal{D}_\nu^{\mu-1} u|_{\partial Q \setminus K} = 0 \qquad (\mu = 1, \dots, m). \tag{0.20}$$

Here $Q \subset \mathbb{R}^n$ is a bounded domain with boundary $\partial Q \in C^\infty$ or $Q = (0,d) \times G$, $G \subset \mathbb{R}^{n-1}$ is a bounded domain (with boundary $\partial G \in C^\infty$ if $n \ge 3$); $K = \emptyset$ if $\partial Q \in C^\infty$, $K = (\{0\} \times \partial G) \cup (\{d\} \times \partial G)$ if $Q = (0,d) \times G$; $\mathcal{D}^\alpha = \mathcal{D}_1^{\alpha_1} \cdots \mathcal{D}_n^{\alpha_n}$, $\mathcal{D}_j = -i(\partial/\partial x_j)$, $\alpha = (\alpha_1, \dots, \alpha_n)$, $|\alpha| = \alpha_1 + \ldots + \alpha_n$; $R_{\alpha\beta Q} = P_Q R_{\alpha\beta} I_Q$, I_Q is the operator of extension of functions by zero outside Q, P_Q is the operator of restriction to Q,

$$R_{\alpha\beta} u(x) = \sum_{h \in \mathcal{M}} a_{\alpha\beta h}(x) u(x+h) \tag{0.21}$$

are difference operators; $\mathcal{M} \subset \mathbb{R}^n$ is a finite set of vectors with integer coordinates, $a_{\alpha\beta h} \in C^\infty(\mathbb{R}^n)$, ν is the inner unit normal vector to ∂Q at x.

The equation (0.19) is said to be strongly elliptic in $\overline{Q}$ if

$$\operatorname{Re}(\mathcal{A}_R u, u)_{L_2(Q)} \ge c_1 \|u\|^2_{W^m(Q)} - c_2 \|u\|^2_{L_2(Q)} \qquad (u \in \dot{C}^\infty(Q)), \tag{0.22}$$

where $c_1 > 0$, $c_2 \ge 0$ do not depend on u; $\dot{C}^\infty(Q)$ is a set of infinitely differentiable functions in Q with compact support in Q.

Section 7 deals with auxiliary results. In this section we study the properties of open components Q_r of the set $Q \setminus (\bigcup_{h \in M} (\partial Q + h))$. Here M is the additive group generated by the set $\mathcal{M}$. We can divide the set $\mathcal{R} = \{Q_r\}$ into disjoint classes in the following way: subdomains Q_{r_1}, Q_{r_2} belong to the same class if there is an $h \in M$ such that $Q_{r_2} = Q_{r_1} + h$. We denote $r = (s, l)$, where $s = 1, 2, \dots$ is the number of a class, l is the number of an element in the sth class ($l = 1, \dots, N = N(s)$). In Section 8, we study the properties of operators $R_{\alpha\beta} \colon L_2(\mathbb{R}^n) \to L_2(\mathbb{R}^n)$ and $R_{\alpha\beta Q} \colon \mathring{W}^m(Q) \to W^m(Q)$, where $\mathring{W}^m(Q) = \{u \in W^m(Q) : \mathcal{D}_\nu^{\mu-1} u|_{\partial Q \setminus K} = 0 \ (\mu = 1, \dots, m)\}$. We prove that a nondegenerate difference operator $R_{\alpha\beta Q}$ with constant coefficients maps $\mathring{W}^m(Q)$ onto $W^m_\gamma(Q)$ continuously and in a one-to-one manner (see Theorem 8.1). Here $W^m_\gamma(Q)$ is a subspace of functions from $W^m(Q)$ with nonlocal conditions on the shifts of ∂Q.

In Section 9, we obtain both necessary and sufficient conditions of strong ellipticity in algebraic form. Let $x \in \overline{Q}_{s1}$ be an arbitrary point. Consider all points $x^i \in \overline{Q}$ such that $x^i - x \in M$. The set $\{x^i\} \in \overline{Q}$ consists of a finite number of points $I = I(s, x)$ ($I \ge N(s)$). We shall number the points x^i so that $x^i = x + h_{si}$

for $i = 1, \dots, N$, $x^1 = x$, where h_{si} satisfies the condition $Q_{si} = Q_{s1} + h_{si}$. We introduce the $I \times I$-matrices $A_{\alpha\beta s}(x)$ with elements $a_{ij}^{\alpha\beta s}(x)$ by the formula

$$a_{ij}^{\alpha\beta s}(x) = \begin{cases} a_{\alpha\beta h}(x^i), & h = x^j - x^i \in \mathcal{M}, \\ 0, & x^j - x^i \notin \mathcal{M}. \end{cases} \tag{0.23}$$

We consider the matrices $R_{\alpha\beta s}(x)$ of order $N \times N$ obtained from $A_{\alpha\beta s}(x)$ by deleting the last $I - N$ rows and columns. If the matrices

$$\sum_{|\alpha|=|\beta|=m} (A_{\alpha\beta s}(x) + A^*_{\alpha\beta s}(x))\xi^{\alpha+\beta}$$

are positive definite for $s = 1, 2, \dots,$ $x \in \overline{Q}_{s1}$, $0 \neq \xi \in \mathbb{R}^n$, then the equation (0.19) is strongly elliptic in $\overline{Q}$ (see Theorem 9.2), where $\xi^\alpha = \xi_1^{\alpha_1} \dots \xi_n^{\alpha_n}$. If the equation (0.19) is strongly elliptic in $\overline{Q}$, then the matrices $\sum_{|\alpha|=|\beta|=m}(R_{\alpha\beta s}(x) + R^*_{\alpha\beta s}(x))\xi^{\alpha+\beta}$ are positive definite for $s = 1, 2, \dots,$ $x \in \overline{Q}_{s1}$, $0 \neq \xi \in \mathbb{R}^n$ (see Theorem 9.1). It is also proved that the necessary and sufficient conditions of strong ellipticity are the same for a dense set of domains Q (see Theorem 9.3). The sufficient condition of strong ellipticity, using the symbol of a differential-difference operator, is much weaker. In contrast to a strongly elliptic differential equation, the symbol of a strongly elliptic differential-difference equation is quasi-polynomial and can change its sign.

In Section 10, we consider the unbounded strongly elliptic operator $\mathcal{A}_R$: $L_2(Q) \to L_2(Q)$ with domain $\mathcal{D}(\mathcal{A}_R) = \{u \in \mathring{W}^m(Q) : \mathcal{A}_R u \in L_2(Q)\}$ acting in the space of distributions $\mathcal{D}'(Q)$ by the formula (0.19). We prove that the sectorial operator $\mathcal{A}_R$ is Fredholm, and the spectrum $\sigma(\mathcal{A}_R)$ is discrete.

Section 11 deals with the smoothness of generalized solutions in subdomains Q_{sl}. A function u is called a generalized solution of the boundary value problem (0.19), (0.20) if $u \in \mathcal{D}(\mathcal{A}_R)$ and $\mathcal{A}_R u = f_0$. It is proved that, if $f_0 \in L_2(Q) \cap W^k(Q_{sl})$ $(s = 1, 2, \dots,$ $l = 1, \dots, N(s))$, then a generalized solution of the boundary value problem (0.19), (0.20) $u \in W^{k+2m}(Q_{sl} \setminus \overline{\mathcal{K}^\varepsilon})$ for each $\varepsilon > 0$ $(s = 1, 2, \dots,$ $l = 1, \dots, N(s))$, where

$$\mathcal{K} = \bigcup_{h_1, h_2 \in M} \{\overline{Q} \cap (\partial Q + h_1) \cap \overline{[(\partial Q + h_2) \setminus (\partial Q + h_1)]}\},$$

$\overline{\mathcal{K}^\varepsilon} = \{x \in \mathbb{R}^n : \rho(x, \mathcal{K}) \leq \varepsilon\}$. Generally speaking, this statement is not valid for $\varepsilon = 0$ (see Example 11.2). *We note that, unlike elliptic differential equations, the smoothness of the generalized solutions of elliptic differential-difference equations can be broken in the domain Q even for infinitely differentiable right-hand sides of the equations. Moreover, the smoothness of the solutions can be violated near the set $\mathcal{K}$.*

In Section 12, we establish necessary and sufficient conditions for the conservation of smoothness of solutions on the boundary of adjacent subdomains.

In Section 13, we examine an elliptic differential equation of order $2m$, with nonlocal conditions relating traces of the unknown function and its derivatives on some parts of the boundary, to a linear combination of traces on the same parts of the boundary displaced towards the interior of the domain. The smoothness of generalized solutions of such problems can be violated near the boundary. Therefore, methods based on the smoothness of solutions (see Section 21) cannot be used to investigate the solvability and the spectrum. Theorems concerning the Fredholm property, the discreteness of the spectrum, and the completeness of the eigenfunction system of the problem under consideration are proved by reducing it to a boundary value problem for a strongly elliptic differential-difference equation.

In order to illustrate the results of Chapter II, we consider the boundary value problems (0.7), (0.8) and (0.3), (0.4). A set $\mathcal{R}$ consists of one class of subdomains $Q_{11} = (0,1) \times (0,1)$, $Q_{12} = (1,2) \times (0,1)$. Further,

$$R_1 = \begin{pmatrix} 1 & \gamma_1 \\ \gamma_2 & 1 \end{pmatrix}.$$

The matrix $R_1 + R_1^*$ is positive definite if and only if $|\gamma_1 + \gamma_2| < 2$. Therefore, the inequality

$$(-\Delta R_Q u, u)_{L_2(Q)} \geq c_1 \|u\|^2_{W^1(Q)} \qquad (u \in \dot{C}^\infty(Q)) \tag{0.24}$$

holds if and only if $|\gamma_1 + \gamma_2| < 2$. The condition $|\gamma_1 + \gamma_2| < 2$ implies that the spectrum $\sigma(\mathcal{A}_R)$ is discrete, and $\sigma(\mathcal{A}_R) \subset \{\lambda \in \mathbb{C} : \operatorname{Re} \lambda > 0\}$. Hence, the boundary value problem (0.7), (0.8) has a unique generalized solution. If $\gamma = \gamma_1 = \gamma_2$, $|\gamma| < 1$, then the operator $\mathcal{A}_R : L_2(Q) \to L_2(Q)$ is self-adjoint, and $\sigma(\mathcal{A}_R) \subset (0, \infty)$. We note that a symbol of $\mathcal{A}_R$ equals $(\xi_1^2 + \xi_2^2)(1 + 2\gamma \cos \xi_1)$. Clearly, it can change sign for $1/2 < |\gamma| < 1$.

Let $|\gamma_1 + \gamma_2| < 2$. If $u \in \mathcal{D}(\mathcal{A}_R)$ is a generalized solution of the boundary value problem (0.7), (0.8), then $R_Q u \in W^2(Q)$ and $u \in W^2(Q_{1l})$ $(l = 1,2)$. In this case, the smoothness of generalized solutions is preserved near a set $\mathcal{K} = \bigcup (i,j)$ $(i = 0,1,2,\ j = 0,1)$. However, if $\gamma_1^2 + \gamma_2^2 \neq 0$, then there is $f_0 \in L_2(Q)$ such that $u \notin W^2(Q)$. For example, let $\gamma_1 = \gamma_2 = \gamma$, $|\gamma| < 1$, and let $f_0 = \pi^2 \sin \pi x_2$. Then there is a unique generalized solution of the boundary value problem (0.7), (0.8) given by

$$u(x) = \begin{cases} \dfrac{\sin \pi x_2}{1+\gamma} \left(\dfrac{\gamma \cosh \pi x_1 - \cosh \pi (x_1 - 1)}{\cosh \pi - \gamma} + 1 \right), & x \in Q_{11}, \\ \dfrac{\sin \pi x_2}{1+\gamma} \left(\dfrac{\gamma \cosh \pi (x_1 - 2) - \cosh \pi (x_1 - 1)}{\cosh \pi - \gamma} + 1 \right), & x \in Q_{12}. \end{cases}$$

Clearly, $u \notin W^2(Q)$ (see Fig. 0.5).

We now apply the results concerning spectrum $\sigma(\mathcal{A}_R)$ to the study of spectrum $\sigma(\mathcal{A}_\gamma)$. If $|\gamma_1 + \gamma_2| < 2$, then the boundary value problem (0.3), (0.4) has a unique generalized solution. In [1], A. V. Bitsadze and A. A. Samarskiĭ proved the existence and uniqueness of a solution of problem (0.3), (0.4) for $\gamma_1 = 0$, $\gamma_2 = 1$ using the maximum principle and the potential theory. It is easy to see that, generally speaking,

$$(\mathcal{A}_\gamma u, v)_{L_2(Q)} \neq (u, \mathcal{A}_\gamma v)_{L_2(Q)} \qquad (u, v \in W^1_\gamma(Q) \cap W^2(Q)).$$

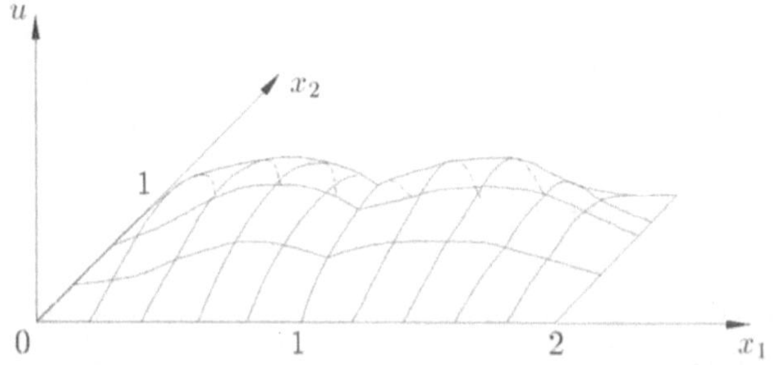

Fig. 0.5

Hence, the operator $\mathcal{A}_\gamma: L_2(Q) \to L_2(Q)$ is not self-adjoint. However, if $\gamma = \gamma_1 = \gamma_2$, $|\gamma| < 1$, then $\sigma(\mathcal{A}_\gamma)$ is real and discrete.

In Chapter III, we consider applications of elliptic differential-difference equations to the mechanics of a deformable body.

Section 14 deals with the elastic model of some sandwich plate with goffered filler. We reduce this model to a variational problem with shifts of argument.

In Section 15, we consider the above variational problem and a corresponding boundary value problem for a strongly elliptic system of differential-difference equations. We prove the existence and the uniqueness of a generalized solution of this problem and the discreteness of the spectrum of the appropriate operator. The convergence of the Ritz method is stated.

Section 16 is devoted to the study of smoothness of solutions.

Section 17 illustrates the results of Sections 14–16 in one dimension.

In Chapter IV, we consider the differential-difference operator $A_R = AR_Q$ with domain $\mathcal{D}(A_R) = \dot{C}^\infty(Q)$. Here $R_Q = P_Q R I_Q$, $R: L_2(\mathbb{R}^n) \to L_2(\mathbb{R}^n)$ is the difference operator given by

$$Ru(x) = \sum_{h \in \mathcal{M}} a_h u(x+h), \tag{0.25}$$

$\mathcal{M}$ is a finite set of vectors with integer coordinates such that, if $h \in \mathcal{M}$, then $-h \in \mathcal{M}$, $a_h = a_{-h}$, a_h are real numbers,

$$A = \sum_{|\alpha|,|\beta| \le m} \mathcal{D}^\alpha a_{\alpha\beta}(x) \mathcal{D}^\beta, \tag{0.26}$$

$a_{\alpha\beta} = a_{\beta\alpha} \in C^\infty(\mathbb{R}^n)$ are real-valued functions, $a_{\alpha\beta}(x) = a_{\alpha\beta}(x+h)$ $(x, x+h \in \overline{Q}$, $h \in M)$.

We assume that

$$\sum_{|\alpha|,|\beta|=m} a_{\alpha\beta}(x)\xi^{\alpha+\beta} > 0 \qquad (x \in \overline{Q},\ 0 \ne \xi \in \mathbb{R}^n);$$

there is s_1 such that $\det R_{s_1} = 0$, and if $m = 1$ the matrices R_s are non-negative, if $m > 1$ the matrices A_s are non-negative ($s = 1, 2, \dots$). Here $A_s = A_s(x)$ are the matrices of order $I \times I$ ($I = I(s,x)$, $s = 1, 2, \dots$, $x \in \overline{Q}_{s1}$) with the elements

$$a^s_{ij}(x) = \begin{cases} a_h, & h = x^j - x^i \in \mathcal{M}, \\ 0, & x^j - x^i \notin \mathcal{M}; \end{cases} \tag{0.27}$$

the $N \times N$-matrices R_s are obtained from A_s by deleting the last $I - N$ rows and columns (cf. (0.23)).

In Section 18, we construct a self-adjoint Friedrichs extension $\mathcal{A}_R$ of the operator A_R.

Section 19 is devoted to the study of the spectrum of this operator. It is proved that the spectrum $\sigma(\mathcal{A}_R)$ consists of isolated eigenvalues $\lambda_0 = 0$ of infinite multiplicity and $\lambda_s > 0$ of finite multiplicity.

In Section 20, we consider the smoothness of solutions. It is shown that the kernel $\mathcal{N}(R_Q) \subset \mathcal{N}(\mathcal{A}_R)$. Therefore, the equation with degeneration $\mathcal{A}_R u = f_0$ can have solutions $u \in \mathcal{D}(\mathcal{A}_R)$ from $L_2(Q)$, not even in $W^1(Q)$. We prove that, if $f_0 \in L_2(Q) \cap W^k(Q_{sl})$ ($s = 1, 2, \dots$, $l = 1, \dots, N(s)$), then $P^R u \in W^{k+2m}(Q_{sl} \setminus \overline{\mathcal{K}^\varepsilon})$ for each $\varepsilon > 0$ ($s = 1, 2, \dots$, $l = 1, \dots, N(s)$), where $P^R: L_2(Q) \to L_2(Q)$ is the orthogonal projection operator on the range $\mathcal{R}(R_Q)$.

To illustrate the results of Chapter IV, we consider the boundary value problems (0.7), (0.8) and (0.5), (0.6). Let $\gamma_1 = \gamma_2 = 1$. Then the difference operator (0.9) will take the form

$$Ru(x) = u(x_1, x_2) + u(x_1 + 1, x_2) + u(x_1 - 1, x_2).$$

Hence,

$$R_1 = \begin{pmatrix} 1 & 1 \\ 1 & 1 \end{pmatrix}.$$

Clearly, $\det R_1 = 0$, $R_1 \geq 0$. The Friedrichs extension $\mathcal{A}_R$ is given by the formulas

$$\mathcal{A}_R u = -\Delta R_Q u \qquad (u \in \mathcal{D}(\mathcal{A}_R) = \{u \in L_2(Q) : R_Q u \in W^2(Q), \\ R_Q u|_{x_2=0} = R_Q u|_{x_2=1} = 0\}). \tag{0.28}$$

The kernel $\mathcal{N}(\mathcal{A}_R) = \mathcal{N}(R_Q)$ consists of functions $u \in L_2(Q)$ such that

$$u(x_1, x_2) = -u(x_1 + 1, x_2) \qquad (0 \leq x_1 \leq 1,\ 0 \leq x_2 \leq 1).$$

The range $\mathcal{R}(\mathcal{A}_R) = \mathcal{R}(R_Q)$ consists of functions $w \in L_2(Q)$ such that

$$w(x_1, x_2) = w(x_1 + 1, x_2) \qquad (0 \leq x_1 \leq 1,\ 0 \leq x_2 \leq 1).$$

We denote by $\mathcal{A}^0_R$ the restriction of $\mathcal{A}_R$ to $\mathcal{D}(\mathcal{A}_R) \cap \mathcal{R}(R_Q)$. The operator $\mathcal{A}^0_R$: $\mathcal{R}(R_Q) \to \mathcal{R}(R_Q)$ has a bounded inverse defined on $\mathcal{R}(R_Q)$, and $\mathcal{D}(\mathcal{A}^0_R) = \{u \in W^2(Q) : u \in \mathcal{R}(R_Q),\ u|_{x_2=0} = u|_{x_2=1} = 0\}$ (see Example 20.4). Therefore, for

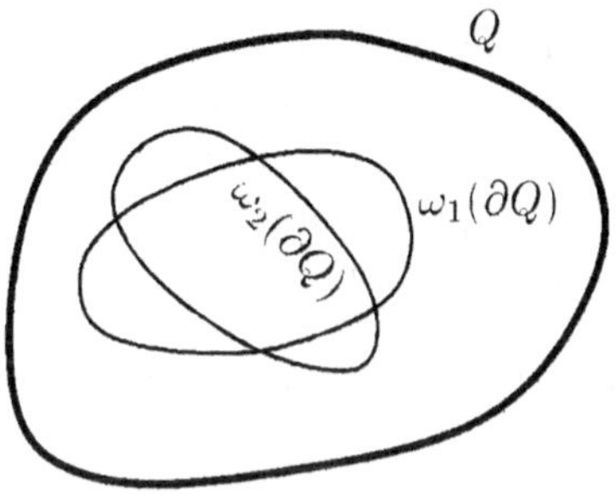

Fig. 0.6

every $f_0 \in \mathcal{R}(R_Q)$ there exists a unique solution of the problem (0.5), (0.6) $w \in W^2(Q)$, and this solution has the form $w = 2(\mathcal{A}_R^0)^{-1} f_0$.

Chapter V is devoted to elliptic differential equations of order $2m$ with nonlocal boundary conditions. The results are applied to the boundary value problems for elliptic differential-difference equations and to the theory of multidimensional diffusion processes.

In Section 21, we consider the equation

$$\sum_{\beta+|\alpha|\le 2m} a_{\alpha\beta}(x) q^\beta \mathcal{D}^\alpha u(x) = f_0(x) \qquad (x \in Q) \tag{0.29}$$

with nonlocal boundary conditions

$$\sum_{s=0}^{S} \sum_{\beta+|\alpha|\le m_\mu} b_{\mu s\alpha\beta}(x) q^\beta (\mathcal{D}^\alpha u)(\omega_s(x))|_{\partial Q} = f_\mu(x) \qquad (x \in \partial Q,\ \mu = 1,\dots,m). \tag{0.30}$$

Here $a_{\alpha\beta}, b_{\mu s\alpha\beta} \in C^\infty(\mathbb{R}^n)$ are complex-valued functions; $n \ge 1$; $Q \subset \mathbb{R}^n$ is a bounded domain with boundary $\partial Q \in C^\infty$ (a bounded open interval if $n = 1$), q is a complex parameter, ω_s are infinitely differentiable nondegenerate transformations mapping some neighborhood γ of the boundary ∂Q onto the set $\omega_s(\gamma)$ so that $\overline{\omega_s(\gamma)} \subset Q$ if $s > 0$ (see Fig. 0.6), $\omega_0(x) \equiv x$; $S > 0$ is an integer. We assume that the operators $\sum_{\beta+|\alpha|=2m} a_{\alpha\beta}(x) q^\beta \mathcal{D}^\alpha$, $\sum_{\beta+|\alpha|=m_\mu} b_{\mu 0\alpha\beta}(x) q^\beta \mathcal{D}^\alpha$ ($\mu = 1,\dots,m$) satisfy the conditions of ellipticity with a parameter for $q \in \theta = \{q \in \mathbb{C} : \varphi_1 \le \arg q \le \varphi_2\}$ (see conditions 21.1, 21.2). We obtain a priori estimates of solutions and the existence of a unique solution for sufficiently large values of a parameter $q \in \theta$. These results are generalized to a case of abstract nonlocal terms.

Section 22 deals with an elliptic equation of the second order in a cylinder $Q = (0,d) \times G$ with the following nonlocal conditions: a trace of a solution for $x_1 = 0, d$ is a linear combination of traces for $x_1 = d_i$ ($0 < d_i < d$, $i = 1,\dots,m$), with a trace on the lateral surface of the cylinder equal to zero. The methods of Section 21 enable it to be proved that a corresponding operator is Fredholm, and for every $0 < \varepsilon < \pi$ all points of a spectrum, except possibly for a finite number, belong to the angle of a complex plane $|\arg\lambda| < \varepsilon$.

In Section 23, we consider an elliptic differential-difference equation

$$AR_Q u = f_0(x) \qquad (x \in Q = (0,d) \times G) \tag{0.31}$$

with boundary condition

$$u|_{\partial Q} = 0. \tag{0.32}$$

Here A is an elliptic differential operator of the second order given by (0.1), R_Q is a nondegenerate operator having shifts along the axis x_1. Generally speaking, the operator AR_Q is not strongly elliptic. Combining the results of Section 22 and Theorem 8.1 on the isomorphism for a nondegenerate difference operator, we establish the Fredholm property, discreteness of spectrum and smoothness of generalized solutions in subdomains Q_{sl}.

Again we consider the boundary value problems (0.3), (0.4) and (0.7), (0.8). The operator $\mathcal{A}_\gamma\colon L_2(Q) \to L_2(Q)$ is Fredholm, and $\operatorname{ind}\mathcal{A}_\gamma = 0$ for all $\gamma_1, \gamma_2 \in \mathbb{R}$. Moreover, the spectrum $\sigma(\mathcal{A}_\gamma)$ is discrete, and for each $0 < \varepsilon$ there is $q_\varepsilon > 0$ such that $\sigma(\mathcal{A}_\gamma) \subset \{\lambda \in \mathbb{C} : |\arg\lambda| < \varepsilon\} \cup \{\lambda \in \mathbb{C} : |\lambda| < q_\varepsilon\}$. Using the above results, we obtain the following statements for the problem (0.7), (0.8). If $\gamma_1\gamma_2 \neq 1$, then $\mathcal{A}_R$ is a Fredholm operator, and $\operatorname{ind}\mathcal{A}_R = 0$. Furthermore, if $u \in \mathcal{D}(\mathcal{A}_R)$ is a generalized solution of the boundary value problem (0.7), (0.8), then $u \in W^2(Q_{1l})$ $(l = 1, 2)$.

In Section 24, we apply the results of Section 21 to the theory of multidimensional diffusion processes in both transversal and non-transversal cases. Using the Hille–Yosida theorem, we can reduce the problem of constructing Feller semigroups to the study of an elliptic differential equation of the second order with homogeneous nonlocal boundary conditions. In the case when nonlocal conditions have the form (0.30), we prove that the closure of the corresponding elliptic operator in $C(\overline{Q})$ is the infinitesimal generator of a Feller semigroup.

In Section 25, we consider an interesting example of a nonlocal elliptic problem satisfying necessary conditions of existence of a Feller semigroup. We prove that the closure of the elliptic operator under consideration is not an infinitesimal generator of a Feller semigroup. The proof is based on the theory of nonlocal elliptic problems, the theory of boundary value problems for elliptic differential-difference equations and the theory of weighted spaces. Combining the above methods, we construct an operator with a negative index. This contradicts the Hille–Yosida theorem.

We note that after the basic part (Chapters I, II), a reader can study Chapters III–V almost independently.

The Appendices A–C are devoted to the elements of functional analysis, the theory of Sobolev spaces, and the theory of elliptic differential equations. The book contains a List of Symbols and an Index. Bibliographical references are discussed in the Notes at the end of each chapter.

The numbering of theorems, lemmas, and formulas has two parts. The first is the section number, the second is the number of the theorem, lemma or formula in that section. For example, Theorem 1.2 is the second theorem in Section 1; (B.1) is the first formula in Appendix B.

This book is mainly based on the investigations of the author and on the courses he gave to students at the Moscow State Aviation Institute. It is addressed to graduate students and mathematicians with interests in functional differential equations and partial differential equations. It could also be useful to specialists in the fields of control theory, theory of diffusion processes, and elasticity theory.

I hope that this book will be of interest to specialists in nonlinear optics. For elliptic functional differential equations arising in the theory of turbulent laser beams, see M. A. Vorontsov, J. Ricklin, and G. Carhart [1] and A. L. Skubachevskiĭ [21].

This book is an introduction to the theory of boundary value problems for elliptic functional differential equations. For the most part, the problems considered here require only well-known mathematical tools. In the general case ($\Gamma_2 \neq \emptyset$, $\overline{\omega(\Gamma_1)} \cap \overline{\Gamma_1} \neq \emptyset$) the solutions of elliptic problems with nonlocal boundary conditions can have power singularities near certain $(n-2)$-dimensional manifolds. It is therefore natural to consider nonlocal elliptic problems in weighted spaces. This theory and related problems for elliptic differential-difference equations will be published in a separate book by the same author, *Nonlocal Boundary Value Problems.*

Chapter I

Boundary Value Problems for Functional Differential Equations in One Dimension

The main purpose of the chapter is to demonstrate some of the methods used in this book in the simplest one-dimensional case. Therefore, we consider here equations of the second order with boundary conditions containing only the values of the unknown function.

Section 1 is devoted to the solvability and the spectrum of ordinary differential equations with abstract boundary conditions. These results are used in Sections 3, 4, 5, and 25. Section 2 deals with the properties of difference operators. In this section we study the connection between ordinary differential equations with nonlocal boundary conditions and boundary value problems for differential-difference equations. In Section 3, we study the solvability and the smoothness of generalized solutions of boundary value problems for differential-difference equations in the interval $(0, d)$. Unlike differential equations, the smoothness of solutions of differential-difference equations can be violated in this interval. In Section 4, we obtain the necessary and sufficient conditions providing the existence of smooth solutions. In Section 5, we consider applications of the results obtained in Sections 3 and 4 to control systems with delay. Section 6 is devoted to the boundary value problem for differential-difference equations with degeneration.

1 Ordinary Differential Equations with Nonlocal Boundary Conditions

Formulation of Nonlocal Boundary Value Problem

Let us consider the equation

$$Au - \lambda u = -a_0(t)u''(t) + A_1 u - \lambda u = f_0(t) \qquad (t \in (0, d)) \qquad (1.1)$$

with *nonlocal boundary conditions*

$$\begin{aligned} B_1u &= u|_{t=0} + (B_{11}u)|_{t=0} + B_{12}u = f_1, \\ B_2u &= u|_{t=d} + (B_{21}u)|_{t=0} + B_{22}u = f_2. \end{aligned} \tag{1.2}$$

Here $a_0 \in C[0,d]$ is a real-valued function, $a_0(t) \geq k > 0$ $(0 \leq t \leq d)$, $f_0 \in L_2(0,d)$ is a complex-valued function, $f_\mu \in \mathbb{C}$ $(\mu = 1,2)$, $\lambda \in \mathbb{C}$ is a spectral parameter, and the operators A_1, $B_{\mu 1}$, $B_{\mu 2}$ satisfy the following condition:

1.1. $A_1 \colon W^1(0,d) \to L_2(0,d)$, $B_{\mu 1} \colon L_2(0,d) \to L_2(0,d)$, $B_{\mu 2} \colon L_2(0,d) \to \mathbb{C}$ *are linear bounded operators, and there exists* $\sigma > 0$ *such that for any* $u \in W^k(0,d)$ $(k = 0,2)$

$$\|B_{\mu 1}u\|_{W^k(0,d)} \leq c\|u\|_{W^k(\sigma,d-\sigma)} \qquad (\mu = 1,2). \tag{1.3}$$

Now we study an example of such nonlocal conditions.

Example 1.1. We consider the differential equation

$$-a_0(t)u''(t) + a_1(t)u'(t) + a_2(t)u(t) - \lambda u(t) = f_0(t) \qquad (t \in (0,d)) \tag{1.4}$$

with nonlocal boundary conditions

$$\begin{aligned} u(0) + \sum_{j=1}^{m} b_{1j}u(d_j) + \int_0^d e_1(t)u(t)\,dt = f_1, \\ u(d) + \sum_{j=1}^{m} b_{2j}u(d_j) + \int_0^d e_2(t)u(t)\,dt = f_2. \end{aligned} \tag{1.5}$$

Here $a_i, e_\mu \in C[0,d]$ are real-valued functions $(i = 0,1,2,\ \mu = 1,2)$, $a_0(t) \geq k > 0$ $(0 \leq t \leq d)$, $f_0 \in L_2(0,d)$ is a complex-valued function, $b_{\mu j}, f_\mu \in \mathbb{C}$; $\lambda \in \mathbb{C}$ is a spectral parameter, $0 = d_0 < d_1 < \ldots < d_m < d_{m+1} = d$.

In contrast to well-known boundary conditions, nonlocal conditions (1.5) contain the values of the unknown function not only at the ends of $(0,d)$, but also at the interior points of the interval.

We introduce the operators A_0, A_1, $B_{\mu 1}$, $B_{\mu 2}$ $(\mu = 1,2)$ given by

$$A_0u = -a_0(t)u''(t), \qquad \mathcal{D}(A_0) = W^2(0,d), \tag{1.6}$$

$$A_1u = a_1(t)u'(t) + a_2(t)u(t), \qquad \mathcal{D}(A_1) = W^1(0,d), \tag{1.7}$$

$$B_{11}u = \sum_{j=1}^{m} b_{1j}u(t+d_j)\eta(t), \qquad \mathcal{D}(B_{11}) = L_2(0,d), \tag{1.8}$$

$$B_{21}u = \sum_{j=1}^{m} b_{2j}u(t-d+d_j)\eta(t-d), \qquad \mathcal{D}(B_{21}) = L_2(0,d), \tag{1.9}$$

$$B_{\mu 2}u = \int_0^d e_\mu(t)u(t)\,dt, \qquad \mathcal{D}(B_{\mu 2}) = L_2(0,d) \quad (\mu = 1,2). \tag{1.10}$$

Here $\eta \in \dot{C}^\infty(\mathbb{R})$, $\eta(t) = 1$ $(|t| < \delta)$, $\eta(t) = 0$ $(|t| > 2\delta)$, $0 < 2\delta < \min(d_{j+1} - d_j)$ $(j = 0, \ldots, m)$. Thus, the values of $(B_{11}u)(t)$ and $(B_{21}u)(t)$ do not depend on the values of functions $u(t + d_j)$ and $u(t - d + d_j)$, when $|t| > 2\delta$ and $|t - d| > 2\delta$, respectively. Now we can rewrite the nonlocal boundary value problem (1.4), (1.5) for $u \in W^2(0, d)$ in the form (1.1), (1.2). Clearly, the condition 1.1 holds for $\sigma = \min\{(d_1 - 2\delta), (d - d_m - 2\delta)\}$.

Solvability and Spectrum

We introduce the bounded linear operators $\mathcal{L}, \mathcal{L}_0, \mathcal{L}_\tau: W^2(0, d) \to \mathcal{W}(0, d) = L_2(0, d) \times \mathbb{C} \times \mathbb{C}$ by the formulas $\mathcal{L}u = (Au - \lambda u, B_1u, B_2u)$, $\mathcal{L}_0 u = (A_0 u - \lambda u, u(0), u(d))$, $\mathcal{L}_\tau u = \mathcal{L}_0 u + \tau(\mathcal{L} - \mathcal{L}_0)u$, where $0 \le \tau \le 1$. Evidently, $\mathcal{L}_\tau = \mathcal{L}_0$ for $\tau = 0$, $\mathcal{L}_\tau = \mathcal{L}$ for $\tau = 1$. We use the norms

$$\begin{aligned} |||u|||_{W^2(0,d)} &= \{\|u\|^2_{W^2(0,d)} + |\lambda|^2 \|u\|_{L_2(0,d)}\}^{1/2}, \\ |||f|||_{\mathcal{W}(0,d)} &= \{\|f_0\|^2_{L_2(0,d)} + |\lambda|^{3/2}(|f_1|^2 + |f_2|^2)\}^{1/2}, \end{aligned}$$

depending on a parameter λ (see Appendix C), where $f = (f_0, f_1, f_2)$, $|\lambda| \ge 1$.

Definition 1.1. A function $u \in W^2(0, d)$ is called a *generalized solution* of the problem (1.1), (1.2) if

$$\mathcal{L}u = f. \tag{1.11}$$

The main purpose of this section is to prove the following:

Theorem 1.1. *Let condition* 1.1 *be fulfilled.*

Then we have:

(a) $\mathcal{L}(\lambda): W^2(0, d) \to \mathcal{W}(0, d)$ *is a Fredholm operator, and* $\operatorname{ind} \mathcal{L}(\lambda) = 0$ *for all* $\lambda \in \mathbb{C}$.

(b) *For any* $\varepsilon > 0$, *there exists* $q_1 > 1$ *such that for* $\lambda \in \Omega_{\varepsilon, q_1} = \{\lambda \in \mathbb{C} : |\arg \lambda| \ge \varepsilon,\ |\lambda| \ge q_1\}$ *the operator* $\mathcal{L}(\lambda)$ *has a bounded inverse* $\mathcal{L}^{-1}(\lambda)$: $\mathcal{W}(0, d) \to W^2(0, d)$.

We introduce the unbounded operator $\mathcal{A}_B: L_2(0, d) \to L_2(0, d)$ defined by

$$\mathcal{A}_B u = Au, \quad \mathcal{D}(\mathcal{A}_B) = W^2_B(0, d) = \{u \in W^2(0, d) : B_\mu u = 0,\ \mu = 1, 2\}.$$

Granting Theorem 1.1 for the moment, we shall prove the following statement.

Theorem 1.2. *Let condition* 1.1 *hold.*

Then we have:

(a) $\mathcal{A}_B: L_2(0, d) \to L_2(0, d)$ *is a Fredholm operator, and* $\operatorname{ind} \mathcal{A}_B = 0$.

(b) *The spectrum* $\sigma(\mathcal{A}_B)$ *is discrete.*

(c) *For* $\lambda \notin \sigma(\mathcal{A}_B)$, *the resolvent* $R(\lambda, \mathcal{A}_B): L_2(0, d) \to L_2(0, d)$ *is a compact operator.*

(d) *For every* $0 < \varepsilon < \pi$, *all points of the spectrum* $\sigma(\mathcal{A}_B)$, *except, possibly, for a finite number of them, belong to the sector of complex plane defined by* $|\arg \lambda| < \varepsilon$.

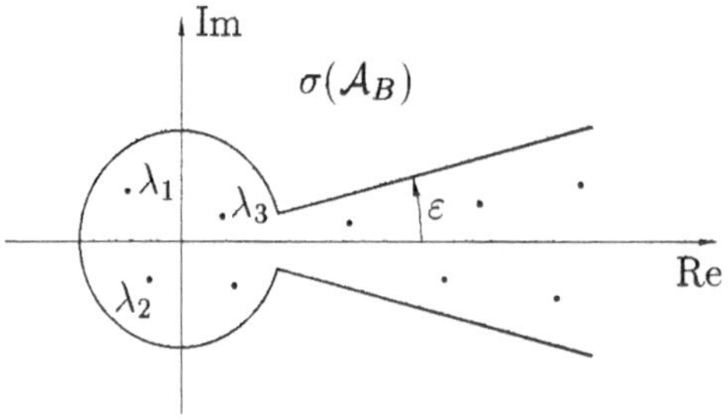

Fig. I.1

Proof. Let $\mu \in \Omega_{\varepsilon,q_1}$. Then, by Theorem 1.1, there exists a bounded operator $(\mathcal{A}_B - \mu I)^{-1}\colon L_2(0,d) \to W_B^2(0,d)$. Hence, from the compactness of the imbedding operator of $W^2(0,d)$ into $L_2(0,d)$ it follows that operator $(\mathcal{A}_B - \mu I)^{-1}\colon L_2(0,d) \to L_2(0,d)$ is compact. Thus, by Theorem A.8, the spectrum $\sigma(\mathcal{A}_B)$ consists of isolated eigenvalues of finite multiplicity, and for every $\varepsilon > 0$, all points of the spectrum $\sigma(\mathcal{A}_B)$, except, possibly, for a finite number of them, belong to the sector $|\arg\lambda| < \varepsilon$ (see Fig. I.1). The statement (a) follows from Theorem A.1. □

Let us consider an example of the nonlocal problem (1.1), (1.2), which cannot be represented in the form (1.4), (1.5).

Example 1.2. We consider the problem

$$-u''(t) + a_2(t)u(t) - \lambda u(t) = f_0(t) \qquad (t \in (0,d)), \tag{1.12}$$

$$u(0) + (B_{11}u)|_{t=0} = f_1, \qquad u(d) = f_2, \tag{1.13}$$

where $B_{11}u = F^{-1}(\arctan\xi \cdot F(\eta u))$, $\eta = \eta(t) \in \dot{C}^\infty(\mathbb{R})$, $\eta(t) = 1$ $(t \in [2\sigma, d - 2\sigma])$, $\eta(t) = 0$ $(t \notin (\sigma, d-\sigma))$, $F(v) = (Fv)(\xi)$ is the Fourier transform of v with respect to t, $F^{-1}(w) = (F^{-1}w)(t)$ is the inverse Fourier transform of w with respect to ξ, $\xi \in \mathbb{R}$.

Proof of Theorem 1.1

First we prove the following a priori estimate.

Lemma 1.1. *Let condition* 1.1 *be fulfilled.*

Then, for every $\varepsilon > 0$, *there exists* $q_1 > 1$ *such that, for* $\lambda \in \Omega_{\varepsilon,q_1}$ *and* $0 \le \tau \le 1$, *we have the estimate*

$$c_1 |||\mathcal{L}_\tau u|||_{\mathcal{W}(0,d)} \le |||u|||_{W^2(0,d)} \le c_2 |||\mathcal{L}_\tau u|||_{\mathcal{W}(0,d)} \quad (u \in W^2(0,d)), \tag{1.14}$$

where constants c_1, c_2 *do not depend on* λ, τ, *and* u.

Proof. Let us prove the right side of the inequality (1.14). Let $\mathcal{L}_\tau u = f$. Then

$$\mathcal{L}_0 u = f + \Phi, \tag{1.15}$$

where $\Phi = (-\tau A_1 u, -\tau(B_{11}u)|_{t=0} - \tau B_{12}u, -\tau(B_{21}u)|_{t=d} - \tau B_{22}u)$.

By virtue of Theorem C.11, there exists $q_0 = q_0(\varepsilon) > 1$ such that for $\lambda \in \Omega_{\varepsilon,q_0}$ the solution of the "local" problem (1.15) is estimated as

$$|||u|||_{W^2(0,d)} \le k_1 |||f + \Phi|||_{\mathcal{W}(0,d)}. \tag{1.16}$$

By virtue of condition 1.1 and the inequality (B.30), we obtain

$$\|A_1 u\|_{L_2(0,d)} \le k_2 |\lambda|^{-1/2} |||u|||_{W^2(0,d)}, \tag{1.17}$$

$$|\lambda|^{3/4} |\tau B_{\mu 2} u| \le k_3 |\lambda|^{3/4} \|u\|_{L_2(0,d)} \le k_3 |\lambda|^{-1/4} |||u|||_{W^2(0,d)}. \tag{1.18}$$

From inequalities (B.31), (B.30), and inequality (1.3) it follows that

$$\begin{aligned} |\lambda|^{3/4} \left|\tau(B_{11}u)|_{t=0}\right| &\le k_4 |\lambda|^{1/2} (\|B_{11}u\|_{W^1(0,d)} + |\lambda|^{1/2} \|B_{11}u\|_{L_2(0,d)}) \\ &\le k_5 (\|B_{11}u\|_{W^2(0,d)} + |\lambda| \cdot \|B_{11}u\|_{L_2(0,d)}) \\ &\le c k_5 (\|u\|_{W^2(\sigma,d-\sigma)} + |\lambda| \cdot \|u\|_{L_2(\sigma,d-\sigma)}). \end{aligned}$$

Let us introduce a function $\xi \in \dot{C}^\infty(0,d)$ such that $\xi(t) = 1$ $(t \in [\sigma, d-\sigma])$, $\xi(t) = 0$ $(t \notin (\sigma/2, d-\sigma/2))$. From the last inequality, Theorem C.11, Leibniz' formula and inequality (B.29) we have:

$$\begin{aligned} |\lambda|^{3/4} \left|\tau(B_{11}u)|_{t=0}\right| &\le k_6 |||u|||_{W^2(\sigma,d-\sigma)} \le k_6 |||\xi u|||_{W^2(0,d)} \\ &\le k_7 \|(A_0 - \lambda I)(\xi u)\|_{L_2(0,d)} \\ &\le k_8 (\|(A_0 + \tau A_1 - \lambda I)u\|_{L_2(0,d)} + \|u\|_{W^1(0,d)}) \\ &\le k_9 (\|(A_0 + \tau A_1 - \lambda I)u\|_{L_2(0,d)} \\ &\qquad + |\lambda|^{-1/2} |||u|||_{W^2(0,d)}). \end{aligned} \tag{1.19}$$

Similarly,

$$\begin{aligned} &|\lambda|^{3/4} \left|\tau(B_{21}u)|_{t=d}\right| \\ &\qquad \le k_9 (\|(A_0 + \tau A_1 - \lambda I)u\|_{L_2(0,d)} + |\lambda|^{-1/2} |||u|||_{W^2(0,d)}). \end{aligned} \tag{1.20}$$

From inequalities (1.16)–(1.20) it follows that

$$\begin{aligned} |||u|||_{W^2(0,d)} &\le k_1 \{(1 + 2k_9) |||\mathcal{L}_\tau u|||_{\mathcal{W}(0,d)} \\ &\qquad + (2k_3 |\lambda|^{-1/4} + (k_2 + 2k_9)|\lambda|^{-1/2}) |||u|||_{W^2(0,d)}\}. \end{aligned}$$

Hence, choosing $q_1 > q_0$ such that $k_1(2k_3 q_1^{-1/4} + (k_2 + 2k_9) q_1^{-1/2}) < 1/2$, we obtain:

$$|||u|||_{W^2(0,d)} \le 2k_1 (1 + 2k_9) |||\mathcal{L}_\tau u|||_{\mathcal{W}(0,d)}.$$

The left side of inequality (1.14) follows from condition 1.1 and inequalities (1.17)–(1.19). □

Now we restate a version of the well-known method concerning continuation along a parameter for our current use (see O. A. Ladyzhenskaya, N. N. Ural'tseva [1], Chapter III, Section 1, Lemma 1.1).

Lemma 1.2. *Let H_1, H_2 be Hilbert spaces, and let $L, L_0 \colon H_1 \to H_2$ be linear bounded operators. Assume that an operator L_0 has a bounded inverse $L_0^{-1} \colon H_2 \to H_1$, and that, for any $0 \le \tau \le 1$,*

$$C_1 \|L_\tau u\|_{H_2} \le \|u\|_{H_1} \le C_2 \|L_\tau u\|_{H_2} \qquad (u \in H_1), \tag{1.21}$$

where $L_\tau = L_0 + \tau(L - L_0)$, $C_1, C_2 > 0$ do not depend on τ and u.

Then the operator L has a bounded inverse $L^{-1} \colon H_2 \to H_1$.

Proof. By a well-known inverse operator theorem, the operator $L_\tau = L_0(I + \tau L_0^{-1}(L - L_0))$ has a bounded inverse $L_\tau^{-1} \colon H_2 \to H_1$ for $0 \le \tau \le \tau_1 = C_1/4C_2$ with the norm $\|L_\tau^{-1}\| \le C_2$. Then again the operator $L_\tau = L_{\tau_1}(I + (\tau - \tau_1) L_{\tau_1}^{-1}(L - L_0))$ has a bounded inverse $L_\tau^{-1} \colon H_2 \to H_1$ for $\tau_1 \le \tau \le 2\tau_1$ with the norm $\|L_\tau^{-1}\| \le C_2$. Repeating this procedure, after a finite number of steps we prove the existence of a bounded inverse operator $L^{-1} \colon H_2 \to H_1$. □

Combining Theorem C.11 and Lemmas 1.1, 1.2, we find that the statement (b) of Theorem 1.1 holds. Thus, by virtue of the compactness of the imbedding operator of $W^2(0,d)$ into $L_2(0,d)$ and Theorem A.1, we obtain the statement (a). The proof of Theorem 1.1 is now complete. □

Some Generalizations

Now we consider a generalization of the problem (1.4), (1.5) to a system of two differential equations with nonlocal boundary conditions. The results of this subsection will be applied to the theory of multidimensional diffusion processes (see Section 25).

We consider a system of differential equations

$$Au_i = -u_i''(t) + q^2 u_i(t) = f_{i0}(t) \qquad (t \in (a_i, b_i),\ i = 1, 2) \tag{1.22}$$

with nonlocal boundary conditions

$$\begin{aligned} B_{11}u &= u_1(a_1) - \gamma_{11} u_2(d_{21}) &= f_{11}, \\ B_{12}u &= u_1(b_1) - \gamma_{12} u_2(d_{22}) &= f_{12}, \\ B_{21}u &= u_2(a_2) - \gamma_{21} u_1(d_{11}) &= f_{21}, \\ B_{22}u &= u_2(b_2) - \gamma_{22} u_1(d_{12}) &= f_{22}. \end{aligned} \tag{1.23}$$

Here $f_{i0} \in L_2(a_i, b_i)$ are complex-valued functions, $\gamma_{ij}, f_{ij} \in \mathbb{C}$, $q \in \mathbb{C}$ is a parameter, $a_i < d_{ij} < b_i$ $(i, j = 1, 2)$.

Denote $W^{2,2}(a,b) = \prod_{i=1}^{2} W^2(a_i,b_i)$, $L_2^2(a,b) = \prod_{i=1}^{2} L_2(a_i,b_i)$, $\mathcal{W}^2(a,b) = \prod_{i=1}^{2} \mathcal{W}(a_i,b_i)$ with the norms

$$|||u|||_{W^{2,2}(a,b)} = \left\{ \sum_{i=1}^{2} |||u_i|||^2_{W^2(a_i,b_i)} \right\}^{1/2},$$

$$|||v|||_{L_2^2(a,b)} = \left\{ \sum_{i=1}^{2} \|v_i\|^2_{L_2(a_i,b_i)} \right\}^{1/2},$$

$$|||f|||_{\mathcal{W}^2(a,b)} = \left\{ \sum_{i=1}^{2} |||f_i|||^2_{\mathcal{W}(a_i,b_i)} \right\}^{1/2},$$

where $u = (u_1, u_2)$, $v = (v_1, v_2)$, $f = (f_1, f_2)$, $f_i = (f_{i0}, f_{i1}, f_{i2})$.

We introduce the bounded linear operators $\mathcal{L}(q), \mathcal{L}_0(q), \mathcal{L}_\tau(q) \colon W^{2,2}(a,b) \to \mathcal{W}^2(a,b)$ by the formulas $\mathcal{L}u = \{Au_i, B_{i1}u, B_{i2}u\}$, $\mathcal{L}_0 u = \{Au_i, u_i(a_i), u_i(b_i)\}$, $\mathcal{L}_\tau u = \mathcal{L}_0 u + \tau(\mathcal{L} - \mathcal{L}_0)u$.

The operator $\mathcal{L}_0$ corresponds to a pair of independent boundary value problems for ordinary differential equations. We set $q^2 = -\lambda$. Then, by virtue of Theorem C.11, we obtain:

Lemma 1.3. *The operator $\mathcal{L}_0(q) : W^{2,2}(a,b) \to \mathcal{W}^2(a,b)$ is Fredholm, and $\operatorname{ind} \mathcal{L}_0(q) = 0$ for all $q \in \mathbb{C}$. For every $\varepsilon > 0$, there exists $q_0 > 0$ such that, for $q \in \theta_{\varepsilon,q_0} = \{q \in \mathbb{C} : |q| \geq q_0, |\arg q| \leq (\pi - \varepsilon)/2\} \cup \{q \in \mathbb{C} : |q| \geq q_0, |\arg q - \pi| \leq (\pi - \varepsilon)/2\}$, the operator $\mathcal{L}_0(q)$ has a bounded inverse $\mathcal{L}_0^{-1}(q) \colon \mathcal{W}^2(a,b) \to W^{2,2}(a,b)$, and each $u \in W^{2,2}(a,b)$ satisfies inequality*

$$c_3 |||\mathcal{L}_0 u|||_{\mathcal{W}^2(a,b)} \leq |||u|||_{W^{2,2}(a,b)} \leq c_4 |||\mathcal{L}_0 u|||_{\mathcal{W}^2(a,b)}, \tag{1.24}$$

where constants $c_3, c_4 > 0$ do not depend on q and u.

Repeating the proof of Lemma 1.1, by virtue of Lemma 1.3, we have:

Lemma 1.4. *For every $\varepsilon > 0$, there is $q_1 > 1$ such that, for $\lambda \in \theta_{\varepsilon,q_1}$ and $0 \leq \tau \leq 1$, we have the estimate*

$$c_5 |||\mathcal{L}_\tau u|||_{\mathcal{W}^2(a,b)} \leq |||u|||_{W^{2,2}(a,b)} \leq c_6 |||\mathcal{L}_\tau u|||_{\mathcal{W}^2(a,b)}, \tag{1.25}$$

where constants $c_5, c_6 > 0$ do not depend on q and u.

Combining Lemmas 1.2–1.4 and Theorem A.9, we obtain:

Theorem 1.3.

(a) *$\mathcal{L}(q) \colon W^{2,2}(a,b) \to \mathcal{W}^2(a,b)$ is a Fredholm operator, and $\operatorname{ind} \mathcal{L}(q) = 0$ for all $q \in \mathbb{C}$.*

(b) *For any $\varepsilon > 0$, there exists $q_1 > 1$ such that for $q \in \theta_{\varepsilon,q_1}$ the operator $\mathcal{L}(q)$ has a bounded inverse $\mathcal{L}^{-1}(q) \colon \mathcal{W}^2(a,b) \to W^{2,2}(a,b)$.*

(c) *The operator function $\mathcal{L}^{-1}(q) \colon \mathcal{W}^2(a,b) \to W^{2,2}(a,b)$ is a finitely meromorphic Fredholm operator function in $\mathbb{C}$.*

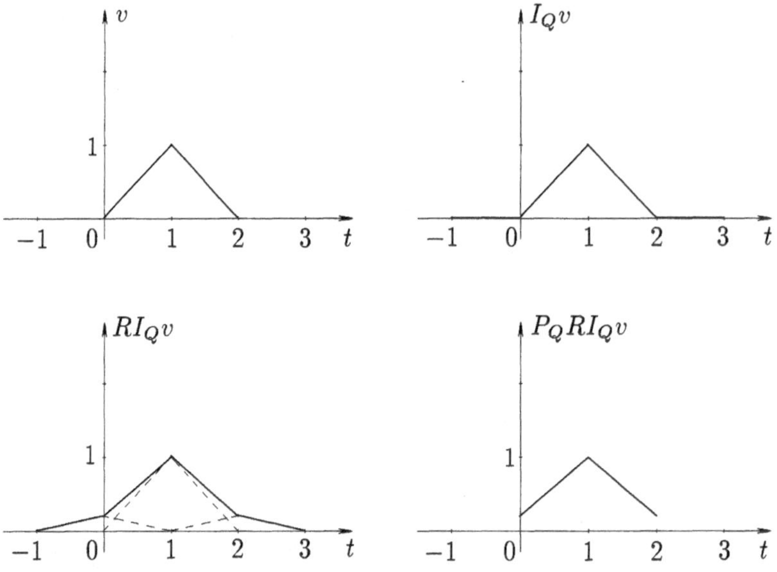

Fig. I.2

2 Difference Operators in One Dimension

Difference Operators in $L_2(\mathbb{R})$

We consider the *difference* operator $R: L_2(\mathbb{R}) \to L_2(\mathbb{R})$ defined by the formula

$$(Rv)(t) = \sum_{j=-m}^{m} b_j v(t+j). \tag{2.1}$$

Here b_j are real numbers.
We introduce the operators

$$I_Q: L_2(0,d) \to L_2(\mathbb{R}), \quad P_Q: L_2(\mathbb{R}) \to L_2(0,d), \quad R_Q: L_2(0,d) \to L_2(0,d)$$

by the formulas

$$(I_Q v)(t) = v(t) \qquad (t \in (0,d)), \qquad v(t) = 0 \quad (t \notin (0,d)), \tag{2.2}$$

$$(P_Q v)(t) = v(t) \qquad (t \in (0,d)), \tag{2.3}$$

$$R_Q = P_Q R I_Q, \tag{2.4}$$

where $Q = (0,d)$ (see Fig. I.2). These operators will be used in the study of boundary value problems for differential-difference equations (see Section 3). The shifts $t \to t+j$ can map the points of $[0,d]$ into the set $[-m,0] \cup [d, d+m]$. Hence, we must consider the boundary conditions for the differential-difference equation not only at the points $\{0\}, \{d\}$, but also on the set $[-m, 0] \cup [d, d+m]$. In order to satisfy homogeneous boundary conditions, we introduce the operator I_Q. The use of the operator P_Q is necessary to consider the equation not on the whole axis $\mathbb{R}$, but on the interval $(0, d)$ only.

Lemma 2.1. *$I_Q^* = P_Q$, $P_Q^* = I_Q$, i.e., for all $u \in L_2(0,d)$, $v \in L_2(\mathbb{R})$,*

$$(I_Q u, v)_{L_2(\mathbb{R})} = (u, P_Q v)_{L_2(0,d)}.$$

The proof follows from (2.2) and (2.3).

From Lemma 2.1 we obtain:

Lemma 2.2. *The operators*

$$R: L_2(\mathbb{R}) \to L_2(\mathbb{R}), \qquad R_Q: L_2(0,d) \to L_2(0,d)$$

are bounded, $R^* v(t) = \sum_{j=-m}^{m} b_j v(t-j)$, $R_Q^* = P_Q R^* I_Q$.

Let $A: H \to H$ be a symmetric linear operator, where H is a Hilbert space. The operator $A: H \to H$ is said to be *positive* (or *non-negative*) if $(Ax, x) > 0$ (or $(Ax, x) \geq 0$) for all $0 \neq x \in \mathcal{D}(A)$. The operator $A: H \to H$ is said to be *positive definite* if $(Ax, x) \geq c(x, x)$ for all $x \in \mathcal{D}(A)$, where $c > 0$.

Lemma 2.3. *Let R be a self-adjoint, positive operator.*

Then R_Q is a self-adjoint, positive operator.

Proof. By virtue of Lemmas 2.1 and 2.2 and the positiveness of operator R, we have

$$(R_Q v, v)_{L_2(0,d)} = (R I_Q v, I_Q v)_{L_2(\mathbb{R})} > 0 \quad \text{for } 0 \neq v \in L_2(0,d). \qquad \square$$

We denote $R(\xi) = b_0 + 2\sum_{j=1}^{m} b_j \cos j\xi$ $(\xi \in \mathbb{R})$.

Lemma 2.4. *If R given by (2.1) is self-adjoint, then it is positive if and only if*

$$0 \leq R(\xi) \not\equiv 0 \qquad (\xi \in \mathbb{R}). \tag{2.5}$$

Proof. By the Plancherel Theorem,

$$(Ru, u)_{L_2(\mathbb{R})} = \int_{-\infty}^{+\infty} (Ru)(t)\overline{u(t)}\, dt = \int_{-\infty}^{+\infty} R(\xi)|\widehat{u}(\xi)|^2\, d\xi$$

for all $u(t) \in L_2(\mathbb{R})$, where

$$\widehat{u}(\xi) = \frac{1}{\sqrt{2\pi}} \int_{-\infty}^{+\infty} u(t) \exp(-it\xi)\, dt$$

is the Fourier transform of u. Thus, the necessity is evident. Since $R(\xi) \not\equiv 0$ is an analytical function, the set $\{\xi \in \mathbb{R} : R(\xi) = 0\}$ can consist only of isolated points. Hence, $\text{mes}\{\xi \in \mathbb{R} : R(\xi) = 0\} = 0$. The sufficiency follows from this. $\square$

Difference Operators in $L_2(0,d)$

Let $d=N+\theta$, where $0<\theta\le 1$, N is a natural number.

Without loss of generality we assume that $m=N$. In fact, if $m<N$, then we can suppose that $b_j=0$ for $|j|>m$. If $m>N$, then the operator R_Q does not depend on the coefficients b_j for $|j|>N$.

If $0<\theta<1$, we denote $Q_{1k}=(k-1,k-1+\theta)$ $(k=1,\dots,N+1)$ and $Q_{2k}=(k-1+\theta,k)$ $(k=1,\dots,N)$. If $\theta=1$, we denote $Q_{1k}=(k-1,k)$ $(k=1,\dots,N+1)$. Thus, there are two classes of intervals Q_{1k} and Q_{2k} if $0<\theta<1$, and there is only one class of intervals Q_{1k} if $\theta=1$. Every two intervals of the same class can be obtained from another by a shift.

Let $P_s\colon L_2(0,d)\to L_2(\bigcup_k Q_{sk})$ be the operator of orthogonal projection onto $L_2(\bigcup_k Q_{sk})$, where $L_2(\bigcup_k Q_{sk})=\{u\in L_2(0,d):u(t)=0 \text{ for } t\in(0,d)\setminus\bigcup_k Q_{sk}\}$, P_1 is the identity operator if $\theta=1$. Evidently,

$$L_2(0,d)=\bigoplus_s L_2\Big(\bigcup_k Q_{sk}\Big). \tag{2.6}$$

Lemma 2.5. *$L_2(\bigcup_k Q_{sk})$ is an invariant subspace of the operator R_Q.*

We obtain the proof from the following property of the intervals Q_{sk}. For every interval Q_{sk} and integer j, either $Q_{sm}=Q_{sk}+j$ or $Q_{sk}+j\subset\mathbb{R}\setminus(0,d)$.

We introduce an isomorphism of the Hilbert spaces

$$U_s\colon L_2\Big(\bigcup_k Q_{sk}\Big)\to L_2^M(Q_{s1})$$

by the formula

$$(U_s v)_k(t)=v(t+k-1)\qquad (t\in Q_{s1},\ k=1,\dots,M), \tag{2.7}$$

where $L_2^M(Q_{s1})=\prod_{k=1}^M L_2(Q_{s1})$, $M=N+1$ if $s=1$, $M=N$ if $s=2$ (see Fig. I.3).

Let us denote by R_1 the matrix of order $(N+1)\times(N+1)$ with the elements

$$r_{ik}=b_{k-i}\qquad (i,k=1,\dots,N+1). \tag{2.8}$$

Denote by R_2 the matrix of order $N\times N$ obtained from R_1 by deleting the last column and the last row. In other words,

$$R_1=\begin{pmatrix} b_0 & b_1 & \dots & b_N\\ b_{-1} & b_0 & \dots & b_{N-1}\\ \vdots & \vdots & \ddots & \vdots\\ b_{-N} & b_{-N+1} & \dots & b_0\end{pmatrix},$$

$$R_2=\begin{pmatrix} b_0 & b_1 & \dots & b_{N-1}\\ b_{-1} & b_0 & \dots & b_{N-2}\\ \vdots & \vdots & \ddots & \vdots\\ b_{-N+1} & b_{-N+2} & \dots & b_0\end{pmatrix}.$$

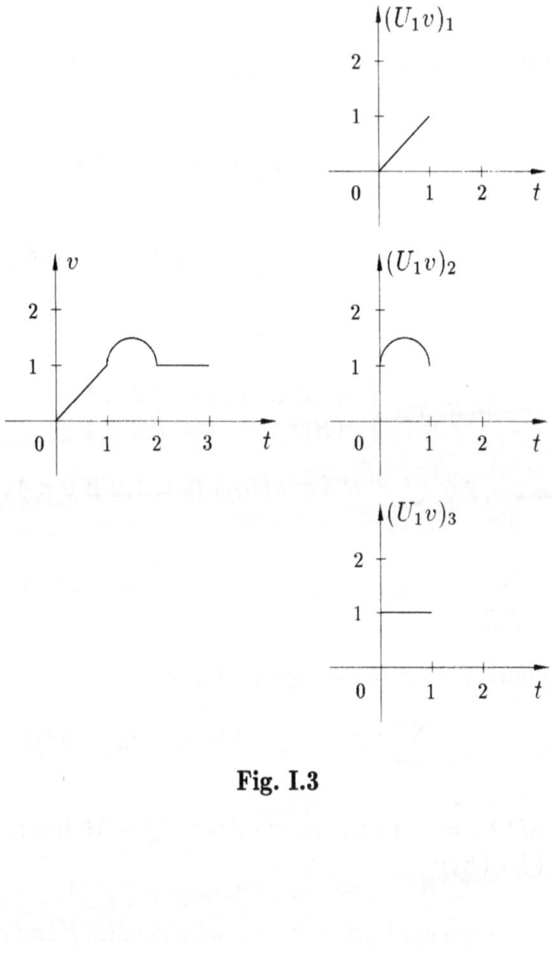

Fig. I.3

We introduce the operator

$$R_{Qs} = U_s R_Q U_s^{-1} \colon L_2^M(Q_{s1}) \to L_2^M(Q_{s1}). \tag{2.9}$$

Here and everywhere in this section $s = 1, 2$ if $\theta < 1$ and $s = 1$ if $\theta = 1$.

Lemma 2.6. *The operator R_{Qs} is the operator of multiplication by the matrix R_s.*

Proof. Let $V(t) \in L_2^M(Q_{s1})$ and $v(t) = U_s^{-1}V(t) \in L_2(\bigcup_k Q_{sk})$. Then from (2.1), (2.4), and (2.7) we have

$$\begin{aligned}(R_{Qs}V)_i(t) = (U_s R_Q v)_i(t) &= \sum_{j=-i+1}^{M-i} b_j v(t+i-1+j) \\ &= \sum_{k=1}^{M} b_{k-i} v(t+k-1) \\ &= \sum_{k=1}^{M} b_{k-i} V_k(t) \qquad (t \in Q_{s1}). \quad \square \end{aligned} \tag{2.10}$$

Lemma 2.7.

$$\sigma(R_Q) = \begin{cases} \sigma(R_1) \cup \sigma(R_2) & \textit{if } \theta < 1, \\ \sigma(R_1) & \textit{if } \theta = 1. \end{cases}$$

Proof. By Lemmas 2.5, 2.6, $\bigcup_s \sigma(R_s) \subset \sigma(R_Q)$ ($s = 1, 2$ if $\theta < 1$, $s = 1$ if $\theta = 1$). Let $\lambda \notin \bigcup_s \sigma(R_s)$. Then the operator

$$A_\lambda = \sum_s U_s^{-1}(R_{Qs} - \lambda I)^{-1} U_s P_s \colon\ L_2(0,d) \to L_2(0,d)$$

is bounded. By Lemma 2.5, $P_s R_Q = R_Q P_s$. Hence,

$$A_\lambda(R_Q - \lambda I) = \sum_s U_s^{-1}(R_{Qs} - \lambda I)^{-1} U_s (R_Q - \lambda I) P_s = I.$$

Similarly, $(R_Q - \lambda I)A_\lambda = I$. Thus, the operator $R_Q - \lambda I$ has a bounded inverse operator for $\lambda \notin \bigcup_s \sigma(R_s)$. $\square$

Lemma 2.8. *If R_Q is self-adjoint,then it is positive definite if and only if the matrix R_1 is positive definite.*

A proof follows from Lemma 2.7.
Similarly, we obtain:

Lemma 2.9. *If R_Q is self-adjoint and positive, then it is positive definite.*

Example 2.1. Let $d = 2$, and let $(Rv)(t) = 2v(t) + b(v(t+1) + v(t-1))$, where $b \in \mathbb{R}$.

Then R is self-adjoint and $R(\xi) = 2(1 + b\cos\xi)$. If $|b| \le 1$, then, by virtue of Lemmas 2.3, 2.4, 2.9, the operator $R_Q: L_2(0,2) \to L_2(0,2)$ is positive definite.

In this case we have only one class of intervals $Q_{11} = (0,1)$, $Q_{12} = (1,2)$. Evidently, the matrix

$$R_1 = \begin{pmatrix} 2 & b \\ b & 2 \end{pmatrix}.$$

By Lemma 2.8, the operator R_Q is positive definite if and only if $|b| < 2$.

To demonstrate the methods of this section, we will obtain directly the necessary and sufficient conditions of positive definiteness of the operator $R_Q: L_2(0,2) \to L_2(0,2)$ directly. In fact, passing to a new variable, we have

$$\begin{aligned} (R_Q v, v)_{L_2(0,2)} &= \int_0^1 [2v(t) + bv(t+1)]\overline{v(t)}\,dt + \int_1^2 [2v(t) + bv(t-1)]\overline{v(t)}\,dt \\ &= \int_0^1 \{[2v(t) + v(t+1)]\overline{v(t)} + [bv(t) + 2v(t+1)]\overline{v(t+1)}\}\,dt \\ &= \int_0^1 (R_1 V(t), V(t))\,dt, \end{aligned}$$

where $V(t) = \begin{pmatrix} v(t) \\ v(t+1) \end{pmatrix}$, $(\cdot,\cdot)$ is the inner product in the space $\mathbb{C}^2$.

Difference Operators in $W^k(0,d)$

Lemma 2.10. *The operator R_Q is continuous from $\mathring{W}^k(0,d)$ into $W^k(0,d)$, and, for all $v \in \mathring{W}^k(0,d)$,*

$$(R_Q v)^{(j)} = R_Q v^{(j)} \qquad (j \le k). \tag{2.11}$$

Proof. Clearly, the equality (2.11) is fulfilled for all $v \in \dot{C}^\infty(0,d)$. Thus, by Lemma 2.2, for all $v \in \dot{C}^\infty(0,d)$ we have

$$\begin{aligned} \|(R_Q v)^{(j)}\|_{L_2(0,d)} = \|R_Q v^{(j)}\|_{L_2(0,d)} &\le k_1 \|v^{(j)}\|_{L_2(0,d)} \qquad (j \le k), \\ \|R_Q v\|_{L_2(0,d)} &\le k_1 \|v\|_{L_2(0,d)}. \end{aligned}$$

Hence, for all $v \in \dot{C}^\infty(0,d)$

$$\|R_Q v\|_{W^k(0,d)} \le k_1 \|v\|_{W^k(0,d)}. \tag{2.12}$$

Since the set $\dot{C}^\infty(0,d)$ is dense in the space $\mathring{W}^k(0,d)$, from (2.12) it follows that the operator R_Q is continuous from $\mathring{W}^k(0,d)$ into $W^k(0,d)$. Hence, from the density of $\dot{C}^\infty(0,d)$ in $\mathring{W}^k(0,d)$ and equality (2.11) for $v \in \dot{C}^\infty(0,d)$ we obtain the equality (2.11) for $v \in \mathring{W}^k(0,d)$. □

Lemma 2.11. *Let a function* $a(t) \in C(\mathbb{R})$ *be* 1*-periodic.*
Then $aR_Q = R_Q a$.

Proof. Let $v \in L_2(0,d)$. Then, assuming $t \in Q_{sk}$, passing to the new variable $t' = t - k + 1$ and using formula (2.7) and 1-periodicity of the function $a(t)$, we obtain

$$\begin{aligned}(R_Q(av))(t) &= (R_Q(av))(t'+k-1) = (U_s R_Q P_s(av))_k(t') \\ &= (R_s U_s P_s(av))_k(t') = a(t')(R_s U_s P_s v)_k(t') \\ &= a(t')(R_Q v)(t) = a(t)(R_Q v)(t). \end{aligned}$$ □

Lemma 2.12. *Let the operator* $R_Q: L_2(0,d) \to L_2(0,d)$ *have a bounded inverse, and let* $R_Q v \in W^k(Q_{si})$ *for all* s *and* $i = 1, \dots, M$.
Then $v \in W^k(Q_{sj})$ *and*

$$\|v\|_{W^k(Q_{sj})} \le c \sum_{i=1}^{M} \|R_Q v\|_{W^k(Q_{si})}, \tag{2.13}$$

where $c > 0$ *does not depend on* v.

Proof. Clearly, $(U_s P_s R_Q v)_i \in W^k(Q_{s1})$ $(i = 1, \dots, M)$, i.e., $(R_s U_s P_s v)_i \in W^k(Q_{s1})$ $(i = 1, \dots, M)$. From this we have $(U_s P_s v)_j = (R_s^{-1}(R_s U_s P_s v))_j \in W^k(Q_{s1})$ $(j = 1, \dots, M)$. Thus, $v \in W^k(Q_{sj})$ and the inequality (2.13) is fulfilled. □

Let us denote by $W_\gamma^k(0,d)$ the subspace of functions from $W^k(0,d)$ satisfying conditions

$$u^{(\mu)}(0) = \sum_{i=1}^{N} \gamma_{1i} u^{(\mu)}(i), \qquad u^{(\mu)}(d) = \sum_{i=1}^{N} \gamma_{2i} u^{(\mu)}(d-i), \tag{2.14}$$

where γ_{ji} are real numbers $(j = 1, 2,\ i = 1, \dots, N)$, $\mu = 0, \dots, k-1$, $k \ge 1$.

Theorem 2.1. *Let* $\det R_j \ne 0$ $(j = 1, 2)$.
Then there exist real numbers γ_{ji} $(j = 1, 2,\ i = 1, \dots, N)$ *such that the operator* R_Q *maps* $\mathring{W}^k(0,d)$ *onto* $W_\gamma^k(0,d)$ *continuously and in a one-to-one manner.*

Proof. 1. At first we prove that there exist γ_{ji} such that $R_Q(\mathring{W}^k(0,d)) \subset W_\gamma^k(0,d)$.

We denote by R_1^1 (R_1^2) the matrix, obtained from R_1 by deleting the first (the last) column. Denote the ith row of the matrix R_1^1 (R_1^2) by $e_i(g_i)$. The matrix, obtained from R_1 by deleting the first row and the first column, coincides with the matrix R_2 obtained from R_1 by crossing out the last row and the last column. Thus, condition $\det R_2 \ne 0$ implies that

$$e_1 = \sum_{i=1}^{N} \gamma_{1i} e_{i+1}, \qquad g_{N+1} = \sum_{i=1}^{N} \gamma_{2i} g_{N+1-i}, \tag{2.15}$$

where γ_{1i}, γ_{2i} are real numbers.

By Lemma 2.10, $R_Q(\mathring{W}^k(0,d)) \subset W^k(0,d)$. Thus, (2.7), (2.15), and Lemma 2.5 imply that for $v \in \mathring{W}^k(0,d)$

$$\begin{aligned}(R_Qv)^{(\mu)}(0) &= (U_1P_1R_Qv)_1^{(\mu)}(0) = (R_1U_1P_1v^{(\mu)})_1(0)\\ &= \sum_{i=1}^{N} \gamma_{1i}(R_1U_1P_1v^{(\mu)})_{i+1}(0)\\ &= \sum_{i=1}^{N} \gamma_{1i}(R_Qv)^{(\mu)}(i) \qquad (\mu = 0,\dots,k-1). \qquad (2.16)\end{aligned}$$

Similarly,

$$(R_Qv)^{(\mu)}(d) = \sum_{i=1}^{N} \gamma_{2i}(R_Qv)^{(\mu)}(d-i) \qquad (\mu = 0,\dots,k-1). \qquad (2.17)$$

Hence,

$$R_Q(\mathring{W}^k(0,d)) \subset W^k_\gamma(0,d).$$

2. Now let us prove the inverse inclusion

$$W^k_\gamma(0,d) \subset R_Q(\mathring{W}^k(0,d)).$$

Assume $u \in W^k_\gamma(0,d)$. By virtue of Lemma 2.7, the operator $R_Q\colon L_2(0,d) \to L_2(0,d)$ has a bounded inverse $R_Q^{-1}\colon L_2(0,d) \to L_2(0,d)$. We shall show that $v = R_Q^{-1}u \in \mathring{W}^k(0,d)$.

Without loss of generality, we consider the case $\theta = 1$. In this case $s = 1$, $M = N+1$. By virtue of Lemma 2.12, $v \in W^k(Q_{1p})$ $(p = 1,\dots,N+1)$. Therefore, by virtue of Theorem B.10, it is sufficient to prove that $v^{(\mu)}|_{t=p-0} = v^{(\mu)}|_{t=p+0}$, $v^{(\mu)}|_{t=0} = v^{(\mu)}|_{t=N+1} = 0$ $(p = 1,\dots,N,\ \mu = 0,\dots,k-1)$, where

$$v^{(\mu)}|_{t=p-0} = \lim_{\substack{t\to p,\\ t<p}} v^{(\mu)}(t), \qquad v^{(\mu)}|_{t=p+0} = \lim_{\substack{t\to p,\\ t>p}} v^{(\mu)}(t).$$

Denote $\varphi_p^\mu = v^{(\mu)}|_{t=p+0}$ $(p = 0,\dots,N,\ \mu = 0,\dots,k-1)$, $\psi_j^\mu = v^{(\mu)}|_{t=j-0}$ $(j = 1,\dots,N+1,\ \mu = 0,\dots,k-1)$. Since $R_Qv \in W^k(0,d)$, then

$$(R_Qv)^{(\mu)}|_{t=p-0} = (R_Qv)^{(\mu)}|_{t=p+0} \qquad (p = 1,\dots,N,\ \mu = 0,\dots,k-1).$$

Thus, the functions φ_p^μ and ψ_j^μ satisfy the following conditions, for every $\mu = 0,\dots,k-1$,

$$\sum_{p=1}^{N+1} r_{i+1,p}\varphi_{p-1}^\mu = \sum_{p=1}^{N+1} r_{ip}\psi_p^\mu \qquad (i = 1,\dots,N). \qquad (2.18)$$

Moreover, the function $R_Q v$ satisfies the conditions (2.16), (2.17), which can be rewritten in the form

$$\sum_{p=1}^{N+1} r_{1p}\varphi^{\mu}_{p-1} = \sum_{i=1}^{N} \gamma_{1i} \sum_{p=1}^{N+1} r_{i+1,p}\varphi^{\mu}_{p-1}, \tag{2.19}$$

$$\sum_{p=1}^{N+1} r_{N+1,p}\psi^{\mu}_{p} = \sum_{i=1}^{N} \gamma_{2i} \sum_{p=1}^{N+1} r_{N+1-i,p}\psi^{\mu}_{p}. \tag{2.20}$$

From the conditions (2.15), (2.19), and (2.20) we obtain

$$\left(r_{11} - \sum_{i=1}^{N} \gamma_{1i} r_{i+1,1}\right)\varphi^{\mu}_0 = 0,$$

$$\left(r_{N+1,N+1} - \sum_{i=1}^{N} \gamma_{2i} r_{N+1-i,N+1}\right)\psi^{\mu}_{N+1} = 0.$$

The factor preceding φ^{μ}_0 (ψ^{μ}_{N+1}) is non-zero. Otherwise, by virtue of (2.15), the first (last) row of R_1 is equal to a linear combination of the remaining rows. But this is impossible since $\det R_1 \neq 0$. Hence, $\varphi^{\mu}_0 = \psi^{\mu}_{N+1} = 0$.

Thus, the system (2.18) will have the form

$$\sum_{p=1}^{N} r_{i+1,p+1}\varphi^{\mu}_p = \sum_{p=1}^{N} r_{ip}\psi^{\mu}_p \qquad (i = 1,\dots,N,\ \mu = 0,\dots,k-1).$$

Since $r_{i+1,p+1} = r_{ip}$ $(i,p = 1,\dots,N)$ and $\det R_2 \neq 0$, we obtain $\varphi^{\mu}_p = \psi^{\mu}_p$ $(p = 1,\dots,N,\ \mu = 0,\dots,k-1)$ and have proved that $W^k_{\gamma}(0,d) \subset R_Q(\mathring{W}^k(0,d))$. $\square$

Example 2.2. Let $d = 2$ and $(Rv)(t) = b_0 v(t) + b_1 v(t+1) + b_{-1} v(t-1)$, where $b_j \in \mathbb{R}$.

In this case we have one class of intervals $Q_{11} = (0,1)$, $Q_{12} = (1,2)$. Evidently, the matrix

$$R_1 = \begin{pmatrix} b_0 & b_1 \\ b_{-1} & b_0 \end{pmatrix}, \qquad R_2 = (b_0).$$

Assume that $b_0^2 - b_1 b_{-1} \neq 0$, $b_0 \neq 0$ (see Theorem 2.1). Then $R_Q(\mathring{W}^1(0,2)) = W^1_{\gamma}(0,2)$, where $W^1_{\gamma}(0,2) = \{u \in W^1(0,2) : u(0) = \gamma_{11} u(1),\ u(2) = \gamma_{21} u(1)\}$, where $\gamma_{11} = b_1/b_0$, $\gamma_{21} = b_{-1}/b_0$.

Example 2.3. Let $d = 2\frac{1}{3}$ and $(Rv)(t) = b_0 v(t) + b_1 v(t+1) + b_{-1} v(t-1)$, where $b_j \in \mathbb{R}$.

In this case we have two classes of intervals $Q_{1k} = (k-1, k-1+1/3)$ $(k = 1,2,3)$ and $Q_{2k} = (k-1+1/3, k)$ $(k = 1,2)$. The matrices R_1 and R_2 have the following form

$$R_1 = \begin{pmatrix} b_0 & b_1 & 0 \\ b_{-1} & b_0 & b_1 \\ 0 & b_{-1} & b_0 \end{pmatrix}, \qquad R_2 = \begin{pmatrix} b_0 & b_1 \\ b_{-1} & b_0 \end{pmatrix}.$$

Let $\det R_1 \neq 0$, $\det R_2 \neq 0$, i.e., $b_0(b_0^2 - 2b_1b_{-1}) \neq 0$, $b_0^2 - b_1b_{-1} \neq 0$. Then, by virtue of Theorem 2.1, $R_Q(\mathring{W}^1(0,2\frac{1}{3})) = W_\gamma^1(0,2\frac{1}{3})$, where $W_\gamma^1(0,2\frac{1}{3}) = \{u \in W^1(0,2\frac{1}{3}) : u(0) = \gamma_{11}u(1)+\gamma_{12}u(2),\ u(2\frac{1}{3}) = \gamma_{21}u(1\frac{1}{3})+\gamma_{22}u(\frac{1}{3})\}$, $\gamma_{11} = b_0b_1/\Delta$, $\gamma_{12} = -b_1^2/\Delta$, $\gamma_{21} = b_{-1}b_0/\Delta$, $\gamma_{22} = -b_{-1}^2/\Delta$, $\Delta = b_0^2 - b_1b_{-1}$.

Self-Adjoint Difference Operators

From (2.6), (2.7), and (2.9) we have

$$R_Q = \sum_s U_s^{-1} R_s U_s P_s. \tag{2.21}$$

Lemma 2.13. *If the operator $R: L_2(\mathbb{R}) \to L_2(\mathbb{R})$ is self-adjoint, then the operator $R_Q: L_2(0,d) \to L_2(0,d)$ is self-adjoint.*

The proof follows from Lemma 2.2.

Lemma 2.14. *The operator $R_Q: L_2(0,d) \to L_2(0,d)$ is self-adjoint if and only if the matrix R_1 is symmetric.*

Proof. Using the representation (2.6), the isomorphism $U_s: L_2(\bigcup_k Q_{sk}) \to L_2^M(Q_{s1})$ and the formula (2.21), we obtain

$$\begin{aligned}(R_Qu,v)_{L_2(0,d)} &= \sum_s (U_s^{-1}R_sU_sP_su, P_sv)_{L_2(\bigcup_k Q_{sk})} \\ &= \sum_s (R_sU_sP_su, U_sP_sv)_{L_2^M(Q_{s1})} \\ &= \sum_s (U_sP_su, R_s^*U_sP_sv)_{L_2^M(Q_{s1})} \\ &= \sum_s (P_su, U_s^{-1}R_s^*U_sP_sv)_{L_2(\bigcup_k Q_{sk})} \\ &= \Big(u, \sum_s U_s^{-1}R_s^*U_sP_sv\Big)_{L_2(0,d)},\end{aligned}$$

where R_s^* is the transposed matrix. Thus,

$$R_Q^* = \sum_s U_s^{-1} R_s^* U_s P_s. \tag{2.22}$$

Since the matrix R_2 is obtained from R_1 by crossing out the last column and the last row, Lemma 2.14 follows from (2.22). □

In this subsection we assume that the operator $R_Q: L_2(0,d) \to L_2(0,d)$ is self-adjoint. We denote by R_Q^R the restriction of operator R_Q to $\mathcal{R}(R_Q)$. Let $P^R: L_2(0,d) \to L_2(0,d)$ and $P_s^R: L_2^M(Q_{s1}) \to L_2^M(Q_{s1})$ be the operators of orthogonal projection onto the subspaces $\mathcal{R}(R_Q)$ and $\mathcal{R}(R_{Qs})$, respectively.

Lemma 2.15. $L_2^M(Q_{s1}) = \mathcal{N}(R_{Qs}) \oplus \mathcal{R}(R_{Qs})$.

Proof. By Lemma 2.14, the operator $R_{Qs}: L_2^M(Q_{s1}) \to L_2^M(Q_{s1})$ is self-adjoint. Therefore, it is sufficient to prove that a linear manifold $\mathcal{R}(R_{Qs})$ is closed in the space $L_2^M(Q_{s1})$. Since the matrices R_s are symmetric, we obtain $\mathbb{C}^M = \mathcal{N}(R_s) \oplus \mathcal{R}(R_s)$. Thus, $\mathcal{R}(R_s)$ is the set of vectors $y \in \mathbb{C}^M$ satisfying the equations

$$\sum_{j=1}^{M} \bar{a}_{ij} y_j = 0 \qquad (i = 1, \dots, m),$$

where $m = \dim \mathcal{N}(R_s)$, and the vectors $a_i = (a_{i1}, \dots, a_{iM})$ form an orthogonal basis in $\mathcal{N}(R_s)$. Hence, $\mathcal{R}(R_{Qs})$ is the set of vector-valued functions $Y(t) \in L_2^M(Q_{s1})$ satisfying the equations

$$\sum_{j=1}^{M} \bar{a}_{ij} Y_j(t) = 0 \qquad (i = 1, \dots, m) \tag{2.23}$$

for almost all $t \in Q_{s1}$. From this it follows that $\mathcal{R}(R_{Qs})$ is closed in $L_2^M(Q_{s1})$. □

Since the operator $R_Q: L_2(0,d) \to L_2(0,d)$ is self-adjoint, from Lemma 2.15, the formula (2.21) and Banach's inverse operator theorem, we have Lemma 2.16.

Lemma 2.16. $L_2(0,d) = \mathcal{N}(R_Q) \oplus \mathcal{R}(R_Q)$, *the operator* $R_Q^R: \mathcal{R}(R_Q) \to \mathcal{R}(R_Q)$ *has a bounded inverse, i.e., for every* $v \in L_2(Q)$

$$\|P^R v\|_{L_2(0,d)} \le c \|R_Q v\|_{L_2(0,d)}, \tag{2.24}$$

where $c > 0$ *is a constant.*

Evidently,

$$P_s^R = U_s P^R U_s^{-1}. \tag{2.25}$$

From this it follows that

$$P^R = \sum_s U_s^{-1} P_s^R U_s P_s, \tag{2.26}$$

$$U_s P_s P^R = P_s^R U_s P_s. \tag{2.27}$$

Remark 2.1. The operator $P_s^R: L_2^M(Q_{s1}) \to L_2^M(Q_{s1})$ is the operator of multiplication by some matrix. We also denote this matrix by P_s^R. Multiplication by the matrix P_s^R in the space $\mathbb{C}^M$ is the operator of orthogonal projection onto the range of the matrix R_s.

Example 2.4. Let $d = 2$ and $(Rv)(t) = v(t) + v(t+1) + v(t-1)$. In this case we have one class of intervals $Q_{11} = (0,1)$, $Q_{12} = (1,2)$. Evidently, the matrix

$$R_1 = \begin{pmatrix} 1 & 1 \\ 1 & 1 \end{pmatrix}.$$

By Lemma 2.7, $\sigma(R_Q) = \sigma(R_1) = \{0\} \cup \{2\}$. Evidently, $\mathcal{N}(R_Q) = \{v \in L_2(0,2) : v(t) = -v(t+1) \text{ for } t \in (0,1)\}$, $\mathcal{R}(R_Q) = \{v \in L_2(0,2) : v(t) = v(t+1) \text{ for } t \in (0,1)\}$.

Lemma 2.17. *Let $v \in L_2(0,d)$, and let $R_Q v \in W^k(Q_{si})$ for all s and $i = 1,\dots,M$.*

Then $P^R v \in W^k(Q_{sj})$ and

$$\|P^R v\|_{W^k(Q_{sj})} \le c \sum_{i=1}^{M} \|R_Q v\|_{W^k(Q_{si})}, \tag{2.28}$$

where $c > 0$ does not depend on v.

Proof. Evidently, $(U_s P_s R_Q v)_i \in W^k(Q_{s1})$ $(i = 1,\dots,M)$. From Lemmas 2.16, 2.5, and formula (2.9) we have

$$U_s P_s R_Q v = U_s P_s R_Q P^R v = R_s U_s P_s P^R v.$$

The operator of multiplication by the matrix R_s in $\mathbb{C}^M$ maps the image of the matrix R_s onto itself in a one-to-one manner. Therefore, there exists a matrix $\widehat{R}_s$ such that for every $y \in \mathcal{R}(R_s)$ we have $\widehat{R}_s R_s y = y$. From this, by virtue of formula (2.27) and Remark 2.1, we have

$$U_s P_s P^R v = \widehat{R}_s (R_s U_s P_s P^R v) = \widehat{R}_s (U_s P_s R_Q v). \tag{2.29}$$

Hence, $P^R v \in W^k(Q_{sj})$, and the inequality (2.28) is fulfilled. □

3 The Boundary Value Problem for the Differential-Difference Equation

Solvability

We consider the *differential-difference equation*

$$-(Rv)''(t) + A_1 v - \lambda v(t) = f_0(t) \qquad (t \in (0,d)) \tag{3.1}$$

with homogeneous boundary conditions

$$v(t) = 0 \qquad (t \in [-N,0] \cup [d, d+N]). \tag{3.2}$$

Here $R\colon L_2(\mathbb{R}) \to L_2(\mathbb{R})$ is the difference operator defined by

$$(Rv)(t) = \sum_{j=-N}^{N} b_j v(t+j),$$

$b_j \in \mathbb{R}$, $A_1 \colon \mathring{W}^1(0,d) \to L_2(0,d)$ is a linear bounded operator, $d = N+\theta$, $0 < \theta \leq 1$, N is a natural number, $f_0 \in L_2(0,d)$ is a complex-valued function, $\lambda \in \mathbb{C}$ is a spectral parameter. One can easily reduce a differential-difference equation with non-homogeneous boundary conditions to a differential-difference equation with homogeneous boundary conditions (see proofs of Theorems 3.7, 3.8). Therefore, without loss of generality, we can study the equation (3.1) with homogeneous boundary conditions (3.2). Since the shifts $t \to t+j$ can map the points of $[0,d]$ into the set $[-N,0] \cup [d,d+N]$, we consider the boundary conditions for the equation (3.1) not only at the ends of the interval $(0,d)$, but also on the set $[-N,0] \cup [d,d+N]$.

We introduce the operators $I_Q \colon L_2(0,d) \to L_2(\mathbb{R})$, $P_Q \colon L_2(\mathbb{R}) \to L_2(0,d)$ and $R_Q \colon L_2(0,d) \to L_2(0,d)$ by formulas (2.2)–(2.4). The operator I_Q allows us to consider the homogeneous boundary conditions (3.2). We use the operator P_Q to study the equation (3.1) not on the whole axis $\mathbb{R}$, but on the interval $(0,d)$.

Let $\mathcal{A}_R \colon L_2(0,d) \to L_2(0,d)$ be the unbounded operator given by

$$\left.\begin{aligned} \mathcal{D}(\mathcal{A}_R) &= \{v \in \mathring{W}^1(0,d) : R_Q v \in W^2(0,d)\}, \\ \mathcal{A}_R v &= -(R_Q v)''(t) + A_1 v \quad (v \in \mathcal{D}(\mathcal{A}_R)). \end{aligned}\right\} \tag{3.3}$$

Definition 3.1. A function $v \in \mathcal{D}(\mathcal{A}_R)$ is called a *generalized solution* of the problem (3.1), (3.2) if

$$\mathcal{A}_R v - \lambda v = f_0. \tag{3.4}$$

This definition is equivalent to the following:

Definition 3.2. A function $v \in \mathring{W}^1(0,d)$ is called a *generalized solution* of the problem (3.1), (3.2) if, for all $w \in \mathring{W}^1(0,d)$,

$$\int_0^d \{(R_Q v)'(t)\overline{w'(t)} + (A_1 v)(t)\overline{w(t)} - \lambda v(t)\overline{w(t)}\}\, dt = \int_0^d f_0(t)\overline{w(t)}\, dt. \tag{3.5}$$

In fact, if a function $v \in \mathcal{D}(\mathcal{A}_R)$ is a generalized solution of the boundary value problem (3.1), (3.2) in the sense of Definition 3.1, then multiplying (3.4) by $\overline{w(t)}$ and integrating by parts, we obtain (3.5).

Now let $v \in \mathring{W}^1(0,d)$ be a generalized solution of the problem (3.1), (3.2) in the sense of Definition 3.2. Then from (3.5), using the rule of differentiation of distributions, we have for all $w \in \dot{C}^\infty(0,d)$

$$\langle -(R_Q v)'', \overline{w} \rangle + \int_0^d (A_1 v - \lambda v)\overline{w}\, dt = \int_0^d f_0 \overline{w}\, dt. \tag{3.6}$$

Since $A_1 v - \lambda v \in L_2(0,d)$, $f_0 \in L_2(0,d)$, from (3.6) it follows that $(R_Q v)'' \in L_2(0,d)$ and the equality (3.4) is fulfilled.

Let us consider examples of the operator A_1.

Example 3.1. Let

$$A_1 v = a_1(t)(R_{1Q} v)'(t) + a_2(t)(R_{2Q} v)(t), \tag{3.7}$$

where $a_1, a_2 \in C^\infty[0,d]$ are real-valued functions, $R_{iQ} = P_Q R_i I_Q$,

$$(R_i v)(t) = \sum_{j=-N}^{N} b_{ij} v(t+j) \qquad (i = 1, 2),$$

$b_{ij} \in \mathbb{R}$. Then, by Lemmas 2.2, 2.10, the operator $A_1 \colon \mathring{W}^1(0,d) \to L_2(0,d)$ is bounded.

Example 3.2. Let

$$A_1 v = \sum_{j=1}^{J} a_{1j}(t) v'(\sigma_j(t)) + \sum_{j=1}^{J} a_{2j}(t) v(\sigma_j(t))$$
$$(v(\sigma_j(t)) = 0 \text{ for } \sigma_j(t) \notin [0,d]),$$

where $a_{ij} \in C^\infty[0,d]$, $\sigma_j, \gamma_j \in C^\infty(\mathbb{R})$ $(j = 1, \dots, J)$ are monotone real-valued functions such that γ_j is the inverse function for σ_j, $v'(\sigma_j(t)) = v'(x)|_{x=\sigma_j(t)}$. Evidently, the operator $A_1 \colon \mathring{W}^1(0,d) \to L_2(0,d)$ is bounded.

For the boundary value problem (3.1), (3.2), in which an operator A_1 has the form of Example 3.1, we shall demonstrate methods of determining generalized solutions (see Examples 3.3, 3.4).

To prove the Fredholm property of the operator $\mathcal{A}_R$, we shall reduce the problem (3.1), (3.2) to a differential equation with nonlocal boundary conditions. Then we shall apply the results of Section 1.

Theorem 3.1. *Assume that* $\det R_s \neq 0$ $(s = 1, 2)$.

Then the unbounded operator $\mathcal{A}_R \colon L_2(0,d) \to L_2(0,d)$ *is Fredholm, and* $\operatorname{ind} \mathcal{A}_R = 0$.

Proof. By Theorem 2.1, there exist numbers γ_{ji} $(j = 1, 2,\ i = 1, \dots, N)$ such that the operator R_Q maps $\mathring{W}^1(0,d)$ onto $W^1_\gamma(0,d)$ continuously and in a one-to-one manner. Here $W^1_\gamma(0,d)$ is the subspace of functions from $W^1(0,d)$ satisfying the conditions

$$u(0) = \sum_{i=1}^{N} \gamma_{1i} u(i), \qquad u(d) = \sum_{i=1}^{N} \gamma_{2i} u(d-i). \tag{3.8}$$

Thus,

$$\mathcal{A}_R = \mathcal{A}_\gamma R_Q,$$

where $\mathcal{A}_\gamma \colon W^2(0,d) \cap W^1_\gamma(0,d) \to L_2(0,d)$ is the bounded operator given by

$$\mathcal{A}_\gamma u = -u'' + A_1 R_Q^{-1} u \qquad (u \in W^2(0,d) \cap W^1_\gamma(0,d)).$$

By virtue of Theorem 1.2, the operator $\mathcal{A}_\gamma$ is Fredholm, and $\operatorname{ind} \mathcal{A}_\gamma = 0$. Hence, from Theorem A.1 it follows that $\mathcal{A}_R$ is a Fredholm operator, and $\operatorname{ind} \mathcal{A}_R = 0$. □

From Lemma 2.7 it follows that:
for $\theta < 1$, an operator $R_Q: L_2(0,d) \to L_2(0,d)$ is invertible if and only if $\det R_1 \neq 0$ and $\det R_2 \neq 0$; for $\theta = 1$, an operator $R_Q: L_2(0,d) \to L_2(0,d)$ is invertible if and only if $\det R_1 \neq 0$.

The proof of Lemma 2.15 implies that, if $0 \in \sigma(R_Q)$, then $\dim \mathcal{N}(\mathcal{A}_R) = \infty$. Hence, the operator $\mathcal{A}_R$ is not Fredholm. Thus, for $\theta < 1$, the operator $\mathcal{A}_R$ is Fredholm if and only if $0 \notin \sigma(R_Q)$. For $\theta = 1$, in Theorem 3.1 we suppose in addition that $\det R_2 \neq 0$.

Problem 3.1. *Assume that* $\theta = 1$, $0 \notin \sigma(R_Q)$, *but* $\det R_2 = 0$. *Is the operator* $\mathcal{A}_R$ *Fredholm?*

Remark 3.1. If the operator R_Q is self-adjoint, and $0 \notin \sigma(R_Q)$, then the operator $\mathcal{A}_R$ is Fredholm (see A. G. Kamenskiĭ [1]).

Smoothness of Generalized Solutions

In contrast to ordinary differential equations, the smoothness of generalized solutions of boundary value problems for differential-difference equations can be violated on the interval $(0,d)$.

Theorem 3.2. *Assume* $\det R_s \neq 0$ ($s = 1,2$ *if* $\theta < 1$, $s = 1$ *if* $\theta = 1$). *Let* $f_0 \in W^k(0,d)$, *and let an operator* A_1 *have the form* (3.7) *if* $k \geq 1$. *Let* $v(t)$ *be a generalized solution of the boundary value problem* (3.1), (3.2).

Then $v \in W^{k+2}(j-1,j)$ $(j = 1,\dots,N+1)$ *if* $\theta = 1$, *and* $v \in W^{k+2}(j-1, j-1+\theta)$ $(j = 1,\dots,N+1)$, $v \in W^{k+2}(j-1+\theta, j)$ $(j = 1,\dots,N)$ *if* $\theta < 1$.

The proof follows from Definition 3.1 and Lemmas 2.7, 2.12.

Let us consider the example, in which the smoothness of the generalized solution is violated even for an infinitely differentiable right-hand side of the equation. This example also shows how to find a generalized solution of the problem (3.1), (3.2) by means of reducing to a differential equation with nonlocal boundary conditions.

Example 3.3. We consider the boundary value problem

$$-(Rv)''(t) = 1 \qquad (t \in (0,2)), \tag{3.9}$$

$$v(t) = 0 \qquad (t \in [-1,0] \cup [2,3]), \tag{3.10}$$

where $(Rv)(t) = v(t) + bv(t+1) + bv(t-1)$, $|b| \neq 1$. Then $R_1 = \begin{pmatrix} 1 & b \\ b & 1 \end{pmatrix}$ and $\det R_1 \neq 0$ (see Example 2.2). By virtue of Theorem 2.1, the operator R_Q maps $\mathring{W}^1(0,2)$ onto $W^1_\gamma(0,2)$ continuously and in a one-to-one manner, where $W^1_\gamma(0,2)$ is the subspace of functions $u \in W^1(0,2)$ such that

$$u(0) = bu(1) = u(2). \tag{3.11}$$

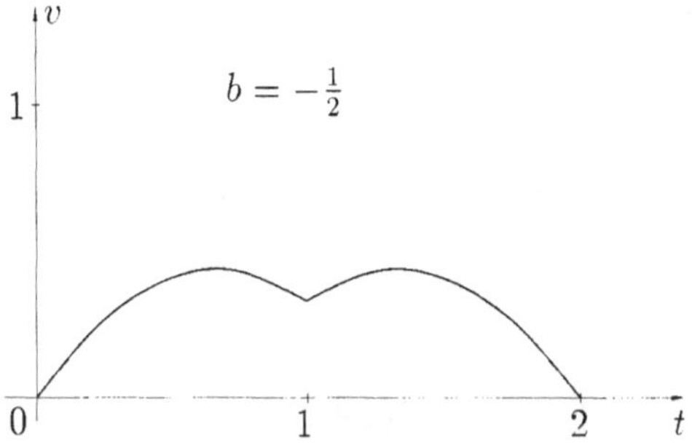

Fig. I.4

Thus, the boundary value problem (3.9), (3.10) is equivalent to the equation

$$-u''(t) = 1 \tag{3.12}$$

with nonlocal conditions (3.11). Substituting a general solution of the equation (3.12) into (3.11), we see that there exists a unique generalized solution of the problem (3.12), (3.11)

$$u(t) = -t^2/2 + t + b/(2(1-b)).$$

Therefore, we can find a generalized solution of the problem (3.9), (3.10) in the following way: $v(t) = R_Q^{-1}u(t)$. Since

$$R_1^{-1} = \frac{1}{1-b^2}\begin{pmatrix} 1 & -b \\ -b & 1 \end{pmatrix},$$

then $R_Q^{-1} = P_Q R' I_Q$, where $(R'w)(t) = [w(t) - bw(t+1) - bw(t-1)]/(1-b^2)$.

Thus, there exists a unique generalized solution of the problem (3.9), (3.10)

$$v(t) = \begin{cases} -\dfrac{t^2}{2(1+b)} + \dfrac{t}{1-b^2} & (t \in (0,1)), \\ -\dfrac{t^2}{2(1+b)} + \dfrac{1-2b}{1-b^2}t + \dfrac{2b}{1-b^2} & (t \in (1,2)) \end{cases} \tag{3.13}$$

(see Fig. I.4).

Evidently, the derivative $v'(t)$ has a discontinuity at the point $t = 1$ if $b \neq 0$, i.e., $v \in \mathring{W}^1(0,2) \setminus W^2(0,2)$.

We now demonstrate another way of determining a generalized solution of a boundary value problem for a differential-difference equation.

Example 3.4. We again consider the boundary value problem (3.9), (3.10).

Let us introduce the new functions

$$\begin{aligned} v_1(t) &= v(t) \qquad (t \in (0,1)), \\ v_2(t) &= v(t+1) \qquad (t \in (0,1)). \end{aligned}$$

Then from (3.9) and (3.10) it follows that

$$\begin{aligned} -v_1''(t) - bv_2''(t) &= 1 \quad (t \in (0,1)), \\ -bv_1''(t) - v_2''(t) &= 1 \quad (t \in (0,1)). \end{aligned} \tag{3.14}$$

A general solution of the system (3.14) has the form:

$$\begin{aligned} v_1(t) &= -t^2/(2(1+b)) + c_1 t + c_2, \\ v_2(t) &= -t^2/(2(1+b)) + c_3 t + c_4. \end{aligned} \tag{3.15}$$

Since $v \in \mathring{W}^1(0,2)$ and $\mathring{W}^1(0,2) \subset C[0,2]$, we obtain

$$v(0) = v(2) = 0, \qquad v(1-0) = v(1+0).$$

The condition $R_Q v \in W^2(0,2)$ implies that

$$(R_Q v)'(1-0) = (R_Q v)'(1+0).$$

We can rewrite these conditions in the following form:

$$\begin{aligned} v_1(0) &= 0, \\ v_2(1) &= 0, \\ v_1(1) &= v_2(0), \\ v_1'(1) + bv_2'(1) &= bv_1'(0) + v_2'(0). \end{aligned} \tag{3.16}$$

From the system (3.16) we find constants $c_1, \dots, c_4$. Substituting these constants into (3.15), we have

$$\begin{aligned} v_1(t) &= -\frac{t^2}{2(1+b)} + \frac{t}{1-b^2} && (t \in (0,1)), \\ v_2(t) &= -\frac{t^2}{2(1+b)} + \frac{-bt}{1-b^2} + \frac{1}{2(1-b)} && (t \in (0,1)). \end{aligned} \tag{3.17}$$

From (3.17) we obtain (3.13).

Spectrum

We introduce the sesquilinear form $b_R[u,v]$ with domain $\mathcal{D}(b_R) = \mathring{W}^1(0,d)$ by the formula

$$b_R[u,v] = \int_0^d \{(R_Q u)'(t)\overline{v'(t)} + (A_1 u)(t)\overline{v(t)}\}\, dt. \tag{3.18}$$

Lemma 3.1. *Let the matrix $R_1 + R_1^*$ be positive definite.*

Then there exist constants $c_1 > 0$ and $c_2 \geq 0$ such that

$$\operatorname{Re} b_R[u] \geq c_1 \|u\|^2_{\mathring{W}^1(0,d)} - c_2 \|u\|^2_{L_2(0,d)} \qquad (u \in \mathring{W}^1(0,d)). \tag{3.19}$$

Proof. Since $A_1: \mathring{W}^1(0,d) \to L_2(0,d)$ is a bounded operator, from Lemmas 2.8, 2.10, and the Schwarz inequality, it follows that

$$\begin{aligned}\operatorname{Re} b_R[u] &= \frac{1}{2}((R_Q + R_Q^*)u', u')_{L_2(0,d)} + \operatorname{Re}(A_1 u, u)_{L_2(0,d)} \\ &\geq k_1 \|u'\|^2_{L_2(0,d)} - k_2 \|u\|_{\mathring{W}^1(0,d)} \|u\|_{L_2(0,d)}.\end{aligned}$$

Hence, using the inequality

$$ab \leq \varepsilon a^2 + \varepsilon^{-1} b^2 \qquad (a, b, \varepsilon \in \mathbb{R})$$

and the equivalent norm in $\mathring{W}^1(0,d)$, we obtain

$$\operatorname{Re} b_R[u] \geq (k_3 - k_2\varepsilon)\|u\|^2_{\mathring{W}^1(0,d)} - k_2\varepsilon^{-1}\|u\|^2_{L_2(0,d)}.$$

We now choose ε such that $\varepsilon < k_3/2k_2$. □

In order to study an adjoint operator $\mathcal{A}_R^*: L_2(0,d) \to L_2(0,d)$, we assume that the operator A_1 satisfies the following condition:

3.1. *There exists a linear bounded operator* $A_1^+: \mathring{W}^1(0,d) \to L_2(0,d)$ *such that, for all* $v, w \in \mathring{W}^1(0,d)$,

$$(A_1 v, w)_{L_2(0,d)} = (v, A_1^+ w)_{L_2(0,d)}. \tag{3.20}$$

Example 3.5. Let A_1 be the operator defined in Example 3.1. We introduce the operator $A_1^+ w = -R_{1Q}^*(a_1 w)' + R_{2Q}^*(a_2 w)$. By Lemma 2.2, the operator $A_1^+: \mathring{W}^1(0,d) \to L_2(0,d)$ is bounded and satisfies the identity (3.20).

Example 3.6. Let A_1 be an operator defined in Example 3.2. We introduce the operator

$$A_1^+ w = -\frac{d}{dx}\sum_{j=1}^{J} a_{1j}(\gamma_j(x))|\gamma_j'(x)|w(\gamma_j(x)) + \sum_{j=1}^{J} a_{2j}(\gamma_j(x))|\gamma_j'(x)|w(\gamma_j(x)),$$

where $w(\gamma_j(x)) = 0$ for $\gamma_j(x) \notin [0,d]$. Clearly, the operator $A_1^+: \mathring{W}^1(0,d) \to L_2(0,d)$ is bounded. Changing the variables $x = \sigma_j(t)$ and integrating by parts, we obtain (3.20).

Together with $\mathcal{A}_R$ we also consider the unbounded operator $\mathcal{A}_R^+: L_2(0,d) \to L_2(0,d)$ given by

$$\left.\begin{aligned}\mathcal{A}_R^+ u(t) &= -(R_Q^* u)''(t) + A_1^+ u, \\ \mathcal{D}(\mathcal{A}_R^+) &= \{u \in \mathring{W}^1(0,d) : R_Q^* u \in W^2(0,d)\}.\end{aligned}\right\} \tag{3.21}$$

Theorem 3.3. *Let the matrix* $R_1 + R_1^*$ *be positive definite.*

Then we have:

(a) *$\mathcal{A}_R: L_2(0,d) \to L_2(0,d)$ is the m-sectorial operator associated with the form b_R.*
(b) *The operator $\mathcal{A}_R: L_2(0,d) \to L_2(0,d)$ is Fredholm, and* $\operatorname{ind} \mathcal{A}_R = 0$.
(c) *The spectrum $\sigma(\mathcal{A}_R)$ is discrete, and $\sigma(\mathcal{A}_R) \subset \{\lambda \in \mathbb{C} : \operatorname{Re}\lambda > -c_2\}$, where $c_2 \geq 0$ is a constant in Lemma* 3.1.
(d) *If $\lambda \notin \sigma(\mathcal{A}_R)$, then the resolvent $R(\lambda, \mathcal{A}_R): L_2(0,d) \to L_2(0,d)$ is a compact operator.*
(e) *If the condition* 3.1 *holds, then $\mathcal{A}_R^* = \mathcal{A}_R^+$, $(\mathcal{A}_R^+)^* = \mathcal{A}_R$.*

Proof. Lemmas 2.10, 3.1 imply that b_R is a bounded sectorial form on $\mathring{W}^1(0,d)$ with a vertex $\gamma = -c_2$. We denote by B_{b_R} the m-sectorial operator associated with b_R. Since Definitions 3.1, 3.2 are equivalent, by virtue of Theorem A.10, we obtain $\mathcal{A}_R = B_{b_R}$. Similarly, using the condition 3.1, we have $\mathcal{A}_R^+ = B_{b_R^*}$. Now the statements of Theorem 3.3 follow from Theorems A.14, A.15 and compactness of the imbedding operator from $\mathring{W}^1(0,d)$ into $L_2(0,d)$. □

Theorem 3.4. *Let the matrix $R_1 + R_1^*$ be positive definite, and let*

$$\operatorname{Re}(A_1 u, u)_{L_2(0,d)} \geq 0 \qquad (u \in \mathring{W}^1(0,d)). \tag{3.22}$$

Then there exists a unique generalized solution of the boundary value problem (3.1), (3.2).

Proof. By virtue of (3.22), we can set $c_2 = 0$ in Lemma 3.1. Thus, by Theorem 3.3, $\sigma(\mathcal{A}_R) \subset \{\lambda \in \mathbb{C} : \operatorname{Re}\lambda > 0\}$, i.e., $0 \notin \sigma(\mathcal{A}_R)$. □

Theorem 3.5. *Let the matrix R_1 be symmetric and positive definite. Assume that the condition* 3.1 *holds, and $A_1 = A_1^+$.*

Then the operator $\mathcal{A}_R: L_2(0,d) \to L_2(0,d)$ is self-adjoint, the spectrum $\sigma(\mathcal{A}_R)$ consists of real isolated eigenvalues of finite multiplicity $-c_2 < \lambda_1 \leq \lambda_2 \leq \ldots \leq \lambda_s \leq \ldots$. There exists an orthonormal basis $\{v_s\}$ in $L_2(0,d)$ consisting of eigenfunctions of $\mathcal{A}_R$. Moreover, the functions $\{v_s/\sqrt{\lambda_s + c_2}\}$ form an orthonormal basis in $\mathring{W}^1(0,d)$ with inner product given by

$$(u,v)'_{\mathring{W}^1(0,d)} = b_R[u,v] + c_2(u,v)_{L_2(0,d)}. \tag{3.23}$$

A proof follows from Lemmas 2.14, 3.1, and Theorem A.16.

Remark 3.2. Let $A_R: L_2(0,d) \to L_2(0,d)$ be an unbounded operator given by

$$\left.\begin{aligned} &\mathcal{D}(A_R) = \dot{C}^\infty(0,d), \\ &A_R v = -(R_Q v)''(t) + A_1 v \quad (v \in \mathcal{D}(A_R)). \end{aligned}\right\} \tag{3.24}$$

Then $\mathcal{A}_R$ is a Friedrichs extension of A_R.

Theorem 3.6. *Let the matrix R_1 be symmetric, and let* $\det R_s \neq 0$ $(s = 1, 2)$. *Assume that the condition* 3.1 *is fulfilled, and* $A_1 = A_1^+$.

Then we have:

(a) *The operator* $\mathcal{A}_R \colon L_2(0,d) \to L_2(0,d)$ *is self-adjoint.*

(b) *The spectrum* $\sigma(\mathcal{A}_R)$ *is discrete.*

(c) *For* $\lambda \notin \sigma(\mathcal{A}_R)$, *the resolvent* $R(\lambda, \mathcal{A}_R) \colon L_2(0,d) \to L_2(0,d)$ *is a compact operator.*

Proof. By Theorem 2.1, the operator R_Q maps $\mathring{W}^1(0,d)$ onto $W_\gamma^1(0,d)$ continuously and in a one-to-one manner, where $W_\gamma^1(0,d)$ is the subspace of functions from $W^1(0,d)$ satisfying the conditions (3.8). From Theorem 1.2 it follows that there exists $\mu \in \mathbb{R}$ such that the operator $\mathcal{A}_\gamma - \mu I \colon W^2(0,d) \cap W_\gamma^1(0,d) \to L_2(0,d)$ has a bounded inverse $(\mathcal{A}_\gamma - \mu I)^{-1} \colon L_2(0,d) \to W^2(0,d) \cap W_\gamma^1(0,d)$, where $\mathcal{A}_\gamma u = -u'' + A_1 R_Q^{-1} u$. Hence, the unbounded operator $\mathcal{A}_R - \mu R_Q \colon L_2(0,d) \to L_2(0,d)$ has a bounded inverse $(\mathcal{A}_R - \mu R_Q)^{-1} \colon L_2(0,d) \to \mathring{W}^1(0,d)$. By Theorem A.8 concerning an operator with compact resolvent, the operator $\mathcal{A}_R - \mu R_Q$ has a discrete spectrum. On the other hand, using Lemmas 2.14, 2.10 and integrating by parts, we have

$$((\mathcal{A}_R - \mu R_Q)u, v)_{L_2(0,d)} = (u, (\mathcal{A}_R - \mu R_Q)v)_{L_2(0,d)}$$

for $u, v \in \mathcal{D}(\mathcal{A}_R)$. Hence, by virtue of Theorem A.6, the operator $\mathcal{A}_R - \mu R_Q$ is self-adjoint. Thus, by virtue of Lemmas 2.2 and 2.14, the operator $\mathcal{A}_R = (\mathcal{A}_R - \mu R_Q) + \mu R_Q$ is also self-adjoint. Let $\lambda \in \mathbb{C} \setminus \mathbb{R}$. Then there exist the bounded operators $(\mathcal{A}_R - \lambda I)^{-1} \colon L_2(0,d) \to L_2(0,d)$ and $(\mathcal{A}_R - \mu R_Q)^{-1} \colon L_2(0,d) \to \mathring{W}^1(0,d)$. From this it follows that

$$(\mathcal{A}_R - \lambda I)^{-1} = (\mathcal{A}_R - \mu R_Q)^{-1} (I + (\mu R_Q - \lambda I)(\mathcal{A}_R - \mu R_Q)^{-1})^{-1}.$$

Hence, $(\mathcal{A}_R - \lambda I)^{-1} \colon L_2(0,d) \to L_2(0,d)$ is a compact operator. Now it suffices to apply Theorem A.8 again. □

Problem 3.2. *Let* $\det R_s \neq 0$ $(s = 1, 2)$, *and let condition* 3.1 *be fulfilled. Is the spectrum* $\sigma(\mathcal{A}_R)$ *discrete?*

Problems 3.1 and 3.2 are closely associated with the following problem.

Problem 3.3. *Study the solvability and the spectrum of a differential-difference equation of order* m *with general boundary conditions.*

As the following example shows, the spectrum $\sigma(\mathcal{A}_R)$ is not in general semi-bounded.

Example 3.7. Let $\mathcal{A}_R v = -(R_Q v)''$, where $d = 2$, $(Rv)(t) = v(t) + 2v(t+1) + 2v(t-1)$. Then $R_1 = \begin{pmatrix} 1 & 2 \\ 2 & 1 \end{pmatrix}$. Thus, the conditions of Theorem 3.6 are fulfilled.

Denote by $\omega_n(t)$ the functions such that $\omega_n(t) \in \dot{C}^\infty(\mathbb{R}^n)$, $\operatorname{supp}\omega_n \subset (0,1)$, $\|\omega_n\|_{L_2(0,1)} = 1$, $\|\omega_n'\|_{L_2(0,1)} \to +\infty$ as $n \to +\infty$. Define the functions u_n and v_n by the formulas

$$\begin{aligned} u_n(t) &= \quad v_n(t) = \omega_n(t) && (t \in [0,1]), \\ u_n(t) &= -v_n(t) = \omega_n(t-1) && (t \in [1,2]). \end{aligned}$$

Then $(\mathcal{A}_R u_n, u_n)_{L_2(0,2)} = 6\|\omega_n'\|^2_{L_2(0,1)} \to +\infty$ as $n \to +\infty$, $(\mathcal{A}_R v_n, v_n)_{L_2(0,2)} = -2\|\omega_n'\|^2_{L_2(0,1)} \to -\infty$ as $n \to +\infty$, $\|u_n\|^2_{L_2(0,2)} = \|v_n\|^2_{L_2(0,2)} = 2$. Hence, by virtue of a spectral theorem, $\sigma(\mathcal{A}_R)$ is not semi-bounded.

Remark 3.3. The functional-differential operators A_1, defined in Examples 3.1 and 3.2, are bounded from $\mathring{W}^1(0,d)$ into $L_2(0,d)$ and satisfy the condition 3.1. Thus, if the corresponding conditions for the operator R_Q are fulfilled, then Theorems 3.1–3.3 are valid.

Let us consider when the conditions of Theorems 3.4–3.6 hold in the case of Example 3.1.

Example 3.8. Let $A_1 v = a_2(t)(R_{2Q}v)(t)$, $R_{2Q} = P_Q R_2 I_Q$, $(R_2 v)(t) = \sum_{j=-N}^{N} b_{2j}\, v(t+j)$, $b_{2j} \in \mathbb{R}$, $a_2(t) \in C^\infty(\mathbb{R})$ is a 1-periodic non-negative function. Assume that the matrix $R_{21} + R_{21}^*$ is non-negative, where R_{21} corresponds to the operator R_{2Q}.

Then the condition (3.22) of Theorem 3.4 is fulfilled.

In fact, by Lemmas 2.11, 2.7, we have

$$(a_2(R_{2Q} + R_{2Q}^*)v, v)_{L_2(0,d)} \geq 0 \qquad (v \in L_2(0,d)).$$

Example 3.9. Let A_1 be a differential-difference operator defined in Example 3.1, where $a_1(t)$ is a real constant, $a_2(t)$ is a 1-periodic function. Denote by R_{11} and R_{21} the matrices corresponding to the operators R_{1Q} and R_{2Q}, respectively. Assume that R_{21} is symmetric and R_{11} is skew-symmetric.

Then the operator A_1 satisfies the conditions of Theorems 3.5 and 3.6.

In fact, by Lemmas 2.10, 2.11, and 2.14,

$$(A_1 u, v)_{L_2(0,d)} = (u, A_1 v)_{L_2(0,d)} \qquad (u, v \in \mathring{W}^1(0,d)).$$

Non-Integer Shifts

Theorems 3.1–3.6 can be generalized if the shifts in the difference operator are commensurable. When the shifts are incommensurable, Theorem 3.2 concerning the smoothness of generalized solutions and some other results are not true.

Example 3.10. We consider the boundary value problem

$$-(v - \varepsilon R v)''(t) = f_0(t) \qquad (t \in (0,\pi)), \tag{3.25}$$

$$v(t) = 0 \qquad (t \in [-1,0] \cup [\pi, \pi+1]), \tag{3.26}$$

where $(Rv)(t) = v(t-1) + v(t+\tau)$, τ is an irrational number such that $0.9 < \tau < 1$ and, for all integers p, q, $p - q\tau \neq \pi$, $0 < \varepsilon < 1/4$ is sufficiently small, $f_0 \in L_2(0,\pi)$.

1. As in Definition 3.1, a function $v \in \mathring{W}^1(0,\pi)$ is called a generalized solution of the problem (3.25), (3.26) if $v - \varepsilon R_Q v \in W^2(0,\pi)$ and

$$-(v - \varepsilon R_Q v)''(t) = f(t) \qquad (t \in (0,\pi)). \tag{3.27}$$

This definition is equivalent to the following:

A function $v \in \mathring{W}^1(0,\pi)$ is called a generalized solution of the problem (3.25), (3.26) if

$$(v' - \varepsilon(R_Q v)', \varphi')_{L_2(0,\pi)} = (f_0, \varphi)_{L_2(0,\pi)} \tag{3.28}$$

for all $\varphi \in \mathring{W}^1(0,\pi)$.

Since the operator $R_Q = P_Q R I_Q \colon L_2(0,\pi) \to L_2(0,\pi)$ is bounded, by virtue of the Riesz theorem concerning a general form of linear functional in a Hilbert space, there exist bounded operators $B \colon \mathring{W}^1(0,\pi) \to \mathring{W}^1(0,\pi)$ and $G \colon L_2(0,\pi) \to \mathring{W}^1(0,\pi)$ such that

$$\begin{aligned} ((R_Q v)', \varphi')_{L_2(0,\pi)} &= (Bv, \varphi)'_{\mathring{W}^1(0,\pi)}, \\ (f_0, \varphi)_{L_2(0,\pi)} &= (Gf_0, \varphi)'_{\mathring{W}^1(0,\pi)}. \end{aligned}$$

Here $(u, w)'_{\mathring{W}^1(0,\pi)} = (u', w')_{L_2(0,\pi)}$ is the equivalent inner product in the space $\mathring{W}^1(0,\pi)$.

Hence, the identity (3.28) will have the form

$$(v - \varepsilon Bv, \varphi)'_{\mathring{W}^1(0,\pi)} = (Gf_0, \varphi)'_{\mathring{W}^1(0,\pi)}.$$

From this we obtain

$$v - \varepsilon Bv = Gf_0. \tag{3.29}$$

If $\varepsilon < \|B\|^{-1}$, then the operator $I - \varepsilon B$ has a bounded inverse. Thus, the boundary value problem (3.25), (3.26) has a unique generalized solution $v = (I - \varepsilon B)^{-1} G f_0$.

2. Let $v \in \mathring{W}^1(0,\pi)$ be a generalized solution of the boundary value problem (3.25), (3.26), where $f_0(t) \equiv 1$. As in Lemma 2.10, $(R_Q v)' = R_Q v'$. Therefore, from (3.27) it follows that

$$(I - \varepsilon R_Q) v'(t) = \Phi_\varepsilon(t) \qquad (t \in (0,\pi)), \tag{3.30}$$

where $\Phi_\varepsilon(t) = -t + c_\varepsilon$, c_ε is some constant.

The operator $R_Q \colon L \to L$ is bounded and $\|R_Q\|_L \le 2$, where $L = L_2(0,\pi)$ or $L = L_\infty(0,\pi)$. Therefore, $\varepsilon \|R_Q\|_L \le 1/2$ and

$$v' = \sum_{j=0}^{\infty} (\varepsilon R_Q)^j \Phi_\varepsilon, \tag{3.31}$$

where the series (3.31) converges in L.

Evidently, either $|\Phi_\varepsilon(0+0)| \geq \pi/2$, or $|\Phi_\varepsilon(\pi-0)| \geq \pi/2$, where

$$\varphi(a+0) = \lim_{\substack{t\to a,\\ t>a}} \varphi(t), \quad \varphi(a-0) = \lim_{\substack{t\to a,\\ t<a}} \varphi(t), \qquad a \in \mathbb{R}.$$

Without loss of generality, we assume that

$$|\Phi_\varepsilon(0+0)| \geq \pi/2.$$

Then

$$|(I_Q\Phi_\varepsilon)(0+0) - (I_Q\Phi_\varepsilon)(0-0)| \geq \pi/2. \tag{3.32}$$

Let us denote $h(p,q) = p - q\tau$, where $p, q \geq 0$ are integers. We define the set M consisting of zero and numbers $h(p,q) \in (0,\pi)$ such that either $h(p,q) = g+1$ for some $g \in M$, or $h(p,q) = l - \tau$ for some $l \in M$. Since τ is an irrational number,

$$h(p_1,q_1) \neq h(p_2,q_2) \qquad ((p_1,q_1) \neq (p_2,q_2)). \tag{3.33}$$

We shall prove that the derivative $v'(t)$ is not continuous at the points of the set $M \cap (0,\pi)$.

By definition of the set M, the shift of argument $-p+q\tau$ ($p-q\tau \in M$, $p+q=j$) transforms the discontinuity of the function $I_Q\Phi_\varepsilon$ at the point 0 into a discontinuity of the function $R_Q^j\Phi_\varepsilon$ at the point $p-q\tau \in (0,\pi)$. By virtue of (3.33), the shift of argument $-p_1+q_1\tau$ of the function $R_Q^j\Phi_\varepsilon$ ($p_1+q_1=j$, $(p_1,q_1) \neq (p,q)$) cannot eliminate the discontinuity of this function at the point $p-q\tau$. At the same time the shifts of the point π do not belong to the set M and cannot eliminate the discontinuity of the function $R_Q^j\Phi_\varepsilon$ at the points of the set M.

But there is also a possibility of obtaining discontinuities of the function $R_Q^i\Phi_\varepsilon (i>j)$ at the point $p-q\tau \in M$. Generally speaking, $(R_Q^{i-j}\Phi_\varepsilon)(0+0) \neq 0$, i.e., $(I_QR_Q^{i-j}\Phi_\varepsilon)(0+0) - (I_QR_Q^{i-j}\Phi_\varepsilon)(0-0) \neq 0$. The shift of argument $-p+q\tau$ ($p-q\tau \in M$, $p+q=j$) transforms the discontinuity of the function $I_QR_Q^{i-j}\Phi_\varepsilon$ at the point 0 into a discontinuity of the function $R_Q^i\Phi_\varepsilon$ at the point $p-q\tau \in (0,\pi)$.

Let us consider an arbitrary point $p-q\tau \in M$. We shall estimate the discontinuities of the functions $(\varepsilon R_Q)^k\Phi_\varepsilon$ at the point $p-q\tau$. Denote $\Delta_{pq}F = F(p-q\tau+0) - F(p-q\tau-0)$. We can rewrite (3.31) in the following form

$$v' = \sum_{k=0}^{J} (\varepsilon R_Q)^k \Phi_\varepsilon + \sum_{k=J+1}^{\infty} (\varepsilon R_Q)^k \Phi_\varepsilon, \tag{3.34}$$

where $J > p+q$ will be chosen later. Then

$$\Delta_{pq}\left(\sum_{k=0}^{J} (\varepsilon R_Q)^k \Phi_\varepsilon\right) = S_1 + S_2, \tag{3.35}$$

where $S_1 = \varepsilon^{p+q} n_{pq}\Delta_{00}(I_Q\Phi_\varepsilon)$, $S_2 = \sum_{k=1}^{J-p-q} \varepsilon^{p+q+k} n_{pq}\Delta_{00}(I_Q R_Q^k \Phi_\varepsilon)$, n_{pq} is the number of all sequences $(0,0),(r_1,s_1),\dots,(r_{p+q},s_{p+q})$ such that $r_l - s_l\tau \in M$ $(l = 1,\dots,p+q)$, $r_l \le r_{l+1}$, $s_l \le s_{l+1}$, $s_{l+1} - s_l + r_{l+1} - r_l = 1$, $r_{p+q} = p$, $s_{p+q} = q$.

By virtue of (3.32), $|S_1| \ge \varepsilon^{p+q} n_{pq}\pi/2$. Evidently,

$$|S_2| \le \varepsilon^{p+q} n_{pq} \|\Phi_\varepsilon\|_{L_\infty(0,\pi)} \sum_{k=1}^{\infty} \varepsilon^k 2^k \le \varepsilon^{p+q+1} n_{pq} 4 \|\Phi_\varepsilon\|_{L_\infty(0,\pi)}.$$

We show that $\|\Phi_\varepsilon\|_{L_\infty(0,\pi)} \le c$ for all $0 < \varepsilon < \min\{\|B\|^{-1}/2, 1/4\}$.

To obtain this, we prove that $\|\Phi_\varepsilon\|_{L_2(0,\pi)} \le c_1$. In fact, by virtue of (3.29), (3.30), we have

$$\begin{aligned}\|\Phi_\varepsilon\|_{L_2(0,\pi)} &\le \|I - \varepsilon R_Q\| \cdot \|v'\|_{L_2(0,\pi)} \le \frac{3}{2}\|v\|'_{\mathring{W}^1(0,\pi)} \\ &\le \frac{3}{2}\|(I - \varepsilon B)^{-1}\| \cdot \|Gf_0\|'_{\mathring{W}^1(0,\pi)} \le \frac{3}{2} \cdot 2\|G\| \cdot \|f_0\|_{L_2(0,\pi)}.\end{aligned}$$

Thus, we can choose $\varepsilon > 0$ so that $\varepsilon 4 \|\Phi_\varepsilon\|_{L_\infty(0,\pi)} \le \pi/4$. Then

$$\left|\Delta_{pq}\left(\sum_{k=0}^{J} (\varepsilon R_Q)^k \Phi_\varepsilon\right)\right| \ge \varepsilon^{p+q} n_{pq}\pi/4, \tag{3.36}$$

$$\begin{aligned}\left\|\sum_{k=J+1}^{\infty} (\varepsilon R_Q)^k \Phi_\varepsilon\right\|_{L_\infty(0,\pi)} &\le \frac{(\varepsilon 2)^{J+1}\|\Phi_\varepsilon\|_{L_\infty(0,\pi)}}{1 - 2\varepsilon} \\ &\le 2^{-J}\varepsilon 4 \|\Phi_\varepsilon\|_{L_\infty(0,\pi)} \le 2^{-J}\pi/4. \end{aligned}\tag{3.37}$$

Since the right-hand side of (3.36) does not depend on J, and the right-hand side of (3.37) does not depend on ε, we can choose J so that $2^{-J} \le \varepsilon^{p+q} n_{pq}/2$. Hence, the function $v'(t)$ is not continuous at the point $p - q\tau \in M$.

3. Now let us show that the set M is dense in $[0,\pi]$. First, we prove that for every $\delta > 0$ there exists $h(p,q) \in M$ such that $0 < h(p,q) < \delta$.

To show this, we use the following well-known

Theorem. *For any irrational number $0 < \tau < 1$ and arbitrary $\delta > 0$ there exist integers p, $q > 0$ such that $p < q$ and*

$$0 < p/q - \tau < q^{-2}, \qquad q^{-1} < \delta.$$

For a proof, see V. I. Arnol'd [1].

By virtue of this Theorem, for every $0 < \delta < \tau$ we can find integers $p, q > 0$ such that

$$0 < p - q\tau < \delta. \tag{3.38}$$

For each integer $s > 0$, we define $r = r(s)$ as follows: 1) $r(1) = 1$; 2) for $h(s, r(s)) \in M$ such that $0 < h(s, r(s)) < \tau$, we set $r(s+1) = r(s) + 1$ if $s + 1 - (r(s)+1)\tau < \tau$, and $r(s+1) = r(s) + 2$ if $s + 1 - (r(s)+1)\tau > \tau$. Clearly, $h(1,1) \in M$. Hence, $h(s, r(s)) \in M$ for every s. Therefore, by virtue of the definition of $r(s)$ and the inequality (3.38), $r(p) = q$. Thus, $h(p,q) \in M$ and $0 < h(p,q) = p - q\tau < \delta$.

From this it follows that, for all $j = 0, 1, 2, 3$ and natural n such that $nh(p,q) < 1$, $j + nh(p,q) < \pi$, we have $j + nh(p,q) \in M$. Thus, the set M is dense in $[0, \pi]$.

We have proved that a generalized solution of the boundary value problem (3.25), (3.26) for $f_0(t) \equiv 1$, does not belong to $W^2(a,b)$ for any interval $(a,b) \subset [0,\pi]$.

Non-Homogeneous Boundary Conditions

We consider the equation

$$-(Ry)''(t) + A_1 y = f_0(t) \qquad (t \in (0,d)) \tag{3.39}$$

with non-homogeneous boundary conditions

$$\left.\begin{aligned} y(t) &= f_1(t) \quad (t \in [-N, 0]), \\ y(t) &= f_2(t) \quad (t \in [d, d+N]). \end{aligned}\right\} \tag{3.40}$$

Here the difference operator R is given by

$$(Ry)(t) = \sum_{j=-N}^{N} b_j y(t+j), \tag{3.41}$$

$b_j \in \mathbb{R}$, $A_1 \colon W^1(-N, d+N) \to L_2(0,d)$ is a linear bounded operator, $f = (f_0, f_1, f_2) \in L_2(0,d) \times W^1(-N,0) \times W^1(d, d+N)$.

Example 3.11. Let

$$A_1 y = a_1(t)(R_1 y)'(t) + a_2(t)(R_2 y)(t), \tag{3.42}$$

where $a_1, a_2 \in C^\infty[0,d]$ are real-valued functions,

$$(R_i y)(t) = \sum_{j=-N}^{N} b_{ij} y(t+j) \qquad (i = 1, 2),$$

$b_{ij} \in \mathbb{R}$. Clearly, the operator $A_1 \colon W^1(-N, d+N) \to L_2(0,d)$ is bounded.

Definition 3.3. A function $y \in W^1(-N, d+N)$ is called a *generalized solution* of the problem (3.39), (3.40) if, for all $w \in \mathring{W}^1(0,d)$,

$$b(y,w) = \int_0^d \{(Ry)'(t)\overline{w'(t)} + (A_1 y)(t)\overline{w(t)}\}\, dt = \int_0^d f_0(t)\overline{w(t)}\, dt. \tag{3.43}$$

Theorem 3.7. *Let the matrix $R_1+R_1^*$ be positive definite. Assume the homogeneous boundary value problem* (3.39), (3.40) *has a unique trivial solution.*

Then the non-homogeneous boundary value problem (3.39), (3.40) *has a unique generalized solution for any $f \in L_2(0,d) \times W^1(-N,0) \times W^1(d,d+N)$, and*

$$\|y\|_{W^1(-N,d+N)} \le c(\|f_0\|_{L_2(0,d)} + \|f_1\|_{W^1(-N,0)} + \|f_2\|_{W^1(d,d+N)}), \tag{3.44}$$

where $c>0$ does not depend on f.

Proof. We denote

$$F(t) = \begin{cases} f_1(t) & (t \in [-N,0]), \\ f_2(t) & (t \in [d,d+N]), \\ f_1(0) + (f_2(d) - f_1(0))t/d & (t \in (0,d)). \end{cases}$$

Clearly, $F \in W^1(-N,d+N)$, and

$$\|F\|_{W^1(-N,d+N)} \le k_1(\|f_1\|_{W^1(-N,0)} + \|f_2\|_{W^1(d,d+N)}).$$

We now define the linear bounded operator $G_0\colon W^1(-N,0) \times W^1(d,d+N) \to W^1(-N,d+N)$ by the formula $G_0 f' = F$, where $f' = (f_1,f_2)$. Let $y = G_0 f' + v$. Clearly, $v \in \mathring{W}^1(0,d)$ and $v(t) = 0$ $(t \in [-N,0] \cup [d,d+N])$. Then the integral identity (3.43) will have the form

$$b(G_0 f', w) + b_R[v,w] = (f_0, w)_{L_2(0,d)}, \tag{3.45}$$

where $b_R[v,w]$ is defined by the formula (3.18).

Using Lemma 3.1, we can introduce the equivalent inner product in $\mathring{W}^1(0,d)$ given by

$$(u,w)'_{\mathring{W}^1(0,d)} = p_R[u,w] + c_2(u,w)_{L_2(0,d)},$$

where $p_R = (b_R + b_R^*)/2$, $q_R = (b_R - b_R^*)/2i$, $b_R = p_R + iq_R$. By virtue of the Riesz theorem about a general form of linear functional in a Hilbert space and Lemma 2.10, there are linear bounded operators $K\colon \mathring{W}^1(0,d) \to \mathring{W}^1(0,d)$, $G_1\colon W^1(-N,d+N) \to \mathring{W}^1(0,d)$, and $G_2\colon L_2(0,d) \to \mathring{W}^1(0,d)$ such that

$$\begin{aligned} q_R[u,w] &= (Ku,w)'_{\mathring{W}^1(0,d)} && (u,w \in \mathring{W}^1(0,d)), \\ b(\Phi,w) &= (G_1\Phi,w)'_{\mathring{W}^1(0,d)} && (\Phi \in W^1(-N,d+N),\ w \in \mathring{W}^1(0,d)), \\ (f_0,w)_{L_2(0,d)} &= (G_2 f_0,w)'_{\mathring{W}^1(0,d)} && (f_0 \in L_2(0,d),\ w \in \mathring{W}^1(0,d)). \end{aligned}$$

Therefore, (3.45) can be written as

$$\begin{aligned} (v,w)'_{\mathring{W}^1(0,d)} + (iKv,w)'_{\mathring{W}^1(0,d)} - c_2(G_2 v,w)'_{\mathring{W}^1(0,d)} \\ = -(G_1 G_0 f',w)'_{\mathring{W}^1(0,d)} + (G_2 f_0,w)'_{\mathring{W}^1(0,d)}. \end{aligned}$$

Thus, we obtain the operator equation in $\mathring{W}^1(0,d)$

$$(I+iK)v - c_2G_2v = -G_1G_0f' + G_2f_0.$$

Since the sesquilinear form q_R is symmetric, we have $K = K^*$. Therefore, the operator $I+iK$ has a bounded inverse $(I+iK)^{-1}\colon \mathring{W}^1(0,d) \to \mathring{W}^1(0,d)$. Thus, we obtain

$$v - c_2(I+iK)^{-1}G_2v = (I+iK)^{-1}(-G_1G_0f' + G_2f_0). \tag{3.46}$$

However, the homogeneous boundary value problem (3.39), (3.40) is equivalent to the equation

$$v - c_2(I+iK)^{-1}G_2v = 0. \tag{3.47}$$

By assumption, the homogeneous equation (3.47) has a unique trivial solution. From the compactness of the imbedding of $\mathring{W}^1(0,d)$ into $L_2(0,d)$ it follows that the operator $-c_2(I+iK)^{-1}G_2\colon \mathring{W}^1(0,d) \to \mathring{W}^1(0,d)$ is compact. Hence, the operator $I - c_2(I+iK)^{-1}G_2$ is Fredholm. Therefore, the equation (3.46) has a unique solution $v \in \mathring{W}^1(0,d)$. We have proved that the boundary value problem (3.39), (3.40) has a unique generalized solution $y = G_0f' + v$ for any $f \in L_2(0,d) \times W^1(-N,0) \times W^1(d,d+N)$. □

Theorem 3.8. *Assume that* $\det R_s \neq 0$ $(s = 1,2$ *if* $\theta < 1$, $s = 1$ *if* $\theta = 1)$. *Let* $f \in W^k(0,d) \times W^{k+2}(-N,0) \times W^{k+2}(d,d+N)$, *and let an operator* A_1 *have the form* (3.42) *if* $k \geq 1$. *Let* $y(t)$ *be a generalized solution of the boundary value problem* (3.39), (3.40).

Then $y \in W^{k+2}(j-1,j)$ $(j = 1,\dots,N+1)$ *if* $\theta = 1$, *and* $y \in W^{k+2}(j-1, j-1+\theta)$ $(j = 1,\dots,N+1)$, $y \in W^{k+2}(j-1+\theta, j)$ $(j = 1,\dots,N)$ *if* $\theta < 1$.

Proof. We introduce the function

$$F(t) = \begin{cases} f_1(t) & (t \in [-N,0]), \\ f_2(t) & (t \in [d,d+N]), \\ \eta(t)\sum\limits_{i=0}^{k+1} f_1^{(i)}(0)t^i/i! & \\ \qquad +\eta(t-d)\sum\limits_{i=0}^{k+1} f_2^{(i)}(d)(t-d)^i/i! & (t \in (0,d)), \end{cases}$$

where $\eta \in \dot{C}^\infty(\mathbb{R})$, $\eta(t) = 1$ $(|t| < 1/4)$, $\eta(t) = 0$ $(|t| > 1/3)$. Clearly, $F \in W^{k+2}(-N,d+N)$. Denote $v = y - F$. A function $v \in \mathring{W}^1(0,d)$ is a generalized solution of the equation

$$-(Rv)''(t) + A_1v = f_0(t) + (RF)'' - A_1F \qquad (t \in (0,d)) \tag{3.48}$$

with homogeneous boundary conditions

$$v(t) = 0 \qquad (t \in [-N,0] \cup [d,d+N]). \tag{3.49}$$

Clearly, $f_0 + (RF)'' - A_1F \in W^k(0,d)$. Hence, by virtue of Theorem 3.2, $v \in W^{k+2}(j-1,j)$ $(j=1,\dots,N+1)$ if $\theta = 1$, and $v \in W^{k+2}(j-1,j-1+\theta)$ $(j=1,\dots,N+1)$, $v \in W^{k+2}(j-1+\theta,j)$ $(j=1,\dots,N)$. Since $y = F+v$ and $F \in W^{k+2}(-N,d+N)$, Theorem 3.8 is proved. □

4 Generalized and Classical Solutions

In Section 3, we proved that the smoothness of generalized solutions can be violated even for infinitely differentiable right-hand sides of equations. In this section we shall obtain the conditions on right-hand sides of equations, which provide appropriate smoothness of the solutions.

Statement of Main Result

We consider the differential-difference equation

$$-(Ry)''(t) + A_1y = f_0(t) \qquad (t \in (0,d)) \tag{4.1}$$

with non-homogeneous boundary conditions

$$\left.\begin{aligned} y(t) &= f_1(t) \quad (t \in [-N,0]), \\ y(t) &= f_2(t) \quad (t \in [d,d+N]), \end{aligned}\right\} \tag{4.2}$$

where

$$(Ry)(t) = \sum_{j=-N}^{N} b_j y(t+j),$$

$b_j \in \mathbb{R}$, $d = N+\theta$, $0 < \theta \le 1$, N is a natural number, $A_1 \colon W^{k+1}(-N,d+N) \to W^k(0,d)$ is a linear bounded operator, $f = (f_0,f_1,f_2) \in \mathcal{W}^k(-N,d+N) = W^k(0,d) \times W^{k+2}(-N,0) \times W^{k+2}(d,d+N)$.

Assume that the following condition is fulfilled:

4.1. $\sum_{j\ne 0} b_j^2 \ne 0$.

We denote by G_j^1 (G_j^2) the jth column of the $N\times(N+1)$-matrix obtained from R_1 by deleting the first (last) row $(j=1,\dots,N+1)$. By virtue of condition 4.1, either $G_1^1 \ne 0$, or $G_{N+1}^2 \ne 0$.

We define the linear bounded operator $\mathcal{L}_R \colon W^{k+2}(-N,d+N) \to \mathcal{W}^k(-N, d+N)$ by the formula

$$\mathcal{L}_R y = (-(Ry)'' + A_1 y, y|_{(-N,0)}, y|_{(d,d+N)}).$$

The main purpose of this section is to prove the following:

Theorem 4.1. *Let condition* 4.1 *be fulfilled, and let* $\det R_s \ne 0$ $(s=1,2)$.

Then we have:

(a) *The operator* $\mathcal{L}_R\colon W^{k+2}(-N,d+N)\to\mathcal{W}^k(-N,d+N)$ *is Fredholm.*
(b) *Assume that* $G_1^1\neq 0$, $G_{N+1}^2\neq 0$ *if* $\theta<1$, *and that the columns* G_1^1, G_{N+1}^2 *are linearly independent if* $\theta=1$. *Then* $\operatorname{ind}\mathcal{L}_R=-2(k+1)$.
(c) *Assume that either* $G_1^1=0$, *or* $G_{N+1}^2=0$ *if* $\theta<1$, *and that the columns* G_1^1, G_{N+1}^2 *are linearly dependent if* $\theta=1$. *Then* $\operatorname{ind}\mathcal{L}_R=-(k+1)$.

Theorem 4.1 implies that the boundary value problem (4.1), (4.2) has a generalized solution from $W^{k+2}(-N,d+N)$ if and only if f is orthogonal to $\dim\mathcal{N}(\mathcal{L}_R)-\operatorname{ind}\mathcal{L}_R$ linearly independent functions in $\mathcal{W}^k(-N,d+N)$.

To prove Theorem 4.1, we first study the connection between the smoothness of a generalized solution and the boundary values of its derivatives. Then we obtain necessary and sufficient conditions of existence of smooth solutions for a differential-difference equation with homogeneous boundary conditions. Finally, we prove Theorem 4.1 reducing a non-homogeneous boundary value problem for a differential-difference equation to a differential-difference equation with homogeneous boundary conditions.

Boundary Conditions for Smooth Solutions

Let us consider the differential-difference equation

$$(Rv)''(t)=f_0(t)\qquad (t\in(0,d)) \tag{4.3}$$

with homogeneous boundary conditions

$$v(t)=0\qquad (t\in[-N,0]\cup[d,d+N]), \tag{4.4}$$

where $f_0\in L_2(0,d)$.

We introduce the unbounded operator $\mathcal{A}_R\colon L_2(0,d)\to L_2(0,d)$ with domain $\mathcal{D}(\mathcal{A}_R)=\{v\in\mathring{W}^1(0,d): R_Qv\in W^2(0,d)\}$ by the formula $\mathcal{A}_Rv=-(R_Qv)''$.

Denote by B_{ij} the cofactor of the element r_{ij} of the matrix R_1.

Remark 4.1. From the proof of Theorem 2.1 it follows that the space $W_\gamma^m(0,d)$ coincides with the subspace of functions $u\in W^m(0,d)$ satisfying the conditions

$$\left.\begin{aligned}\sum_i B_{i1}u^{(\mu)}(i-1)&=0,\\ \sum_i B_{i,N+1}u^{(\mu)}(\theta+i-1)&=0\quad(\mu=0,\dots,m-1).\end{aligned}\right\} \tag{4.5}$$

Here and below we sum over $i=1,\dots,N+1$.

Theorem 4.2. *Let condition* 4.1 *be fulfilled, and let* $\det R_s\neq 0$ $(s=1,2)$. *Assume that* $\theta<1$ *and* $m\geq 2$.

Then $\{v\in\mathcal{D}(\mathcal{A}_R): v, R_Qv\in W^m(0,d)\}=\{v\in\mathcal{D}(\mathcal{A}_R): R_Qv\in W^m(0,d),$ $v^{(\mu)}(0+0)=0$ *if* $G_1^1\neq 0$, *and* $v^{(\mu)}(d-0)=0$ *if* $G_{N+1}^2\neq 0$ $(\mu=1,\dots,m-1)\}$.

Proof. Assume $v \in \mathcal{D}(\mathcal{A}_R)$, $R_Q v \in W^m(0,d)$. Let us prove that $v \in W^m(0,d)$ if and only if $v^{(\mu)}(0+0) = 0$ in the case $G_1^1 \neq 0$ and $v^{(\mu)}(d-0) = 0$ in the case $G_{N+1}^2 \neq 0$ $(\mu = 1, \dots, m-1)$.

Since $R_Q v \in W^m(0,d)$, we have

$$(R_Q v)^{(\mu)}|_{t=l-0} = (R_Q v)^{(\mu)}|_{t=l+0} \qquad (l = 1, \dots, N), \tag{4.6}$$

$$(R_Q v)^{(\mu)}|_{t=\theta+l-1-0} = (R_Q v)^{(\mu)}|_{t=\theta+l-1+0} \qquad (l = 1, \dots, N) \tag{4.7}$$

for each $1 \leq \mu \leq m-1$.

By virtue of Lemma 2.6, we can rewrite the relation (4.6) in the form

$$(R_2 U_2 P_2 v^{(\mu)})_l|_{t=1-0} = (R_1 U_1 P_1 v^{(\mu)})_{l+1}|_{t=0+0}. \tag{4.8}$$

Denote $\varphi_l^\mu = (U_2 P_2 v^{(\mu)})_l|_{t=1-0}$ $(l = 1, \dots, N)$, $\psi_l^\mu = (U_1 P_1 v^{(\mu)})_{l+1}|_{t=0+0}$ $(l = 0, \dots, N)$. Then from (4.8) it follows that

$$\sum_{j=1}^{N} r_{lj} \varphi_j^\mu = \sum_{i=1}^{N+1} r_{l+1,i} \psi_{i-1}^\mu \qquad (l = 1, \dots, N). \tag{4.9}$$

Evidently, $r_{l+1,i+1} = r_{li}$. Hence, we obtain from (4.9) that

$$\sum_{j=1}^{N} r_{lj} (\varphi_j^\mu - \psi_j^\mu) = r_{l+1,1} \psi_0^\mu \qquad (l = 1, \dots, N). \tag{4.10}$$

The determinant of the system (4.10) $\det R_2 = B_{11} \neq 0$. Hence, the system (4.10) has a unique solution $\varphi_j^\mu - \psi_j^\mu = 0$ $(j = 1, \dots, N)$ if and only if $G_1^1 = 0$ or $\psi_0^\mu = 0$. In other words, $v^{(\mu)}|_{t=l-0} = v^{(\mu)}|_{t=l+0}$ $(l = 1, \dots, N)$ if and only if $G_1^1 = 0$ or $v^{(\mu)}|_{t=0+0} = 0$. Similarly, from (4.7) we obtain that $v^{(\mu)}|_{t=\theta+l-1-0} = v^{(\mu)}|_{t=\theta+l-1+0}$ $(l = 1, \dots, N)$ if and only if $G_{N+1}^2 = 0$ or $v^{(\mu)}|_{t=d-0} = 0$.

Thus, $\{v \in \mathcal{D}(\mathcal{A}_R) : v, R_Q v \in W^m(0,d)\} = \{v \in \mathcal{D}(\mathcal{A}_R) : R_Q v \in W^m(0,d),$ $v^{(\mu)}(0+0) = 0$ if $G_1^1 \neq 0$ and $v^{(\mu)}(d-0) = 0$ if $G_{N+1}^2 \neq 0$ $(\mu = 1, \dots, m-1)\}$. □

Theorem 4.3. *Let condition* 4.1 *be fulfilled, and let* $\det R_s \neq 0$ $(s = 1, 2)$. *Assume* $\theta = 1$, $m \geq 2$.

If the columns G_1^1 *and* G_{N+1}^2 *are linearly independent, then* $\{v \in \mathcal{D}(\mathcal{A}_R) : v, R_Q v \in W^m(0,d)\} = \{v \in \mathcal{D}(\mathcal{A}_R) : R_Q v \in W^m(0,d),\ v^{(\mu)}(0+0) = v^{(\mu)}(d-0) = 0\ (\mu = 1, \dots, m-1)\} = \mathring{W}^m(0,d)$.

If there exist $\alpha_1, \alpha_2 \in \mathbb{R}$ *such that* $\alpha_1 G_1^1 + \alpha_2 G_{N+1}^2 = 0$, $\alpha_1^2 + \alpha_2^2 \neq 0$, *then* $\{v \in \mathcal{D}(\mathcal{A}_R) : v, R_Q v \in W^m(0,d)\} = \{v \in \mathcal{D}(\mathcal{A}_R) : R_Q v \in W^m(0,d),$ $\alpha_1 v^{(\mu)}(d-0) + \alpha_2 v^{(\mu)}(0+0) = 0\ (\mu = 1, \dots, m-1)\}$.

Proof. Assume $v \in \mathcal{D}(\mathcal{A}_R)$, $R_Q v \in W^m(0,d)$.

Since $R_Q v \in W^m(0,d)$, we have

$$(R_Q v)^{(\mu)}|_{t=l-0} = (R_Q v)^{(\mu)}|_{t=l+0} \qquad (l = 1, \dots, N) \tag{4.11}$$

for each $1 \le \mu \le m-1$. By virtue of Lemma 2.6, we can rewrite the relation (4.11) in the form

$$(R_1 U_1 v^{(\mu)})_l|_{t=1-0} = (R_1 U_1 v^{(\mu)})_{l+1}|_{t=0+0} \qquad (l = 1, \dots, N). \tag{4.12}$$

Denote $\varphi_l^\mu = (U_1 v^{(\mu)})_l|_{t=1-0}$ $(l = 1, \dots, N+1)$, $\psi_l^\mu = (U_1 v^{(\mu)})_{l+1}|_{t=0+0}$ $(l = 0, \dots N)$. Then from (4.12) it follows that

$$\sum_{i=1}^{N+1} r_{li}\varphi_i^\mu = \sum_{i=1}^{N+1} r_{l+1,i}\psi_{i-1}^\mu \qquad (l = 1, \dots, N).$$

Evidently, $r_{l+1,i+1} = r_{li}$. Hence, we obtain

$$\sum_{j=1}^{N} r_{lj}(\varphi_j^\mu - \psi_j^\mu) = r_{l+1,1}\psi_0^\mu - r_{l,N+1}\varphi_{N+1}^\mu. \tag{4.13}$$

Since $\det R_2 = B_{11} \neq 0$, the system (4.13) has a unique solution $\varphi_j^\mu - \psi_j^\mu = 0$ $(j = 1, \dots, N)$ if and only if $r_{l+1,1}\psi_0^\mu - r_{l,N+1}\varphi_{N+1}^\mu = 0$ $(l = 1, \dots, N)$. Thus, in the case of the linearly independent columns G_1^1 and G_{N+1}^2, $v^{(\mu)}|_{t=l-0} = v^{(\mu)}|_{t=l+0}$ $(l = 1, \dots, N)$ if and only if $v^{(\mu)}|_{t=0+0} = v^{(\mu)}|_{t=d-0} = 0$. In the case when there exist $\alpha_1, \alpha_2 \in \mathbb{R}$ such that $\alpha_1 G_1^1 + \alpha_2 G_{N+1}^2 = 0$ and $\alpha_1^2 + \alpha_2^2 \neq 0$, $v^{(\mu)}|_{t=l-0} = v^{(\mu)}|_{t=l+0}$ $(l = 1, \dots, N)$ if and only if $\alpha_1 v^{(\mu)}|_{t=d-0} + \alpha_2 v^{(\mu)}|_{t=0+0} = 0$. □

Existence of Smooth Solutions for Homogeneous Boundary Conditions

Let $A_R^k\colon W^{k+2}(0,d) \to W^k(0,d)$ be the bounded operator defined by $\mathcal{D}(A_R^k) = \{v \in \mathcal{D}(\mathcal{A}_R) : v, R_Q v \in W^{k+2}(0,d)\}$, $A_R^k v = \mathcal{A}_R v$ $(v \in \mathcal{D}(A_R^k))$, where $k \ge 0$.

Lemma 4.1. *Let* $\det R_s \neq 0$ $(s = 1, 2)$. *Then* $\dim \mathcal{N}(A_R^k) = 0$.

Proof. Let $v \in \mathcal{N}(A_R^k)$. Then $(R_Q v)''(t) = 0$. Hence, $(R_Q v)(t) = c_1 + c_2 t$. Since $\det R_s \neq 0$ $(s = 1, 2)$, we obtain

$$v(t) = U_s^{-1} R_s^{-1} U_s P_s (c_1 + c_2 t) \qquad \Big(t \in \bigcup_j Q_{sj}\Big). \tag{4.14}$$

Thus, a function $v(t)$ is piecewise linear on the interval $(0,d)$. Therefore, $v \in W^2(0,d) \cap \mathring{W}^1(0,d)$ if and only if $v(t) \equiv 0$. □

Lemma 4.2. *Let condition* 4.1 *be satisfied, and let* $\det R_s \neq 0$ $(s = 1, 2)$. *Assume that* $G_1^1 \neq 0$, $G_{N+1}^2 \neq 0$ *if* $\theta < 1$, *and that the columns* G_1^1, G_{N+1}^2 *are linearly independent if* $\theta = 1$.

Then the operator $A_R^0 \colon W^2(0,d) \to L_2(0,d)$ *is Fredholm, and* $\dim \mathcal{N}(A_R^0) = 0$, $\operatorname{codim} \mathcal{R}(A_R^0) = 2$.

Proof. By virtue of Theorems 4.2, 4.3 and Lemma 4.1, it is sufficient to prove that the equation $\mathcal{A}_R v = f_0$ has a solution such that $v'(0+0) = v'(d-0) = 0$ if and only if

$$(f_0, \varphi_i)_{L_2(0,d)} = 0 \qquad (i = 1, 2), \tag{4.15}$$

where $\varphi_1, \varphi_2 \in L_2(0,d)$ are linearly independent functions.

Let us consider a function $u = R_Q v$. Here v is a solution of the equation $\mathcal{A}_R v = f_0$. By virtue of Remark 4.1 and Theorem 2.1, a function u is a solution of the equation

$$-u''(t) = f_0(t) \tag{4.16}$$

with boundary conditions

$$\sum_i B_{i1} u(i-1) = 0, \qquad \sum_i B_{i,N+1} u(\theta + i - 1). \tag{4.17}$$

On the other hand, if u is a solution of the boundary value problem (4.16), (4.17), then the function $v = R_Q^{-1} u$ is a solution of the problem (4.3), (4.4). The general solution of equation (4.16) has the form

$$u(t) = c_1 + c_2 t - \int_0^t (t - \tau) f_0(\tau)\, d\tau. \tag{4.18}$$

Substituting (4.18) into (4.17), we obtain the equations for constants c_1 and c_2,

$$\begin{aligned} c_1 \sum_i B_{i1} + c_2 \sum_i (i-1) B_{i1} &= \sum_i B_{i1} \int_0^{i-1} (i - 1 - \tau) f_0(\tau)\, d\tau, \\ c_1 \sum_i B_{i,N+1} + c_2 \sum_i (\theta + i - 1) B_{i,N+1} & \\ = \sum_i B_{i,N+1} \int_0^{\theta+i-1} & (\theta + i - 1 - \tau) f_0(\tau)\, d\tau. \end{aligned} \tag{4.19}$$

By virtue of (2.7),

$$(U_1 P_1 v')(t) = R_1^{-1} U_1 P_1 u'(t) \qquad (t \in (0, \theta)), \tag{4.20}$$

where $(U_1 P_1 u')(t) = (u'(t), u'(t+1), \ldots, u'(t+N))$.

From (4.20) and (2.7) it follows that the conditions $v'(0+0) = 0$, $v'(d-0) = 0$ will have the form

$$
\begin{aligned}
v'(0+0) &= (U_1 P_1 v')_1(0+0) \\
&= \sum_i \frac{B_{i1}}{\det R_1}(U_1 P_1 u')_i(0+0) \\
&= \sum_i \frac{B_{i1}}{\det R_1} P_1 u'(i-1+0) \\
&= \sum_i \frac{B_{i1}}{\det R_1}\left(c_2 - \int_0^{i-1} f_0(\tau)\,d\tau\right) = 0, \qquad (4.21)\\
v'(d-0) &= (U_1 P_1 v')_{N+1}(\theta-0) \\
&= \sum_i \frac{B_{i,N+1}}{\det R_1}(U_1 P_1 u')_i(\theta-0) \\
&= \sum_i \frac{B_{i,N+1}}{\det R_1} P_1 u'(\theta+i-1-0) \\
&= \sum_i \frac{B_{i,N+1}}{\det R_1}\left(c_2 - \int_0^{\theta+i-1} f_0(\tau)\,d\tau\right) = 0. \qquad (4.22)
\end{aligned}
$$

Thus, the equation $\mathcal{A}_R v = f_0$ has a solution if and only if the system of linear algebraic equations (4.19) is compatible. A solution $v \in \mathcal{D}(\mathcal{A}_R)$ of the equation $\mathcal{A}_R v = f_0$ belongs to $\mathring{W}^2(0,d)$ if and only if the conditions (4.21), (4.22) are fulfilled.

We consider different cases of the system (4.19).

a) Assume the determinant of the system (4.19) is non-zero. Then the boundary value problem (4.16), (4.17) has a unique solution u for every $f_0 \in L_2(0,d)$. Hence, by Theorem 2.1, the equation $\mathcal{A}_R v = f_0$ has a unique solution for every $f_0 \in L_2(0,d)$. This solution $v \in \mathring{W}^2(0,d)$ if and only if the conditions (4.21), (4.22) hold, where

$$
\begin{aligned}
c_2 &= \Delta_2/\Delta, \\
\Delta_2 &= \Big(\sum_i B_{i1}\Big)\Big(\sum_i B_{i,N+1}\int_0^{\theta+i-1}(\theta+i-1-\tau)f_0(\tau)\,d\tau\Big) \\
&\quad -\Big(\sum_i B_{i,N+1}\Big)\Big(\sum_i B_{i1}\int_0^{i-1}(i-1-\tau)f_0(\tau)\,d\tau\Big), \qquad (4.23)\\
\Delta &= \Big(\sum_i B_{i1}\Big)\Big(\sum_i(\theta+i-1)B_{i,N+1}\Big) \\
&\quad -\Big(\sum_i B_{i,N+1}\Big)\Big(\sum_i(i-1)B_{i1}\Big). \qquad (4.24)
\end{aligned}
$$

Hence, $v \in \mathring{W}^2(0,d)$ if and only if the conditions (4.15) are satisfied, where the functions φ_1, φ_2 can be found from the formulas (4.21), (4.23) and (4.22), (4.23), respectively. In particular, using the equality $B_{11} = B_{N+1,N+1}$, we have, for $N < \tau < d$,

$$\varphi_1(\tau) = \Big(\sum_i B_{i1}\Big)^2 \frac{B_{11}(d-\tau)}{\Delta},$$

$$\varphi_2(\tau) = \Big(\sum_i B_{i1}\Big)\Big(\sum_i B_{i,N+1}\Big)\frac{B_{11}(d-\tau)}{\Delta} - B_{11}.$$

Since $B_{11} \neq 0$, the functions $\varphi_1(\tau)$ and $\varphi_2(\tau)$ are linearly independent if $\sum_i B_{i1} \neq 0$.

Now let $\sum_i B_{i1} = 0$. Then, for $0 < \tau < \theta$,

$$\varphi_1(\tau) = -\sum_{i=2}^{N+1} B_{i1},$$

$$\varphi_2(\tau) = -\Big(\sum_i B_{i,N+1}\Big)^2\Big(\sum_{i=2}^{N+1} B_{i1}(i-1-\tau)\Big)\frac{1}{\Delta} - \Big(\sum_i B_{i,N+1}\Big).$$

Since $\sum_{i=1}^{N+1} B_{i1} = 0$, $B_{11} \neq 0$, we have $\sum_{i=2}^{N+1} B_{i1} \neq 0$. In addition, using the relations $\sum_i B_{i1} = 0$, $\Delta \neq 0$, by virtue of (4.24), we obtain $\sum_i B_{i,N+1} \neq 0$. Hence, the functions $\varphi_1(\tau)$, $\varphi_2(\tau)$ are linearly independent.

b) Let now $\Delta = 0$, but $\sum_i B_{i1} \neq 0$ or $\sum_i B_{i,N+1} \neq 0$. Then the equation $\mathcal{A}_R v = f_0$ has a solution if and only if

$$\Delta_2 = \Big(\sum_i B_{i1}\Big)\Big(\sum_i B_{i,N+1}\int_0^{\theta+i-1}(\theta+i-1-\tau)f_0(\tau)\,d\tau\Big) - \Big(\sum_i B_{i,N+1}\Big)\Big(\sum_i B_{i1}\int_0^{i-1}(i-1-\tau)f_0(\tau)\,d\tau\Big) = 0. \quad (4.25)$$

Thus, in the formula (4.18) a constant c_2 can be arbitrary. A constant c_1 can be found from (4.19) as a function of c_2.

By virtue of (4.21), (4.22), $v \in \mathring{W}^2(0,d)$ if and only if a constant c_2 and a function $f_0(\tau)$ satisfy the conditions

$$\Big(\sum_i B_{i1}\Big)c_2 - \sum_i B_{i1}\int_0^{i-1} f_0(\tau)\,d\tau = 0, \quad (4.26)$$

$$\Big(\sum_i B_{i,N+1}\Big)c_2 - \sum_i B_{i,N+1}\int_0^{\theta+i-1} f_0(\tau)\,d\tau = 0. \quad (4.27)$$

From (4.26), (4.27) we have

$$-\Big(\sum_i B_{i,N+1}\Big)\Big(\sum_i B_{i1}\int_0^{i-1} f_0(\tau)\,d\tau\Big) \\ +\Big(\sum_i B_{i1}\Big)\Big(\sum_i B_{i,N+1}\int_0^{\theta+i-1} f_0(\tau)\,d\tau\Big)=0. \quad (4.28)$$

Thus, the equation $\mathcal{A}_R v = f_0$ has a solution $v \in \mathring{W}^2(0,d)$ if and only if the condition (4.15) is fulfilled, where functions $\varphi_1(\tau)$, $\varphi_2(\tau)$ can be found from the formulas (4.25), (4.28), respectively. In particular, for $N<\tau<d$,

$$\varphi_1(\tau)=\Big(\sum_i B_{i1}\Big)B_{11}(d-\tau), \qquad \varphi_2(\tau)=\Big(\sum_i B_{i1}\Big)B_{11}.$$

If $\sum_i B_{i1}\neq 0$, then $\varphi_1(\tau)$ and $\varphi_2(\tau)$ are linearly independent. If $\sum_i B_{i1}=0$, then, for $0<\tau<\theta$, we have

$$\begin{aligned}\varphi_1(\tau)&=-\Big(\sum_i B_{i,N+1}\Big)\Big(\sum_{i=2}^{N+1} B_{i1}(i-1-\tau)\Big)\\ &=-\Big(\sum_i B_{i,N+1}\Big)\Big(\sum_{i=2}^{N+1} B_{i1}(i-1)+B_{11}\tau\Big),\\ \varphi_2(\tau)&=-\Big(\sum_i B_{i,N+1}\Big)\Big(\sum_{i=2}^{N+1} B_{i1}\Big)=B_{11}\Big(\sum_{i=1}^{N+1} B_{i,N+1}\Big).\end{aligned}$$

Evidently, in this case also the functions $\varphi_1(\tau)$ and $\varphi_2(\tau)$ are linearly independent.

c) Let us prove that the case $\sum_i B_{i1}=\sum_i B_{i,N+1}=0$ is not possible. Assume to the contrary that $\sum_i B_{i1}=\sum_i B_{i,N+1}=0$. Then $\Delta=0$. Hence, the equation $-u''(t)=0$ with boundary conditions (4.17) has a nontrivial solution $u=c_1+c_2t$. In this case, by virtue of Theorem 2.1, the equation $\mathcal{A}_R v=0$ has a nontrivial solution $v(t)=R_Q^{-1}u(t)$ satisfying the conditions:

$$\begin{aligned}v(0)&=(U_1P_1v)_1(0)=\sum_i \frac{B_{i1}}{\det R_1}(U_1P_1u)_i(0)\\ &=\sum_i \frac{B_{i1}}{\det R_1}u(i-1)=\frac{c_2}{\det R_1}\Big(\sum_i B_{i1}(i-1)\Big)=0,\\ v(d)&=(U_1P_1v)_{N+1}(\theta)=\sum_i \frac{B_{i,N+1}}{\det R_1}(U_1P_1u)_i(\theta)\\ &=\sum_i \frac{B_{i,N+1}}{\det R_1}u(\theta+i-1)=\frac{c_2}{\det R_1}\Big(\sum_i B_{i,N+1}(\theta+i-1)\Big)=0.\end{aligned}$$

Thus, we must have either $c_2 = 0$, or $\sum_i B_{i1}(i-1) = \sum_i B_{i,N+1}(i-1+\theta) = 0$.

If $c_2 = 0$, then $u(t) = c_1$. Hence, by virtue of (4.14), $v(t)$ is piecewise constant. Since $v \in \mathring{W}^1(0,d)$, then $v(t) \equiv 0$. We obtain a contradiction.

If $\sum_i B_{i1}(i-1) = \sum_i B_{i,N+1}(i-1+\theta) = 0$, then the rank of the matrix of the system (4.19) is equal to zero. Hence, c_1, c_2 are arbitrary constants. In particular, we can choose $c_2 = 0$. Thus, we have the previous case $u(t) = c_1$. □

Lemma 4.3. *Let condition* 4.1 *be satisfied, and let* $\det R_s \neq 0$ $(s = 1, 2)$. *Assume that* $\theta < 1$, *and that* $G_1^1 = 0$ *or* $G_{N+1}^2 = 0$.

Then the operator $A_R^0 \colon W^2(0,d) \to L_2(0,d)$ *is Fredholm, and* $\dim \mathcal{N}(A_R^0) = 0$, $\operatorname{codim} \mathcal{R}(A_R^0) = 1$.

Proof. Without loss of generality, we shall assume that $G_{N+1}^2 = 0$. Then from the condition 4.1 it follows that $G_1^1 \neq 0$. Hence, by virtue of Theorem 4.2 and Lemma 4.1, it is sufficient to prove that the equation $\mathcal{A}_R v = f_0$ has a solution such that

$$v'(0+0) = 0 \tag{4.29}$$

if and only if

$$(f_0, \varphi)_{L_2(0,d)} = 0, \tag{4.30}$$

where $0 \neq \varphi \in L_2(0,d)$.

According to the proof of Lemma 4.2, a function $v \in \mathcal{D}(\mathcal{A}_R)$ is a solution of the equation $\mathcal{A}_R v = f_0$ if and only if a function $u = R_Q v$ is a solution of the problem (4.16), (4.17). Thus, the equation $\mathcal{A}_R v = f_0$ has a solution if and only if the system of linear algebraic equations (4.19) is compatible. From (4.29), (4.21) it follows that a solution $v \in \mathcal{D}(\mathcal{A}_R)$ of the equation $\mathcal{A}_R v = f_0$ belongs to $\mathcal{D}(A_R^0)$ if and only if the equality (4.26) is fulfilled.

Let us consider the two cases of system (4.19).

a) Assume the determinant Δ of the system (4.19) is non-zero. Then, by Theorem 2.1, the equation $\mathcal{A}_R v = f_0$ has a unique solution for every $f_0 \in L_2(0,d)$. This solution belongs to $\mathcal{D}(A_R^0)$ if and only if the condition (4.26) holds. Substituting (4.23) into the equality (4.26), we can rewrite (4.26) in the form (4.30).

In particular, for $N < \tau < d$, we have

$$\varphi(\tau) = \Big(\sum_i B_{i1}\Big)^2 \frac{B_{11}(d-\tau)}{\Delta}.$$

If $\sum_i B_{i1} \neq 0$, then $\varphi \neq 0$.

Let now $\sum_i B_{i1} = 0$. Then, for $0 < \tau < \theta$, we have

$$\varphi(\tau) = -\sum_{i=2}^{N+1} B_{i1} = B_{11} \neq 0.$$

b) Assume that $\Delta = 0$. Then the equation $\mathcal{A}_R v = f_0$ has a solution if and only if the equality (4.25) is fulfilled. We can rewrite (4.25) in the form of (4.30).

In particular, for $N < \tau < d$,

$$\varphi(\tau) = \Big(\sum_i B_{i1}\Big) B_{11}(d-\tau).$$

We prove that $\sum_i B_{i1} \neq 0$. Then $\varphi \neq 0$. In fact, by assumption, $G^2_{N+1} = 0$, $G^1_1 \neq 0$. Hence,

$$R_1 = \begin{pmatrix} a_0 & 0 & 0 & \dots & 0 \\ a_{-1} & a_0 & 0 & \dots & 0 \\ a_{-2} & a_{-1} & a_0 & \dots & 0 \\ \vdots & \vdots & \vdots & \ddots & \vdots \\ a_{-N} & a_{-N+1} & a_{-N+2} & \dots & a_0 \end{pmatrix}.$$

From this it follows that $B_{i1} = 0$ for $i \neq 1$ and $\sum_i B_{i1} = B_{11} \neq 0$.

A solution $v \in \mathcal{D}(\mathcal{A}_R)$ of the equation $\mathcal{A}_R v = f_0$ belongs to $\mathcal{D}(A^0_R)$ if and only if the equality (4.26) is fulfilled. Since $\sum_i B_{i1} \neq 0$, there exists a unique solution c_2 of the equation (4.26) for each f_0 satisfying (4.30). A constant c_1 can be found from the first equation of the system (4.19). □

Lemma 4.4. *Let condition* 4.1 *be satisfied, and let* $\det R_s \neq 0$ $(s = 1, 2)$. *Assume that* $\theta = 1$, *and that the columns* G^1_1, G^2_{N+1} *are linearly dependent.*

Then the operator $A^0_R\colon W^2(0,d) \to L_2(0,d)$ *is Fredholm, and* $\dim \mathcal{N}(A^0_R) = 0$, $\operatorname{codim} \mathcal{R}(A^0_R) = 1$.

Proof. By virtue of Theorem 4.3 and Lemma 4.1, it is sufficient to prove that the equation $\mathcal{A}_R v = f_0$ has a solution such that

$$\alpha_1 v'(d-0) + \alpha_2 v'(0+0) = 0 \tag{4.31}$$

if and only if

$$(f_0, \varphi)_{L_2(0,d)} = 0, \tag{4.32}$$

where $0 \neq \varphi \in L_2(0,d)$, $\alpha_1^2 + \alpha_2^2 \neq 0$, $\alpha_1 G^1_1 + \alpha_2 G^2_{N+1} = 0$.

According to the proof of Lemma 4.2, a function $v \in \mathcal{D}(\mathcal{A}_R)$ is a solution of the equation $\mathcal{A}_R v = f_0$ if and only if a function $u = R_Q v$ is a solution of the problem (4.16), (4.17). Thus, by virtue of (4.19), the equation $\mathcal{A}_R v = f_0$ has a solution if and only if the system of linear algebraic equations

$$\begin{aligned} c_1 \sum_i B_{i1} + c_2 \sum_i (i-1) B_{i1} &= \sum_i B_{i1} \int_0^{i-1} (i-1-\tau) f_0(\tau)\, d\tau, \\ c_1 \sum_i B_{i,N+1} + c_2 \sum_i i B_{i,N+1} &= \sum_i B_{i,N+1} \int_0^{i} (i-\tau) f_0(\tau)\, d\tau \end{aligned} \tag{4.33}$$

is compatible. From (4.31), (4.21), and (4.22) it follows that a solution $v \in \mathcal{D}(\mathcal{A}_R)$ of the equation $\mathcal{A}_R v = f_0$ belongs to $\mathcal{D}(A_R^0)$ if and only if

$$\alpha_1 \sum_i B_{i,N+1}\left(c_2 - \int_0^i f_0(\tau)\,d\tau\right) + \alpha_2 \sum_i B_{i1}\left(c_2 - \int_0^{i-1} f_0(\tau)\,d\tau\right) = 0. \quad (4.34)$$

Let us consider two cases of the system (4.33).

a) Assume the determinant of the system (4.33) is non-zero. Then, by Theorem 2.1, the equation $\mathcal{A}_R v = f_0$ has a unique solution for every $f_0 \in L_2(0,d)$. This solution belongs to $\mathcal{D}(A_R^0)$ if and only if the condition (4.34) holds. Substituting (4.23) into the equality (4.34) and assuming $\theta = 1$, we can rewrite (4.34) in the form of (4.32).

In particular, for $N < \tau < d$, we have

$$\varphi(\tau) = -\alpha_1 B_{11} + \left(\alpha_1 \sum_i B_{i,N+1} + \alpha_2 \sum_i B_{i1}\right)\left(\sum_i B_{i1}\right)\frac{B_{11}(d-\tau)}{\Delta}.$$

If $\alpha_1 \neq 0$ or $\sum_i B_{i1} \neq 0$, then $\varphi \neq 0$.

Let now $\alpha_1 = 0$, $\sum_i B_{i1} = 0$. Then, for $0 < \tau < 1$, we have

$$\varphi(\tau) = -\alpha_2 \sum_{i=2}^{N+1} B_{i1} = \alpha_2 B_{11} \neq 0.$$

b) Let now $\Delta = 0$, but $\sum_i B_{i1} \neq 0$ or $\sum_i B_{i,N+1} \neq 0$. Then the equation $\mathcal{A}_R v = f_0$ has a solution if and only if

$$\Delta_2 = \left(\sum_i B_{i1}\right)\left(\sum_i B_{i,N+1} \int_0^i (i-\tau) f_0(\tau)\,d\tau\right)$$
$$-\left(\sum_i B_{i,N+1}\right)\left(\sum_i B_{i1} \int_0^{i-1} (i-1-\tau) f_0(\tau)\,d\tau\right) = 0. \quad (4.35)$$

Thus, in the formula (4.18) the constant c_2 can be arbitrary. The constant c_1 can be found from (4.19).

A solution $v \in \mathcal{D}(\mathcal{A}_R)$ of the equation $\mathcal{A}_R v = f_0$ belongs to $\mathcal{D}(A_R^0)$ if and only if (4.34) holds.

Let us prove that in the formula (4.34)

$$\alpha_1 \sum_i B_{i,N+1} + \alpha_2 \sum_i B_{i1} \neq 0. \quad (4.36)$$

Without loss of generality, we assume that $\alpha_2 \neq 0$. Then $\alpha G_1^1 = G_{N+1}^2$, where $\alpha = -\alpha_1/\alpha_2$. Therefore, we can rewrite (4.36) in the following form:

$$\sum_i B_{i1} \neq \alpha \sum_i B_{i,N+1}. \quad (4.37)$$

The matrix R_1 looks like

$$R_1 = \begin{pmatrix} a_0 & \alpha a_{-N} & \alpha a_{-N+1} & \dots & \alpha a_{-1} \\ a_{-1} & a_0 & \alpha a_{-N} & \dots & \alpha a_{-2} \\ a_{-2} & a_{-1} & a_0 & \dots & \alpha a_{-3} \\ \vdots & \vdots & \vdots & \ddots & \vdots \\ a_{-N} & a_{-N+1} & a_{-N+2} & \dots & a_0 \end{pmatrix}.$$

Let us prove that $B_{i1} = \alpha B_{i-1,N+1}$ $(i = 2, \dots, N+1)$. The first row of B_{i1} is equal to the last row of $B_{i-1,N+1}$ multiplied by α. Furthermore, the jth row of B_{i1} is equal to the $(j-1)$th row of $B_{i-1,N+1}$ $(j = 2, \dots, N)$. Hence,

$$B_{i1} = \alpha B_{i-1,N+1} \qquad (i = 2, \dots, N+1). \tag{4.38}$$

Assume $\sum_i B_{i1} = \alpha \sum_i B_{i,N+1}$. Then from (4.38) and the equality $B_{11} = B_{N+1,N+1}$ it follows that $B_{11}(1-\alpha) = 0$, i.e., $\alpha = 1$. Since $\Delta = 0$, from (4.24) we have $\sum_i (i-1)B_{i1} = \sum_i iB_{i,N+1}$, i. e.

$$\sum_{i=2}^{N+1} (i-1)B_{i1} = \sum_{i=2}^{N+1} (i-1)B_{i-1,N+1} + (N+1)B_{11}.$$

Substituting $B_{i1} = B_{i-1,N+1}$ $(i = 2, \dots, N+1)$ into the last equality, we obtain a contradiction $B_{11} = 0$. Thus, we have proved that (4.37) is valid.

A solution $v \in \mathcal{D}(\mathcal{A}_R)$ of the equation $\mathcal{A}_R v = f_0$ belongs to $\mathcal{D}(A_R^0)$ if and only if the equality (4.34) is fulfilled. By virtue of (4.36), we can choose c_2 such that (4.34) holds for each f_0 satisfying (4.35).

Evidently, we can rewrite (4.35) in the form (4.32). If $\sum_i B_{i1} \neq 0$, then, for $N < \tau < d$,

$$\varphi(\tau) = \Big(\sum_i B_{i1}\Big) B_{11}(d - \tau).$$

If $\sum_i B_{i1} = 0$, then, for $0 < \tau < 1$,

$$\varphi(\tau) = -\Big(\sum_i B_{i,N+1}\Big)\Big(\sum_{i=2}^{N+1} B_{i1}(i-1) + B_{11}\tau\Big).$$

Thus, $\varphi \neq 0$.

In the proof of Lemma 4.2 it was stated that the case $\sum_i B_{i1} = \sum_i B_{i,N+1} = 0$ is not possible. □

Theorem 4.4. *Let condition* 4.1 *be fulfilled, and let* $\det R_s \neq 0$ $(s = 1, 2)$. *Assume that* $G_1^1 \neq 0$, $G_{N+1}^2 \neq 0$ *if* $\theta < 1$, *and that the columns* G_1^1, G_{N+1}^2 *are linearly independent if* $\theta = 1$.

Then the operator $A_R^k \colon W^{k+2}(0,d) \to W^k(0,d)$ *is Fredholm, and* $\dim \mathcal{N}(A_R^k) = 0$, $\operatorname{codim} \mathcal{R}(A_R^k) = 2(k+1)$.

Proof. Assume $f_0 \in W^k(0,d)$. From Theorems 4.2, 4.3, 2.1 and Lemma 4.2 it follows that the equation $\mathcal{A}_R v = f_0$ has a solution $v \in \mathcal{D}(A_R^k)$ if and only if

$$(f_0, \varphi_j)_{L_2(0,d)} = 0 \qquad (j = 1, 2),$$
$$\sum_i B_{i1} f_0^{(\mu)}(i-1) = 0, \qquad \sum_i B_{i,N+1} f^{(\mu)}(\theta + i - 1) = 0$$
$$(\mu = 0, \dots, k-1),$$

where $\varphi_j \in L_2(0,d)$ are linearly independent functions.

By the Riesz theorem, there exist functions $\Phi_{\mu j} \in W^k(0,d)$ ($j = 1, 2$, $\mu = -1, 0, \dots, k-1$) such that

$$\begin{aligned}
(f_0, \varphi_j)_{L_2(0,d)} &= (f_0, \Phi_{-1,j})_{W^k(0,d)} \qquad (j = 1, 2),\\
\sum_i B_{i1} f_0^{(\mu)}(i-1) &= (f_0, \Phi_{\mu 1})_{W^k(0,d)},\\
\sum_i B_{i,N+1} f_0^{(\mu)}(\theta + i - 1) &= (f_0, \Phi_{\mu 2})_{W^k(0,d)} \qquad (\mu = 0, \dots, k-1).
\end{aligned}$$

Let us prove that functions $\Phi_{\mu j}$ ($\mu = -1, \dots, k-1$, $j = 1, 2$) are linearly independent. Assume

$$\sum_{\mu=-1}^{k-1} \sum_{j=1,2} c_{\mu j} \Phi_{\mu j} = 0.$$

Then, for all $f \in W^k(0,d)$,

$$\sum_{\mu,j} c_{\mu j} (f, \Phi_{\mu j})_{W^k(0,d)} = 0.$$

Hence,

$$\begin{aligned}
c_{-1,1}(f, \varphi_1)_{L_2(0,d)} &+ c_{-1,2}(f, \varphi_2)_{L_2(0,d)}\\
&+ \sum_{\mu=0}^{k-1} \Bigg(c_{\mu 1} \sum_i B_{i1} f^{(\mu)}(i-1)\\
&\qquad + c_{\mu 2} \sum_i B_{i,N+1} f^{(\mu)}(\theta + i - 1) \Bigg) = 0. \qquad (4.39)
\end{aligned}$$

Since φ_1, φ_2 are linearly independent, we have

$$\begin{vmatrix} (\varphi_1, \varphi_1)_{L_2(0,d)} & (\varphi_1, \varphi_2)_{L_2(0,d)} \\ (\varphi_2, \varphi_1)_{L_2(0,d)} & (\varphi_2, \varphi_2)_{L_2(0,d)} \end{vmatrix} \neq 0.$$

We can choose $f_1, f_2 \in \dot{C}^\infty(\bigcup_{s,l} Q_{sl})$ such that

$$\begin{vmatrix} (f_1, \varphi_1)_{L_2(0,d)} & (f_1, \varphi_2)_{L_2(0,d)} \\ (f_2, \varphi_1)_{L_2(0,d)} & (f_2, \varphi_2)_{L_2(0,d)} \end{vmatrix} \neq 0.$$

Substituting $f = f_1$, $f = f_2$ into (4.39), we obtain

$$\begin{aligned} c_{-1,1}(f_1,\varphi_1)_{L_2(0,d)} + c_{-1,2}(f_1,\varphi_2)_{L_2(0,d)} &= 0, \\ c_{-1,1}(f_2,\varphi_1)_{L_2(0,d)} + c_{-1,2}(f_2,\varphi_2)_{L_2(0,d)} &= 0. \end{aligned} \tag{4.40}$$

Since the determinant of the system (4.40) is non-zero, we have $c_{-1,1} = c_{-1,2} = 0$.

Now let us assume that $f^{(\beta)}(0) = 1$, $f^{(\mu)}(i) = 0$ $(i = 0,\dots,N,\ \mu = 0,\dots,k-1\colon (i,\mu) \neq (0,\beta))$ and $f^{(\mu)}(\theta + i - 1) = 0$ $(i = 1,\dots,N+1,\ \mu = 0,\dots,k-1)$. Then from (4.39) we obtain $c_{\beta 1} = 0$. Similarly, we can prove that $c_{\mu 2} = 0$. Hence, $f_0 \in \mathcal{R}(A_R^k)$ if and only if

$$(f_0,\Phi_{\mu j})_{W^k(0,d)} = 0 \qquad (\mu = -1,\dots,k-1,\ j = 1,2),$$

where $\Phi_{\mu j}$ are linearly independent functions. By Lemma 4.1,

$$\dim \mathcal{N}(A_R^k) = 0. \quad \square$$

Theorem 4.5. *Let condition* 4.1 *be fulfilled, and let* $\det R_s \neq 0$ $(s = 1,2)$. *Assume* $\theta = 1$, *and the columns* G_1^1, G_{N+1}^2 *are linearly dependent.*

Then the operator $A_R^k\colon W^{k+2}(0,d) \to W^k(0,d)$ *is Fredholm, and* $\dim \mathcal{N}(A_R^k) = 0$, $\operatorname{codim} \mathcal{R}(A_R^k) = k+1$.

Proof. Let $f_0 \in W^k(0,d)$. From Theorem 4.3 and Lemma 4.4 it follows that the equation $\mathcal{A}_R v = f_0$ has a solution $v \in \mathcal{D}(A_R^k)$ if and only if

$$(f_0,\varphi)_{L_2(0,d)} = 0, \tag{4.41}$$

$$\alpha_1 v^{(\mu+2)}(d-0) + \alpha_2 v^{(\mu+2)}(0+0) = 0 \qquad (\mu = 0,\dots,k-1), \tag{4.42}$$

where $\varphi \neq 0$, $\alpha_1^2 + \alpha_2^2 \neq 0$, $\alpha_1 G_1^1 + \alpha_2 G_{N+1}^2 = 0$.

Denote $u = R_Q v$. By virtue of Lemma 2.6, we have

$$\begin{aligned} v^{(\mu+2)}(0+0) &= (U_1 v^{(\mu+2)})_1(0+0) = \sum_i \frac{B_{i1}}{\det R_1}(U_1 u^{(\mu+2)})_i(0+0) \\ &= \sum_i \frac{B_{i1}}{\det R_1} u^{(\mu+2)}(i-1) \\ &= -\sum_i \frac{B_{i1}}{\det R_1} f_0^{(\mu)}(i-1), \end{aligned} \tag{4.43}$$

$$\begin{aligned} v^{(\mu+2)}(d-0) &= (U_1 v^{(\mu+2)})_{N+1}(1-0) = \sum_i \frac{B_{i,N+1}}{\det R_1}(U_1 u^{(\mu+2)})_i(1-0) \\ &= \sum_i \frac{B_{i,N+1}}{\det R_1} u^{(\mu+2)}(i) = -\sum_i \frac{B_{i,N+1}}{\det R_1} f_0^{(\mu)}(i). \end{aligned} \tag{4.44}$$

Substituting (4.43) and (4.44) into (4.42), we obtain

$$\alpha_1 \sum_i B_{i,N+1} f_0^{(\mu)}(i) + \alpha_2 \sum_i B_{i1} f_0^{(\mu)}(i-1) = 0 \qquad (\mu = 0,\dots,k-1). \tag{4.45}$$

By virtue of the Riesz theorem, we can rewrite (4.41) and (4.45) in the form

$$(f_0, \Phi_\mu)_{W^k(0,d)} = 0 \qquad (\mu = -1, \dots, k-1),$$

where $\Phi_\mu \in W^k(0,d)$.

As in the proof of Theorem 4.4, we can show that functions Φ_μ $(\mu = -1, \dots, k-1)$ are linearly independent. □

Using Lemma 4.3 and Theorem 4.2, and arguing as in the proof of Theorem 4.5, we obtain:

Theorem 4.6. *Let condition* 4.1 *be fulfilled, and let* $\det R_s \neq 0$ $(s = 1, 2)$. *Assume that* $\theta < 1$, *and that* $G_1^1 = 0$ *or* $G_{N+1}^2 = 0$.

Then the operator $A_R^k \colon W^{k+2}(0,d) \to W^k(0,d)$ *is Fredholm, and* $\dim \mathcal{N}(A_R^k) = 0$, $\operatorname{codim} \mathcal{R}(A_R^k) = k + 1$.

Example 4.1. We consider the boundary value problem

$$-(Rv)''(t) = f_0(t) \qquad (t \in (0,2)), \tag{4.46}$$

$$v(t) = 0 \qquad (t \in [-1,0] \cup [2,3]), \tag{4.47}$$

where $(Rv)(t) = v(t) + bv(t+1) + bv(t-1)$, $|b| \neq 0, 1$ (see Example 3.3).

Then $R_1 = \begin{pmatrix} 1 & b \\ b & 1 \end{pmatrix}$. Hence, the conditions of Lemma 4.4 are satisfied and $\alpha_1 = 1$, $\alpha_2 = -1$. The equations (4.33) will have the form

$$\begin{aligned} c_1(1-b) - c_2 b &= -b \int_0^1 (1-\tau) f_0(\tau)\, d\tau, \\ c_1(1-b) + c_2(-b+2) &= -b \int_0^1 (1-\tau) f_0(\tau)\, d\tau + \int_0^2 (2-\tau) f_0(\tau)\, d\tau. \end{aligned}$$

Evidently, the determinant of this system is equal to $2(1-b) \neq 0$. In Example 4.1, the condition (4.34) has the form

$$\int_0^2 f_0(\tau)\, d\tau = 0. \tag{4.48}$$

Thus, the boundary value problem (4.46), (4.47) has a generalized solution $v \in W^2(0,d)$ if and only if the condition (4.48) holds.

Example 4.2. We consider the boundary value problem

$$-(Rv)''(t) = f_0(t) \qquad (t \in (0,3)), \tag{4.49}$$

$$v(t) = 0 \qquad (t \in [-1,0] \cup [3,4]), \tag{4.50}$$

where $(Rv)(t) = 2v(t) + v(t+1) + v(t-1)$. Then

$$R_1 = \begin{pmatrix} 2 & 1 & 0 \\ 1 & 2 & 1 \\ 0 & 1 & 2 \end{pmatrix}.$$

Evidently, $\det R_1 = 4$, $B_{11} = 3$. The columns $G_1^1 = \begin{pmatrix} 1 \\ 0 \end{pmatrix}$, $G_3^2 = \begin{pmatrix} 0 \\ 1 \end{pmatrix}$ are linearly independent. Hence, the conditions of Lemma 4.2 are satisfied. The equations (4.19) will have the form

$$\begin{aligned} 2c_1 &= -2\int_0^1 (1-\tau)f_0(\tau)\,d\tau + \int_0^2 (2-\tau)f_0(\tau)\,d\tau, \\ 2c_1 + 6c_2 &= \int_0^1 (1-\tau)f_0(\tau)\,d\tau - 2\int_0^2 (2-\tau)f_0(\tau)\,d\tau \\ &\qquad +3\int_0^3 (3-\tau)f_0(\tau)\,d\tau. \end{aligned}$$

Hence,

$$c_2 = \frac{1}{2}\Big\{ \int_0^1 (1-\tau)f_0(\tau)\,d\tau - \int_0^2 (2-\tau)f_0(\tau)\,d\tau + \int_0^3 (3-\tau)f_0(\tau)\,d\tau \Big\}.$$

In Example 4.2, conditions (4.21), (4.22) will have the form

$$\begin{gathered} \int_0^1 (3-\tau)f_0(\tau)\,d\tau - \int_0^2 (3-\tau)f_0(\tau)\,d\tau + \int_0^3 (3-\tau)f_0(\tau)\,d\tau = 0, \\ -\int_0^1 \tau f_0(\tau)\,d\tau + \int_0^2 \tau f_0(\tau)\,d\tau - \int_0^3 \tau f_0(\tau)\,d\tau = 0. \end{gathered}$$

Thus, the boundary value problem (4.49), (4.50) has a generalized solution $v \in W^2(0,d)$ if and only if

$$(f_0, \varphi_j)_{L_2(0,d)} = 0 \qquad (j = 1, 2),$$

where

$$\begin{aligned} \varphi_1(\tau) &= \tau - 3 \quad (\tau \in (0,1) \cup (2,3)), & \varphi_1(\tau) &= 0 \quad (\tau \in (1,2)), \\ \varphi_2(\tau) &= \tau \quad (\tau \in (0,1) \cup (2,3)), & \varphi_2(\tau) &= 0 \quad (\tau \in (1,2)). \end{aligned}$$

Proof of Theorem 4.1

By virtue of the compactness of the imbedding operator of $W^{k+2}(-N, d+N)$ into $W^{k+1}(-N, d+N)$, the operator

$$A_1 : W^{k+2}(-N, d+N) \to W^k(0,d)$$

is compact. Therefore, by Theorem A.2, it suffices to prove Theorem 4.1 for the case $A_1 = 0$.

Let us assume that $A_1 = 0$.

We introduce the function

$$\psi(t) = \begin{cases} f_1(t) & (t \in [-N,0]), \\ f_2(t) & (t \in [d,d+N]), \\ \eta(t) \sum\limits_{i=0}^{k+1} f_1^{(i)}(0)t^i/i! & \\ \qquad +\eta(t-d) \sum\limits_{i=0}^{k+1} f_2^{(i)}(d)(t-d)^i/i! & (t \in (0,d)), \end{cases}$$

where $\eta \in \dot{C}^\infty(\mathbb{R})$, $\eta(t) = 1$ ($|t| < 1/4$), $\eta(t) = 0$ ($|t| > 1/3$). Evidently, $\psi \in W^{k+2}(-N,d+N)$. Denote $w = y - \psi$. Then, we can reduce the boundary value problem (4.1), (4.2) to the equation

$$A_R^k w = f_0 + (R\psi)''. \tag{4.51}$$

By Theorems 4.4–4.6, the equation (4.51) has a solution if and only if

$$(f_0 + (R\psi)'', \varphi_j)_{W^k(0,d)} = 0 \qquad (j = 1, \dots, -\operatorname{ind} A_R^k), \tag{4.52}$$

where $\varphi_j \in W^k(0,d)$ are linearly independent functions. Let us introduce a linear functional

$$\Phi_j(\psi) = ((R\psi)'', \varphi_j)_{W^k(0,d)}.$$

Evidently,

$$|f_1^{(i)}(0)| \le c_1 \|f_1\|_{W^{k+2}(-N,0)}, \qquad |f_2^{(i)}(d)| \le c_2 \|f_2\|_{W^{k+2}(d,d+N)}.$$

Here $i = 0, \dots, k+1$, $c_1, c_2 > 0$. Since $\psi \in W^{k+2}(-N,d+N)$, we have $(R\psi)''(t) = (R\psi'')(t)$ for $t \in (0,d)$. Hence,

$$|\Phi_j(\psi)| \le c\{\|f_1\|^2_{W^{k+2}(-N,0)} + \|f_2\|^2_{W^{k+2}(d,d+N)}\}^{1/2} \|\varphi_j\|_{W^k(0,d)}.$$

Then, by the Riesz theorem we obtain

$$(R\psi'', \varphi_j)_{W^k(0,d)} = (f_1, B_1\varphi_j)_{W^{k+2}(-N,0)} + (f_2, B_2\varphi_j)_{W^{k+2}(d,d+N)},$$

where

$$B_1 \colon W^k(0,d) \to W^{k+2}(-N,0), \qquad B_2 \colon W^k(0,d) \to W^{k+2}(d,d+N)$$

are linear bounded operators.

Hence, the conditions (4.52) will have the form

$$(f, G_j)_{\mathcal{W}^k(-N,d+N)} = 0 \qquad (j = 1, \dots, -\operatorname{ind} A_R^k), \tag{4.53}$$

where vector-valued functions $G_j = (\varphi_j, B_1\varphi_j, B_2\varphi_j)$ are linearly independent, $f = (f_0, f_1, f_2)$. Thus, for $A_1 = 0$ the equation $\mathcal{L}_R v = f$ has a solution $v \in W^{k+2}(-N,d+N)$ for $f \in \mathcal{W}^k(-N,d+N)$ if and only if the conditions (4.53) are fulfilled. Furthermore, by Lemma 4.1, $\dim \mathcal{N}(\mathcal{L}_R) = 0$. □

Null Space

Let us show that the operator $\mathcal{L}_R$ can have nontrivial null space.

Example 4.3. We consider the boundary value problem

$$-(Rv)''(t) + A_1 v = 0 \qquad (t \in (0,d)), \tag{4.54}$$

$$v(t) = 0 \qquad (t \in [-N,0] \cup [d, d+N]), \tag{4.55}$$

where

$$(Rv)(t) = \sum_{j=-N}^{N} b_j v(t+j), \qquad A_1 v = \alpha(t)(R_1 v)'(t),$$

$$\alpha(t) = 2\pi \sum_{j=-N}^{N} \frac{b_j \psi_{[-j,d-j]}}{\psi_{[1/4,d+1/4]} + \psi_{[-1/4,d-1/4]}} \qquad (t \in (0,d)),$$

$$(R_1 v)(t) = v(t+1/4) - v(t-1/4),$$

$$\psi_{[a,b]}(t) = 1 \quad (t \in [a,b]), \qquad \psi_{[a,b]}(t) = 0 \quad (t \notin [a,b]),$$

$d > 1$ is an integer.

Let us prove that the function

$$v(t) = (1 - \cos 2\pi t)\psi_{[0,d]} \in W^2(-N, d+N)$$

is the solution of the boundary value problem (4.54), (4.55). The points of discontinuity of the function $\psi_{[0,d]}$ coincide with roots of the functions $1 - \cos 2\pi t$, $\sin 2\pi t$. Therefore,

$$\begin{aligned} -(Rv)''(t) &= -\Big(\sum_{j=-N}^{N} b_j (1 - \cos(2\pi t + 2\pi j))\psi_{[-j,d-j]} \Big)'' \\ &= -4\pi^2 \cos 2\pi t \sum_{j=-N}^{N} b_j \psi_{[-j,d-j]}. \end{aligned}$$

Similarly,

$$\begin{aligned} (R_1 v)'(t) &= \{(1 - \cos(2\pi t + \pi/2))\psi_{[-1/4,d-1/4]} \\ &\qquad - (1 - \cos(2\pi t - \pi/2))\psi_{[1/4,d+1/4]}\}' \\ &= \{(1 + \sin 2\pi t)\psi_{[-1/4,d-1/4]} - (1 - \sin 2\pi t)\psi_{[1/4,d+1/4]}\}' \\ &= 2\pi \cos 2\pi t (\psi_{[1/4,d+1/4]} + \psi_{[-1/4,d-1/4]}). \end{aligned}$$

Thus, $-(Rv)''(t) + A_1 v = 0$.

Remark 4.2. By virtue of Theorem 4.1, the boundary value problem (4.1), (4.2) has a solution $v \in W^{k+2}(-N, d+N)$ if and only if a right-hand side $f = (f_0, f_1, f_2) \in \mathcal{W}^k(-N, d+N)$ satisfies $(\dim \mathcal{N}(\mathcal{L}_R) - \operatorname{ind} \mathcal{L}_R)$ conditions of orthogonality in the space $\mathcal{W}^k(-N, d+N)$. Evidently, if $v \in W^{k+2}(0,d)$, then $v \in C^{k+1}[0,d]$.

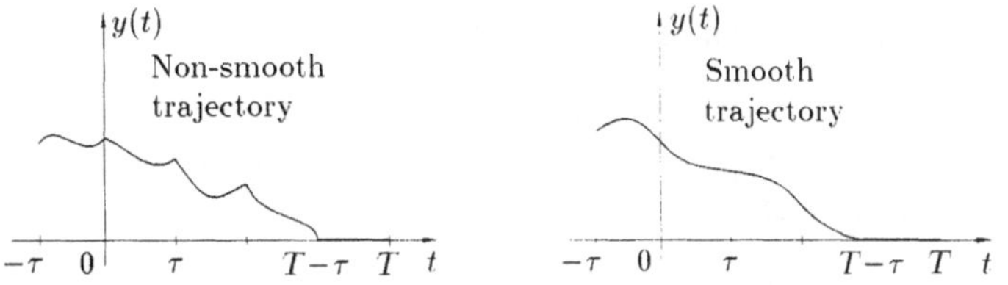

Fig. I.5

Problem 4.1. *To obtain necessary and sufficient conditions of existence of smooth solutions in the case of a system of differential-difference equations.*

This problem has applications to control theory with delay (see Problem 5.1).

5 Applications to Control Systems with Delay

Formulation of Problem

We consider a linear control system with delay described by the equation

$$y'(t) + ay'(t-\tau) + by(t) + cy(t-\tau) = u(t) \qquad (0 < t), \tag{5.1}$$

where a, b, c are real constants, the delay $\tau > 0$ is a constant, and $u(t)$ is a control function (see Fig. 0.1).

A previous history of the system is defined by the initial condition

$$y(t) = \varphi(t) \qquad (t \in [-\tau, 0]), \tag{5.2}$$

where $\varphi(t)$ is a given function.

We shall study the problem of how to reduce the system (5.1), (5.2) to equilibrium. Let us find a control function $u(t)$ $(0 < t < T)$ such that

$$y(t) = 0 \qquad (t \in [T-\tau, T]), \tag{5.3}$$

where $T > 2\tau$ (see Fig. I.5). If we set $u(t) \equiv 0$ $(t > T)$, then the solution of the problem (5.1), (5.3) is $y(t) \equiv 0$ $(t > T)$.

Evidently, a function $y(t)$ satisfying the conditions (5.1)–(5.3) is not unique. Therefore, we also assume that

$$\int_0^T u^2(t)\, dt \to \min . \tag{5.4}$$

Thus, we obtain the variational problem for energy functional

$$J(y) = \int_0^T \{y'(t) + ay'(t-\tau) + by(t) + cy(t-\tau)\}^2\, dt \to \min \tag{5.5}$$

with boundary conditions (5.2), (5.3).

Variational Problem and Boundary Value Problem

Without loss of generality, we assume that $\tau = 1$. Assume $y \in W^1(-1,T)$ is a solution of the variational problem (5.5), (5.2), (5.3), where $\varphi \in W^1(-1,0)$. In this section we consider the spaces of real-valued functions. We denote

$$\widetilde{W} = \{v \in W^1(-1,T) : v(t) = 0 \ (t \in [-1,0] \cup [T-1,T])\}.$$

Let $v \in \widetilde{W}$ be an arbitrary fixed function. Then a function $y+sv \in W^1(-1,T)$ and satisfies the boundary conditions (5.2), (5.3) for each $s \in \mathbb{R}$. Denote $J(y+sv) = F(s)$. Since

$$J(y+sv) \geq J(y) \qquad (s \in \mathbb{R}), \tag{5.6}$$

we have

$$\left.\frac{dF}{ds}\right|_{s=0} = 0. \tag{5.7}$$

From (5.7) we obtain

$$\begin{aligned} b[y,v] = \int_0^T & (y'(t) + ay'(t-1) + by(t) + cy(t-1)) \\ & \times (v'(t) + av'(t-1) + bv(t) + cv(t-1))\, dt = 0. \end{aligned} \tag{5.8}$$

In the terms containing $v(t-1)$ or $v'(t-1)$, we change the variable $\xi = t-1$ and return to the old variable $t = \xi$. Then, since $v \in \widetilde{W}$, it follows from (5.8) that

$$\begin{aligned} \int_0^{T-1} & \{[(1+a^2)y'(t) + ay'(t-1) + ay'(t+1)]v'(t) \\ & + [(ac+b)y(t) + cy(t-1) + aby(t+1)]v'(t) \\ & + [(ac+b)y'(t) + aby'(t-1) + cy'(t+1)]v(t) \\ & + [(b^2+c^2)y(t) + bcy(t-1) + bcy(t+1)]v(t)\}\, dt = 0. \end{aligned}$$

Using the condition $v(0) = v(T-1) = 0$ and integrating by parts, we have

$$\int_0^{T-1} y(t)v'(t)\, dt = -\int_0^{T-1} y'(t)v(t)\, dt.$$

Therefore,

$$\begin{aligned} b[y,v] = \int_0^{T-1} & \{[(1+a^2)y'(t) + ay'(t-1) + ay'(t+1)]v'(t) \\ & +[(ab-c)(y'(t-1) - y'(t+1)) + (b^2+c^2)y(t) \\ & +bc(y(t-1) + y(t+1))]v(t)\}\, dt = 0. \end{aligned} \tag{5.9}$$

From (5.9) it follows that

$$(1+a^2)y'(t) + ay'(t-1) + ay'(t+1) \in W^1(0,T-1). \tag{5.10}$$

Thus, integrating by parts, we have

$$\begin{aligned}
&-[(1+a^2)y'(t)+ay'(t-1)+ay'(t+1)]' \\
&\quad +(ab-c)(y'(t-1)-y'(t+1))+(b^2+c^2)y(t) \\
&\quad +bc(y(t-1)+y(t+1))=0 \qquad (t\in(0,T-1)). \qquad (5.11)
\end{aligned}$$

We define the bounded operator $R\colon L_2(-1,T)\to L_2(0,T-1)$ and A_1: $W^1(-1,T)\to L_2(0,T-1)$ by the formulas

$$\begin{aligned}
(Ry)(t)&=(1+a^2)y(t)+ay(t-1)+ay(t+1) && (t\in(0,T-1)),\\
(A_1y)(t)&=(ab-c)(y'(t-1)-y'(t+1)) \\
&\quad +(b^2+c^2)y(t)+bc(y(t-1)+y(t+1)) && (t\in(0,T-1)).
\end{aligned}$$

Since $(Ry)'(t)=(Ry')(t)$ $(t\in(0,T-1))$ for $y\in W^1(-1,T)$, the equation (5.11) will have the form

$$-(Ry)''(t)+A_1y=0 \qquad (t\in(0,d)). \qquad (5.12)$$

A function $y(t)$ satisfies the following boundary conditions:

$$y(t)=\varphi(t) \quad (t\in[-1,0]), \qquad y(t)=0 \quad (t\in[d,d+1]), \qquad (5.13)$$

where $d=T-1$.

Remark 5.1. Since the difference operator R contains only two shifts of argument ±1, we put the boundary conditions on $[-1.0]\cup[d,d+1]$, unlike in Sections 3, 4.

We have proved that, if a function $y\in W^1(-1,T)$ gives a minimum to the variational problem (5.5), (5.2), (5.3), then $y(t)$ is a generalized solution of the boundary value problem (5.12), (5.13) (see Definition 3.3).

Now let $y\in W^1(-1,T)$ be a generalized solution of the boundary value problem (5.12), (5.13). Then for every $v\in\widetilde{W}$ we have

$$J(y+v)=J(y)+J(v)+2b[y,v]\geq J(y).$$

Thus, we have proved the following statement:

Theorem 5.1. *A function $y\in W^1(-1,T)$ gives a minimum to a variational problem* (5.5), (5.2), (5.3) *if and only if y is a generalized solution of the boundary value problem* (5.12), (5.13).

Solvability of a Boundary Value Problem

In order to prove the existence and uniqueness of generalized solution of a boundary value problem (5.12), (5.13), we first obtain some auxiliary statements.

We denote $J_1(w)=\int_0^T(w'(t)+aw'(t-1))^2dt$, where $w\in\widetilde{W}$.

Lemma 5.1. *There exists a constant* $c_1 > 0$ *such that, for all* $w \in \widetilde{W}$,

$$J_1(w) \geq c_1 \|w\|^2_{W^1(0,T-1)}. \tag{5.14}$$

Proof. Similarly to the above, we obtain

$$\begin{aligned} J_1(w) &= \int_0^{T-1} [(1+a^2)w'(t) + aw'(t-1) + aw'(t+1)]w'(t)\,dt \\ &= (R_Q w', w')_{L_2(0,T-1)}. \end{aligned} \tag{5.15}$$

Here $R: L_2(\mathbb{R}) \to L_2(\mathbb{R})$, $R_Q: L_2(0,T-1) \to L_2(0,T-1)$ are linear bounded operators given by $Rv = (1+a^2)v(t) + av(t+1) + av(t-1)$, $R_Q = P_Q R I_Q$.

For every $a \in \mathbb{R}$ a symbol $R(\xi) = (1+a^2) + 2a\cos\xi \geq 0$ and $R(\xi) \not\equiv 0$ ($\xi \in \mathbb{R}$). Therefore, by virtue of Lemmas 2.4, 2.3, 2.9, the operator R_Q is positive definite. Thus, from (5.15) and from the formula (B.29) we obtain

$$J_1(w) \geq c_1 \|w\|^2_{W^1(0,T-1)}. \quad \square$$

Lemma 5.2. *There exists a constant* $c_2 > 0$ *such that, for all* $w \in \widetilde{W}$,

$$J(w) \geq c_2 \|w\|^2_{W^1(0,T-1)}. \tag{5.16}$$

Proof. Assume the inequality (5.16) does not hold. Then, for each $n > 0$, there exists a function $w_n \in \widetilde{W}$ such that

$$J(w_n) \leq \frac{1}{n} \|w_n\|^2_{W^1(0,T-1)}.$$

Without loss of generality, we assume that $\|w_n\|_{W^1(0,T-1)} = 1$. Thus, we have

$$J(w_n) \leq 1/n. \tag{5.17}$$

On the other hand, from the inequality $(\alpha-\beta)^2 \geq \alpha^2/2 - \beta^2$ ($\alpha, \beta \in \mathbb{R}$) and Lemma 5.1, for every $v \in \widetilde{W}$, we obtain

$$J(v) \geq k_1 \|v\|^2_{W^1(0,T-1)} - k_2 \|v\|^2_{L_2(0,T-1)}. \tag{5.18}$$

By virtue of the compactness of the imbedding operator from $\widetilde{W}$ into $L_2(-1,T)$, there exists a subsequence w_{n_k}, which converges to w_0 in the space $L_2(-1,T)$. Thus, from (5.17), (5.18) it follows that

$$\begin{aligned} k_1 \|w_{n_k} - w_{n_m}\|^2_{W^1(0,T-1)} &\leq k_2 \|w_{n_k} - w_{n_m}\|^2_{L_2(0,T-1)} + J(w_{n_k} - w_{n_m}) \\ &\leq k_2 \|w_{n_k} - w_{n_m}\|^2_{L_2(0,T-1)} + 2/n_k + 2/n_m \to 0 \end{aligned}$$

as $k, m \to \infty$.

Hence, $w_{n_k} \to w_0$ in the space $\widetilde{W}$ and $\|w_0\|_{W^1(0,T-1)} = 1$. Therefore, we have

$$J(w_0) = \int_0^T \{w_0'(t) + aw_0'(t-1) + bw_0(t) + cw_0(t-1)\}^2\, dt = 0,$$

i.e.,

$$w_0'(t) + aw_0'(t-1) + bw_0(t) + cw_0(t-1) = 0 \qquad (t \in (0,T)). \tag{5.19}$$

Since $w_0 \in \widetilde{W}$, a function w_0 satisfies the initial condition

$$w_0(t) = 0 \qquad (t \in [-1,0]). \tag{5.20}$$

Then, if $0 < t \leq 1$, the equation (5.19) will have the form

$$w_0'(t) + bw_0(t) = 0, \tag{5.21}$$

and $w_0(0) = 0$. Hence,

$$w_0(t) = 0 \qquad (t \in [0,1]). \tag{5.22}$$

By virtue of (5.22), the equation (5.19) will have the form (5.21) if $1 < t \leq 2$, and $w_0(1) = 0$. From this it follows that

$$w_0(t) = 0 \qquad (t \in [1,2])$$

and so on. Hence, $w_0(t) \equiv 0$ $(0 \leq t \leq T-1)$. But this is impossible, since $\|w_0\|_{W^1(0,T-1)} = 1$. □

Theorem 5.2. *For every $\varphi \in W^1(-1,0)$, there exists a unique generalized solution of the boundary value problem* (5.12), (5.13) *$y \in W^1(-1,T)$, and*

$$\|y\|_{W^1(-1,T)} \leq c\|\varphi\|_{W^1(-1,0)}, \tag{5.23}$$

where $c > 0$ does not depend on φ.

Proof. By virtue of Theorem 5.1 and Lemma 5.2, if $\varphi = 0$, then the boundary value problem (5.12), (5.13) has a unique trivial solution. On the other hand, in the proof of Lemma 5.1 it was shown that the operator $R_Q \colon L_2(0,T-1) \to L_2(0,T-1)$ is self-adjoint and positive definite. Therefore, by virtue of Lemma 2.8 the matrix

$$R_1 = \begin{pmatrix} 1+a^2 & a & 0 & \dots & 0 \\ a & 1+a^2 & a & \dots & 0 \\ 0 & a & 1+a^2 & \dots & 0 \\ \vdots & \vdots & \vdots & \ddots & \vdots \\ 0 & 0 & 0 & \dots & 1+a^2 \end{pmatrix} \tag{5.24}$$

is positive definite. Hence, by Theorem 3.7, for every $\varphi \in W^1(-1,0)$, there is a unique generalized solution of the problem (5.12), (5.13), and the estimate (5.23) holds. □

Smooth Solutions

Let us now study the smoothness of generalized solutions of the problem (5.12), (5.13).

Denote $d = T - 1$. Let $d = N + \theta$, where $0 < \theta \leq 1$ and N is a natural number.

Since the symmetric matrix R_1 is positive definite, from Theorem 3.8 we obtain

Theorem 5.3. *Let $\varphi \in W^{k+2}(-1,0)$, $k \geq 0$, and let $y(t)$ be a generalized solution of the boundary value problem* (5.12), (5.13).

Then $y \in W^{k+2}(j-1,j)$ $(j = 1,\dots,N+1)$ if $\theta = 1$, and $y \in W^{k+2}(j-1,j-1+\theta)$ $(j = 1,\dots,N+1)$, $y \in W^{k+2}(j-1+\theta,j)$ $(j = 1,\dots,N)$ if $\theta < 1$.

Theorem 5.4. *Let $a \neq 0$, and let $\varphi \in W^{k+2}(-1,0)$, where $k \geq 0$.*

Then a generalized solution of the boundary value problem (5.12), (5.13) *$y \in W^{k+2}(-1,T)$ if and only if*

$$(\varphi, \psi_j)_{W^{k+2}(-1,0)} = 0 \qquad (j = 1,\dots,s), \tag{5.25}$$

where $\psi_1,\dots,\psi_s \in W^{k+2}(-1,0)$ are linearly independent functions, $s = k+1$ if $T = 3$, and $s = 2(k+1)$ if $T > 3$.

Proof. Let us define the bounded operator

$$\mathcal{L}_R: W^{k+2}(-1,T) \to W^k(0,T-1) \times W^{k+2}(-1,0) \times W^{k+2}(T-1,T)$$

by the formula

$$\mathcal{L}_R y = \{-(Ry)''(t) + (A_1 y)(t), y|_{[-1,0]}, y|_{[T-1,T]}\}.$$

By virtue of Theorem 5.2, the boundary value problem (5.12), (5.13) has a unique generalized solution $y \in W^1(-1,T)$. Hence, it is sufficient to prove that $\operatorname{ind}\mathcal{L}_R = -s$, where $s = k+1$ if $d = T-1 = 2$ and $s = 2(k+1)$ if $d = T-1 > 2$.

In fact, if $d = 2$, then a matrix R_1 has the form

$$R_1 = \begin{pmatrix} 1+a^2 & a \\ a & 1+a^2 \end{pmatrix}.$$

Therefore, the columns $G_1^1 = (a)$, $G_{N+1}^2 = (a)$ are linearly dependent, and, by virtue of Theorem 4.1, $\operatorname{ind}\mathcal{L}_R = -(k+1)$.

If $d > 2$, then the columns G_1^1, G_{N+1}^2 have the form

$$G_1^1 = \begin{pmatrix} a \\ 0 \\ \vdots \\ 0 \end{pmatrix}, \qquad G_{N+1}^2 = \begin{pmatrix} 0 \\ \vdots \\ 0 \\ a \end{pmatrix}.$$

Since $a \neq 0$, the columns G_1^1, G_{N+1}^2 are linearly independent. Hence, by virtue of Theorem 4.1, $\operatorname{ind}\mathcal{L}_R = -2(k+1)$. □

Thus, if the initial function $\varphi(t)$ is sufficiently smooth and satisfies some conditions of orthogonality, then a generalized solution $y(t)$ of the problem (5.12), (5.13) has corresponding smoothness.

Problem 5.1. *To extend the results of this section to the case when $y(t)$ is an n-dimensional vector-valued function, $u(t)$ is an n-dimensional vector-valued function, and a, b, c are matrices of order $n \times n$.*

6 The Boundary Value Problem for the Differential-Difference Equation with Degeneration

In Sections 3–5 we considered boundary value problems for differential-difference equations in the case when a difference operator was nondegenerate. Now we study a differential-difference equation with degeneration. We assume that a difference operator is nonnegative. As in Section 3, our approach is based on the properties of difference operators and the construction of a Friedrichs extension of differential-difference operator. A generalization to a multidimensional case will be considered in Chapter IV.

Smoothness of Solutions

We consider the differential-difference equation

$$-(Rv)''(t) = f_0(t) \qquad (t \in (0,d)) \tag{6.1}$$

with homogeneous boundary conditions

$$v(t) = 0 \qquad (t \in [-N,0] \cup [d, d+N]). \tag{6.2}$$

Here $R\colon L_2(\mathbb{R}) \to L_2(\mathbb{R})$ is the difference operator defined by

$$(Rv)(t) = \sum_{j=-N}^{N} b_j v(t+j),$$

$b_j \in \mathbb{R}$, $d = N + \theta$, $0 < \theta \le 1$, N is a natural number, and $f_0 \in L_2(0,d)$ is a complex-valued function.

We assume that the following conditions are fulfilled:

6.1. $b_j = b_{-j}$ $(j = 1, \dots, N)$, $b_0 > 0$.

6.2. $0 \subset \bigcup_s \sigma(R_s)$ $(s = 1$ if $\theta = 1$; $s = 1,2$ if $\theta < 1)$.

6.3. *The matrices R_s are nonnegative* $(s = 1$ if $\theta = 1$; $s = 1,2$ if $\theta < 1)$.

From the condition 6.1 and Lemma 2.14 it follows that the operator R_Q is self-adjoint.

Let us consider the unbounded operator $\mathcal{A}_R: L_2(0,d) \to L_2(0,d)$ defined by

$$\begin{aligned} \mathcal{A}_R v &= -(R_Q v)''(t), \\ \mathcal{D}(\mathcal{A}_R) &= \{v \in L_2(0,d) : R_Q v \in W^2(0,d)\}. \end{aligned} \tag{6.3}$$

Definition 6.1. A function $v \in \mathcal{D}(\mathcal{A}_R)$ is called a *generalized solution* of the problem (6.1), (6.2) if

$$\mathcal{A}_R v = f_0. \tag{6.4}$$

Let $P^R: L_2(0,d) \to L_2(0,d)$ be the operator of orthogonal projection onto the subspace $\mathcal{R}(R_Q)$. From Definition 6.1 and Lemma 2.17 we obtain the following:

Theorem 6.1. *Assume the conditions* 6.1–6.3 *are fulfilled. Let $v(t)$ be a generalized solution of the boundary value problem* (6.1), (6.2), *and let $f_0 \in W^k(0,d)$.*

Then $P^R v \in W^{k+2}(j-1,j)$ $(j = 1,\dots,N+1)$ if $\theta = 1$, and $P^R v \in W^{k+2}(j-1,j-1+\theta)$ $(j = 1,\dots,N+1)$, $P^R v \in W^{k+2}(j-1+\theta,j)$ $(j = 1,\dots,N)$ if $\theta < 1$.

Self-Adjoint Operator

Now we shall prove that the operator $\mathcal{A}_R: L_2(0,d) \to L_2(0,d)$ is self-adjoint.

Lemma 6.1. *Let conditions* 6.1–6.3 *be fulfilled.*

Then $\det R_1 = 0$.

Proof. When $\theta = 1$, Lemma 6.1 follows from the condition 6.2. Now let $\theta < 1$, and let $\det R_2 = 0$. Then, by virtue of the extremal property of eigenvalues and condition 6.3, we have $\det R_1 = 0$. □

Lemma 6.2. *Let conditions* 6.1–6.3 *be fulfilled.*

Then the first row of the matrix R_1 is a linear combination of the remaining N rows.

Proof. Denote by S_k the matrix of order $k \times k$ obtained from the matrix R_1 by deleting the last $N+1-k$ rows and columns. If $k = N+1$, then $S_k = R_1$. By virtue of Lemma 6.1 and symmetry of the matrix R_1, it is sufficient to prove the following statement:

A. Let $\det S_k = 0$. Then there exists a vector $(\alpha_1,\dots,\alpha_k) \in \mathcal{N}(S_k)$ such that $\alpha_1 \neq 0$ $(k = 2,\dots,N+1)$.

We shall prove this statement by induction. Evidently, the statement A holds for $k = 2$.

Now let the statement A be correct for $k = m-1$. Assume that for $k = m$ this statement is not true, i.e., $\det S_m = 0$ and the vectors from $\mathcal{N}(S_m)$ have the following form $(0,\alpha_2,\dots,\alpha_m)$. Then $\det S_{m-1} = 0$ (in the opposite case the first column of S_m is a linear combination of the remaining columns). By virtue of (2.8) and symmetry of the matrix R_1, the elements of the matrix R_1 has the following form $r_{ij} = b_{|j-i|}$. Hence, if $(\alpha_1, \alpha_2, \dots, \alpha_{m-1}, \alpha_m) \in \mathcal{N}(S_m)$, then $(\alpha_m, \alpha_{m-1}, \dots, \alpha_2, \alpha_1) \in \mathcal{N}(S_m)$. In fact, let

$$\sum_{j=1}^{m} \alpha_j b_{|j-i|} = 0 \qquad (i = 1, \dots, m).$$

Then, denoting $j = m+1-p$, $i = m+1-q$, we have

$$0 = \sum_{j=1}^{m} \alpha_j b_{|j-i|} = \sum_{p=1}^{m} \alpha_{m+1-p} b_{|p-q|} \qquad (q = 1, \dots, m).$$

Thus, the elements of subspace $\mathcal{N}(S_m)$ have the form $(0, \alpha_2, \dots, \alpha_{m-1}, 0)$. Hence, $(0, \alpha_2, \dots, \alpha_{m-1}) \in \mathcal{N}(S_{m-1})$. Since $\det S_{m-1} = 0$, then by virtue of the induction hypothesis there exists a vector $(\beta_1, \dots, \beta_{m-1}) \in \mathcal{N}(S_{m-1})$ such that $\beta_1 \neq 0$. Therefore, $\dim \mathcal{N}(S_m) + 1 \leq \dim \mathcal{N}(S_{m-1})$. On the other hand, from condition 6.3 we have $S_m \geq 0$. Hence, by virtue of the extremal property of eigenvalues, $\dim \mathcal{N}(S_{m-1}) \leq \dim \mathcal{N}(S_m)$. We have obtained a contradiction. □

Denote $W_{A_R} = \{v \in L_2(0,d) : R_Q v \in W^1(0,d)\}$.

Lemma 6.3. *Let conditions* 6.1–6.3 *be fulfilled.*

Then, for every $v \in W_{A_R}$, *there exists* $v_0 \in \mathring{W}^1(0,d)$ *such that* $R_Q v = R_Q v_0$.

Proof. Let us consider the case $\theta = 1$. Assume that $v \in L_2(0,d)$, $R_Q v \in W^1(0,d)$. From Lemma 2.17 it follows that $(U_1 P^R v)_k \in W^1(0,1)$. Therefore, we can define the vectors $V' = U_1 P^R v|_{t=0}$, $V'' = U_1 P^R v|_{t=1}$. Since $R_Q v \in W^1(0,d)$, we have

$$(R_1 V')_{k+1} = (R_1 V'')_k \qquad (k = 1, \dots, N). \tag{6.5}$$

1. Let us show that there is $v_1 \in \mathring{W}^1(0,d)$ such that

$$R_Q v_1|_{t=k} = R_Q v|_{t=k} \qquad (k = 0, \dots, N+1). \tag{6.6}$$

For this it is sufficient to prove that there exists a solution $X \in \mathbb{C}^{N+2}$ of the system

$$A_1 X = F$$

such that $X_1 = X_{N+2} = 0$, where

$$F = ((R_1 V')_1, \dots, (R_1 V')_{N+1}, (R_1 V'')_{N+1}),$$

and A_1 is the $(N+2) \times (N+2)$-matrix with the elements

$$a_{ij} = b_{|j-i|} \qquad (i, j = 1, \dots, N+2). \tag{6.7}$$

This problem is equivalent to the following

$$K_1 Y = F, \tag{6.8}$$

where K_1 is the matrix of order $(N+2)\times N$ obtained from the matrix A_1 by deleting the first and the last columns, $Y=(X_2,\dots,X_{N+1})$. Denote by K_2 the extended matrix of the system (6.8). By virtue of Lemma 6.2, the first row of the matrices K_1 and K_2 is a linear combination of the ith rows with coefficients γ_{i-1} $(i=2,\dots,N+1)$. From (6.5) we have

$$F=((R_1V')_1,(R_1V'')_1,\dots,(R_1V'')_{N+1}).$$

Therefore, from Lemma 6.2 and the formula (6.7) it follows that the $(N+2)$th row of the matrices K_1 and K_2 is equal to a linear combination of the ith rows with coefficients γ_{N+2-i} $(i=2,\dots,N+1)$. Thus, $\operatorname{rank} K_1=\operatorname{rank} K_1'$, $\operatorname{rank} K_2=\operatorname{rank} K_2'$, where K_1' and K_2' are matrices of order $N\times N$ and $N\times(N+1)$ obtained from K_1 and K_2 by deleting the first and the last rows. By Lemma 6.2, the last column of the matrix K_2' is a linear combination of the columns of K_1'. Hence, $\operatorname{rank} K_1'=\operatorname{rank} K_2'$. Therefore, the system (6.8) is compatible.

2. Thus, we have proved that there exists a function $v_1\in \mathring{W}^1(0,d)$ satisfying conditions (6.6). From (6.6) and Lemma 2.17 it follows that $R_q(v-v_1)|_{t=k}=0$ $(k=0,\dots,N+1)$ and $P^R(v-v_1)\in W^1(j-1,j)$ $(j=1,\dots,N+1)$. Therefore, by virtue of (2.29), $P^R(v-v_1)|_{t=k}=0$ $(k=0,\dots,N+1)$. Hence, $P^R(v-v_1)\in \mathring{W}^1(0,d)$. Then we have $v_0=P^R(v-v_1)+v_1\in \mathring{W}^1(0,d)$ and $R_Qv_0=R_Qv$.

The proof is analogous for the case $\theta<1$. □

Theorem 6.2. *Let conditions* 6.1–6.3 *be fulfilled.*

Then the operator $\mathcal{A}_R\colon L_2(0,d)\to L_2(0,d)$ *is self-adjoint.*

Proof. 1. Let $u,v\in\mathcal{D}(\mathcal{A}_R)$. By virtue of Lemma 6.3, there exist functions $u_1,v_1\in \mathcal{N}(R_Q)$ such that $u_0=u+u_1\in\mathring{W}^1(0,d)$, $v_0=v+v_1\in\mathring{W}^1(0,d)$. From (2.23) it follows that $(R_Qu)'',(R_Qv)''\in\mathcal{R}(R_Q)$. Therefore, by virtue of Lemmas 2.10, 2.16, integrating by parts, we obtain, for $u,v\in\mathcal{D}(\mathcal{A}_R)$,

$$\begin{aligned}(-(R_Qu)'',v)_{L_2(0,d)}&=(-(R_Qu_0)'',v_0)_{L_2(0,d)}=(R_Qu_0',v_0')_{L_2(0,d)}\\&=(u_0',R_Qv_0')_{L_2(0,d)}=(u_0,-(R_Qv_0)'')_{L_2(0,d)}=(u,-(R_Qv)'')_{L_2(0,d)}.\end{aligned}$$

Hence, $\mathcal{A}_R\subset\mathcal{A}_R^*$.

2. We shall now prove that $\mathcal{A}_R^*\subset\mathcal{A}_R$. Let $v\in\mathcal{D}(\mathcal{A}_R^*)$. Then, for all $u\in \dot{C}^\infty(0,d)\subset\mathcal{D}(\mathcal{A}_R)$, we have

$$(-(R_Qu)'',v)_{L_2(0,d)}=(u,\mathcal{A}_R^*v)_{L_2(0,d)}.$$

By Lemma 2.10, $(R_Qu)''=R_Qu''$. Hence,

$$(u,\mathcal{A}_R^*v)_{L_2(0,d)}=-(u'',R_Qv)_{L_2(0,d)}. \tag{6.9}$$

Since $u\in\dot{C}^\infty(0,d)$ is arbitrary, from the definition of a derivative in the sense of distributions (see Appendix B) and identity (6.9), it follows that $R_Qv\in W^2(0,d)$ and

$$(u,\mathcal{A}_R^*v)_{L_2(0,d)}=(u,-(R_Qv)'')_{L_2(0,d)}.$$

Hence, $v\in\mathcal{D}(\mathcal{A}_R)$ and $\mathcal{A}_R^*v=\mathcal{A}_Rv$. □

A Priori Estimates and Spectrum

Lemma 6.4. *Let conditions* 6.1–6.3 *be fulfilled.*
Then, for all $v \in \mathcal{D}(\mathcal{A}_R)$, *we have*

$$c_1 \| R_Q v\|^2_{W^1(0,d)} \leq (\mathcal{A}_R v + R_Q v, v)_{L_2(0,d)} \leq c_2 \|R_Q v\|^2_{W^1(0,d)}, \tag{6.10}$$

where $c_1, c_2 > 0$ *do not depend on* v.

Proof. 1. By virtue of Lemmas 2.7, 2.16, 2.2, 2.13 and the spectral theorem, for all $u \in L_2(0,d)$,

$$(R_Q u, u)_{L_2(0,d)} \geq k_1 \|P^R u\|^2_{L_2(0,d)} \geq k_2 \|R_Q u\|^2_{L_2(0,d)}. \tag{6.11}$$

Now let $v \in \mathcal{D}(\mathcal{A}_R)$. By Lemma 6.3, there exists $v_1 \in \mathcal{N}(R_Q)$ such that $v_0 = v + v_1 \in \mathring{W}^1(0,d)$. From (2.23) it follows that $(R_Q v)'' \in \mathcal{R}(R_Q)$. Therefore, using Lemma 2.10 and integrating by parts, we have

$$(\mathcal{A}_R v + R_Q v, v)_{L_2(0,d)} = (R_Q v_0', v_0')_{L_2(0,d)} + (R_Q v_0, v_0)_{L_2(0,d)}. \tag{6.12}$$

Substituting $u = v_0'$, $u = v_0$ into (6.11), we obtain

$$(\mathcal{A}_R v + R_Q v, v)_{L_2(0,d)} \geq k_2 (\|R_Q v_0'\|^2_{L_2(0,d)} + \|R_Q v_0\|^2_{L_2(0,d)}).$$

This inequality and Lemma 2.10 give the left part of (6.10).

2. Lemma 2.16 and (6.12) imply that

$$\begin{aligned}
&(\mathcal{A}_R v + R_Q v, v)_{L_2(0,d)} \\
&\quad \leq (\|R_Q v_0'\|_{L_2(0,d)} \|P^R v_0'\|_{L_2(0,d)} + \|R_Q v_0\|_{L_2(0,d)} \|P^R v_0\|_{L_2(0,d)}) \\
&\quad \leq k_3 (\|R_Q v_0'\|^2_{L_2(0,d)} + \|R_Q v_0\|^2_{L_2(0,d)}).
\end{aligned}$$

From this, using Lemma 2.10, we obtain the right-part of (6.10). □

Lemma 6.5. *Let conditions* 6.1–6.3 *be fulfilled.*
Then $\dim \mathcal{N}(R_Q) = \infty$.

Proof. By Lemma 6.1, $\det R_1 = 0$. Hence, the system of linear algebraic equations

$$R_1 X = 0 \tag{6.13}$$

has a nontrivial solution. Thus, the subspace $\mathcal{N}(R_{Q1}) \subset L_2^{N+1}(0,\theta)$ consists of vector-valued functions $\mathcal{Z}(t)$ such that

$$R_1 \mathcal{Z}(t) = 0 \qquad (t \in (0,\theta)). \tag{6.14}$$

Evidently, $\dim \mathcal{N}(R_{Q1}) = \infty$. Therefore, from the imbedding

$$U_1^{-1} \mathcal{N}(R_{Q1}) \subset \mathcal{N}(R_Q)$$

we obtain $\dim \mathcal{N}(R_Q) = \infty$. □

Denote by $\mathcal{A}_R^p$ the restriction of the operator $\mathcal{A}_R + pR_Q$ to the set $\mathcal{D}(\mathcal{A}_R) \cap \mathcal{R}(R_Q)$, where $p \geq 0$.

Lemma 6.6. *Let conditions* 6.1–6.3 *be fulfilled.*

Then the unbounded, self-adjoint operator $\mathcal{A}_R^0 \colon \mathcal{R}(R_Q) \to \mathcal{R}(R_Q)$ *is Fredholm, and* $\operatorname{ind} \mathcal{A}_R^0 = 0$. *The spectrum* $\sigma(\mathcal{A}_R^0)$ *is discrete, real and consists of positive eigenvalues* λ_s. *For* $\lambda \notin \sigma(\mathcal{A}_R^0)$, *the resolvent* $R(\lambda, \mathcal{A}_R^0) \colon \mathcal{R}(R_Q) \to \mathcal{R}(R_Q)$ *is compact.*

Proof. 1. Evidently, $\mathcal{N}(R_Q) \subset \mathcal{N}(\mathcal{A}_R + pR_Q) \subset \mathcal{D}(\mathcal{A}_R)$. Hence, by virtue of Lemma 2.16, $\mathcal{R}(R_Q)$ is the invariant subspace of the operator $\mathcal{A}_R + pR_Q$.

Thus, the operator $\mathcal{A}_R^p \colon \mathcal{R}(R_Q) \to \mathcal{R}(R_Q)$ is self-adjoint.

2. Let us consider the operator $\mathcal{A}_R^1$. By virtue of Lemma 2.16, for all $u, v \in \mathcal{D}(\mathcal{A}_R)$,

$$((\mathcal{A}_R + R_Q)u, v)_{L_2(0,d)} = (-(R_Q u)'' + R_Q u, (R_Q^R)^{-1} R_Q v)_{L_2(0,d)}.$$

We define the sesquilinear symmetric form on

$$R_Q(\mathcal{D}(\mathcal{A}_R)) \times R_Q(\mathcal{D}(\mathcal{A}_R))$$

by the formula

$$b_1[R_Q u, R_Q v] = ((\mathcal{A}_R + R_Q)u, v)_{L_2(0,d)}.$$

Denote by W_R the closure of the set $R_Q(\mathcal{D}(\mathcal{A}_R))$ in $W^1(0,d)$. By virtue of Lemma 6.4, the form $b_1[\cdot, \cdot]$ can be continued to a bounded sesquilinear symmetric form defined on $W_R \times W_R$. Since $\dot{C}^\infty(0,d)$ is dense in $\mathring{W}^1(0,d)$, Lemmas 2.10, 6.3 imply that $W_R = R_Q(W_{\mathcal{A}_R})$. By virtue of the Riesz theorem, for all $w_1, w_2 \in W_R$,

$$b_1[w_1, w_2] = (B_1 w_1, w_2)_{W^1(0,d)},$$

where $B_1 \colon W_R \to W_R$ is a bounded self-adjoint operator.

Take arbitrary $u \in W_{\mathcal{A}_R}$ and $v \in \dot{C}^\infty(0,d)$. By virtue of Lemmas 2.10, 6.3, there is a sequence $\{u_n\} \subset \dot{C}^\infty(0,d)$ such that

$$\lim_{n \to \infty} \|R_Q u_n - R_Q u\|_{W^1(0,d)} = 0.$$

Hence, $(\mathcal{A}_R + R_Q)u_n \to (\mathcal{A}_R + R_Q)u$ in the space $\mathcal{D}'(0,d)$. Thus,

$$\begin{aligned} \langle (\mathcal{A}_R + R_Q)u, \overline{v} \rangle &= \lim_{n \to \infty} ((\mathcal{A}_R + R_Q)u_n, v)_{L_2(0,d)} \\ &= \lim_{n \to \infty} (B_1(R_Q u_n), R_Q v)_{W^1(0,d)} \\ &= (B_1(R_Q u), R_Q v)_{W^1(0,d)}. \end{aligned} \tag{6.15}$$

On the other hand, by virtue of Lemma 2.16, for all $f \in \mathcal{R}(R_Q)$ and $v \in \dot{C}^\infty(0,d)$,

$$(f, v)_{L_2(0,d)} = (f, (R_Q^R)^{-1} R_Q v)_{L_2(0,d)}.$$

Therefore, from the Riesz theorem it follows that there exists a linear bounded operator $B_2\colon L_2(0,d) \to W_R$ such that

$$(f,v)_{L_2(0,d)} = (B_2 f, R_Q v)_{W^1(0,d)}. \tag{6.16}$$

By virtue of (6.15), (6.16), a function $u \in \mathcal{D}(\mathcal{A}_R)$ is a solution of the equation

$$(\mathcal{A}_R + R_Q)u = f \tag{6.17}$$

for $f \in \mathcal{R}(R_Q)$ if and only if $u \in W_{\mathcal{A}_R}$ and for all $v \in \dot{C}^\infty(0,d)$

$$(B_1(R_Q u), R_Q v)_{W^1(0,d)} = (B_2 f, R_Q v)_{W^1(0,d)}. \tag{6.18}$$

From Lemma 6.4 and (6.15) it follows that

$$(B_1(R_Q u), R_Q u)_{W^1(0,d)} \geq c_1 \|R_Q u\|^2_{W^1(0,d)} \tag{6.19}$$

for each $u \in W_{\mathcal{A}_R}$. Hence, the operator B_1 has a bounded inverse $B_1^{-1}\colon W_R \to W_R$ and $\|B_1^{-1}\| \leq 1/c_1$. Lemmas 2.10, 6.3 imply that $R_Q(\dot{C}^\infty(0,d))$ is dense in W_R. Therefore, the identity (6.18) is equivalent to the equation

$$R_Q u = B_1^{-1} B_2 f, \tag{6.20}$$

and

$$\|R_Q u\|_{W^1(0,d)} \leq \frac{\|(R_Q^R)^{-1}\|}{c_1} \|f\|_{L_2(0,d)}. \tag{6.21}$$

It follows from (6.20) and Lemma 2.16 that for every $f \in \mathcal{R}(R_Q)$ the equation

$$\mathcal{A}_R^1 u = f \tag{6.22}$$

has a unique solution $u \in \mathcal{D}(\mathcal{A}_R) \cap \mathcal{R}(R_Q)$, and

$$u = (R_Q^R)^{-1} B_1^{-1} B_2 f. \tag{6.23}$$

3. We now consider the unbounded operator $\mathcal{A}_R^0\colon \mathcal{R}(R_Q) \to \mathcal{R}(R_Q)$. Since the operator $\mathcal{A}_R^0$ is self-adjoint, the spectrum $\sigma(\mathcal{A}_R^0)$ is real. Take $\lambda_0 \in \mathbb{C} \backslash \sigma(\mathcal{A}_R^0)$. We consider the equation

$$\mathcal{A}_R^0 u - \lambda_0 u = f_0, \tag{6.24}$$

where $f_0 \in \mathcal{R}(R_Q)$.

This equation has a unique solution $u \in \mathcal{R}(R_Q)$ and

$$\|u\|_{L_2(0,d)} \leq k_1 \|f_0\|_{L_2(0,d)}. \tag{6.25}$$

We rewrite the equation (6.24) in the form

$$\mathcal{A}_R^1 u = R_Q u + \lambda_0 u + f_0. \tag{6.26}$$

By virtue of (6.25), (6.21) and Lemma 2.2, we obtain

$$\|R_Q u\|_{W^1(0,d)} \le \frac{\|(R_Q^R)^{-1}\|}{c_1}\|R_Q u + \lambda_0 u + f_0\|_{L_2(0,d)} \le k_2\|f_0\|_{L_2(0,d)}.$$

Hence, by virtue of Theorem B.8 and Lemma 2.16, the operator $(\mathcal{A}_R^0 - \lambda_0 I)^{-1}$: $\mathcal{R}(R_Q) \to \mathcal{R}(R_Q)$ is compact. Thus, Theorem A.8 implies that the spectrum $\sigma(\mathcal{A}_R^0)$ is discrete, and, for $\lambda \notin \sigma(\mathcal{A}_R^0)$, the resolvent $R(\lambda, \mathcal{A}_R^0): \mathcal{R}(R_Q) \to \mathcal{R}(R_Q)$ is compact.

From (6.12) it follows that $(\mathcal{A}_R v, v)_{L_2(0,d)} \ge 0$ for all $v \in \mathcal{D}(\mathcal{A}_R)$. Therefore, the eigenvalues λ_s of the operator $\mathcal{A}_R^0$ are positive.

The equality $\mathcal{A}_R^0 R(\lambda_0, \mathcal{A}_R^0) = I + \lambda_0 R(\lambda_0, \mathcal{A}_R^0)$ and Theorem A.1 imply that the operator $\mathcal{A}_R^0: \mathcal{R}(R_Q) \to \mathcal{R}(R_Q)$ is Fredholm, and $\operatorname{ind} \mathcal{A}_R^0 = 0$. □

Lemmas 6.5, 6.6 give us:

Theorem 6.3. *Let conditions* 6.1–6.3 *be fulfilled.*

Then the spectrum $\sigma(\mathcal{A}_R)$ consists of the eigenvalue $\lambda_0 = 0$ and real isolated eigenvalues $\lambda_s > 0$. Moreover, the eigenvalue $\lambda_0 = 0$ has infinite multiplicity, and $\lambda_s > 0$ have finite multiplicities.

Example 6.1. We consider the equation

$$-(Rv)''(t) = f_0(t) \qquad (t \in (0,2)) \tag{6.27}$$

with homogeneous boundary conditions

$$v(t) = 0 \qquad (t \in [-1,0] \cup [2,3]). \tag{6.28}$$

Here $(Rv)(t) = v(t) + v(t+1) + v(t-1)$ (see Example 2.4).

In this case we have one class of intervals $Q_{11} = (0,1)$, $Q_{12} = (1,2)$. The matrix

$$R_1 = \begin{pmatrix} 1 & 1 \\ 1 & 1 \end{pmatrix}.$$

From this it follows that

$$\sigma(R_Q) = \sigma(R_1) = \{0\} \cup \{2\}, \tag{6.29}$$

$$\mathcal{N}(R_Q) = \{v \in L_2(0,2) : v(t) = -v(t+1) \text{ for } t \in (0,1)\}, \tag{6.30}$$

$$\mathcal{R}(R_Q) = \{v \in L_2(0,2) : v(t) = v(t+1) \text{ for } t \in (0,1)\}. \tag{6.31}$$

Hence,

$$\mathcal{D}(\mathcal{A}_R) = \big\{v \in L_2(0,2) : v(t) + v(t+1) \in W^2(0,1),$$
$$(v(t) + v(t+1))^{(i)}|_{t=0} = (v(t) + v(t+1))^{(i)}|_{t=1},\ i = 0,1\big\}. \tag{6.32}$$

Let us consider the equation

$$\mathcal{A}_R v = 0. \tag{6.33}$$

From (6.33) it follows that $R_Q v = at + b$. By virtue of (6.31), $a = 0$. Thus, $\mathcal{N}(\mathcal{A}_R) = \mathcal{N}(R_Q) \oplus \{v(t) \equiv \text{const}\}$.

Now let us consider the problem

$$\mathcal{A}_R v = \lambda v, \tag{6.34}$$

where $\lambda \neq 0$. From (6.34) it follows that

$$\left.\begin{aligned} -v_1''(t) - v_2''(t) &= \lambda v_1(t) \qquad (t \in (0,1)),\\ -v_1''(t) - v_2''(t) &= \lambda v_2(t) \qquad (t \in (0,1)), \end{aligned}\right\} \tag{6.35}$$

where

$$\begin{aligned} v_1(t) &= v(t) \qquad && (t \in (0,1)),\\ v_2(t) &= v(t+1) \qquad && (t \in (0,1)). \end{aligned}$$

The system of equations (6.35) will give us $\lambda(v_1(t) - v_2(t)) = 0$. Hence, a function $v(t) = v_1(t) = v_2(t)$ satisfies the equation

$$-v''(t) = \frac{\lambda}{2} v(t) \qquad (t \in (0,1)). \tag{6.36}$$

By virtue of (6.32), a function $v(t)$ satisfies the following conditions

$$v(0) = v(1), \qquad v'(0) = v'(1). \tag{6.37}$$

The eigenvalues of the problem (6.36), (6.37) are $\lambda_s = 8\pi^2 s^2$ $(s = 1, 2, \ldots)$. For each eigenvalue λ_s, there are two linearly independent eigenfunctions $v_{s1}(t) = \sin 2\pi st$, $v_{s2} = \cos 2\pi st$ corresponding to λ_s. Thus, the spectrum $\sigma(\mathcal{A}_R)$ consists of:

$\lambda_0 = 0$, which corresponds to the infinite dimensional subspace of the eigenfunctions $\mathcal{N}(\mathcal{A}_R) = \mathcal{N}(R_Q) \oplus \{v_0(t) \equiv \text{const}\}$, and

$\lambda_s = 8\pi^2 s^2$ $(s = 1, 2, \ldots)$, which correspond to the linearly independent eigenfunctions $v_{s1} = \sin 2\pi st$, $v_{s2} = \cos 2\pi st$.

Self-Adjoint Perturbations

We define the unbounded operator $A_R \colon L_2(0,d) \to L_2(0,d)$ given by

$$A_R v = -(R_Q v)'' \qquad (v \in \mathcal{D}(A_R) = \dot{C}^\infty(0,d)).$$

Theorem 6.4. *Let conditions* 6.1–6.3 *be fulfilled.*

Then $\mathcal{A}_R \colon L_2(0,d) \to L_2(0,d)$ *is the Friedrichs extension of* A_R.

Proof. Denote by $b_R[u,v]$ $(u,v \in \mathcal{D}(b_R))$ a closure of non-negative symmetric sesquilinear form

$$(\mathcal{A}_R u, v)_{L_2(0,d)} \qquad (u,v \in \dot{C}^\infty(0,d)).$$

Clearly, $\mathcal{D}(b_R) = H_{b_R}$ is a Hilbert space with inner product

$$(u,v)_{b_R} = b_R[u,v] + (u,v)_{L_2(0,d)}$$

(see (A.6)). By Theorem A.10, there exists an m-sectorial operator B_{b_R} associated with the form b_R. The operator B_{b_R} is called a Friedrichs extension of $\mathcal{A}_R$. Theorem A.13 implies that B_{b_R} is self-adjoint and non-negative. We prove that $B_{b_R} = \mathcal{A}_R$.

First we show that $H_{b_R} = W_{\mathcal{A}_R}$. In fact, if $v \in H_{b_R}$, then, by Lemma 6.4, $v \in W_{\mathcal{A}_R}$. Now let $v \in W_{\mathcal{A}_R}$. By virtue of Lemma 6.3, there are $v_0 \in \mathring{W}^1(0,d)$ and $v_1 \in \mathcal{N}(R_Q)$ such that $v = v_0 + v_1$. Since $\dot{C}^\infty(0,d)$ is dense in $\mathring{W}^1(0,d)$, from Lemma 6.4 we obtain $v_0 \in H_{b_R}$. On the other hand, $U_s P_s v_1 \in \mathcal{N}(R_{Qs})$. If rank $R_s = j < M$, then a subspace $\mathcal{N}(R_{Qs})$ consists of vector-valued functions $V \in L_2^M(Q_{s1})$ such that their $i_1,\dots,i_j$th coordinates are linear combinations of the remaining $M-j$ coordinates. Since $\dot{C}^\infty(Q_{s1})$ is dense in $L_2(Q_{s1})$, we conclude that for any $p = 1,2,\dots$ there is $V_{sp} \in \mathcal{N}(R_{Qs}) \cap \dot{C}^{\infty,M}(Q_{s1})$, for which $\sum_s \|V_{sp} - U_s P_s v_1\|^2_{L_2^M(Q_{s1})} < 1/p$. By construction, $u_p = \sum_s U_s^{-1} V_{sp} \in \mathcal{N}(R_Q) \cap \dot{C}^\infty(0,d)$ and $\|u_p - u\|_{L_2(0,d)} \to 0$ as $p \to \infty$. Therefore, $v_1 \in H_{b_R}$. Hence, $v = v_0 + v_1 \in H_{b_R}$. We stated that $H_{b_R} = W_{\mathcal{A}_R}$.

We now prove that $B_{b_R} = \mathcal{A}_R$. Let $u \in \mathcal{D}(B_{b_R})$. Since the sesquilinear form b_R is symmetric, by virtue of Theorem A.10(a) and Lemma 2.14, for all $v \in \dot{C}^\infty(0,d)$

$$\begin{aligned}(B_{b_R} u, v)_{L_2(0,d)} &= b_R[u,v] = \overline{b_R[v,u]} = \overline{(\mathcal{A}_R v, u)}_{L_2(0,d)} \\ &= (u, \mathcal{A}_R v)_{L_2(0,d)} = (-R_Q u, v'')_{L_2(0,d)} = (\mathcal{A}_R u, v)_{L_2(0,d)}.\end{aligned}$$

Hence, $B_{b_R} \subset \mathcal{A}_R$. Now let $u \in \mathcal{D}(\mathcal{A}_R)$. Similarly to the above we have

$$(\mathcal{A}_R u, v)_{L_2(0,d)} = b_R[u,v] \qquad (v \in \dot{C}^\infty(0,d)).$$

Since $H_{b_R} = W_{\mathcal{A}_R}$, then $\dot{C}^\infty(0,d)$ is a core of b_R. Therefore, Theorem A.10(d) implies that $u \in \mathcal{D}(B_{b_R})$ and $B_{b_R} u = \mathcal{A}_R u$. Thus, $\mathcal{A}_R \subset B_{b_R}$. We have proved that $\mathcal{A}_R = B_{b_R}$. □

Theorem 6.5. *Let conditions 6.1–6.3 be fulfilled. Assume*

$$A_1 \colon L_2(0,d) \to L_2(0,d)$$

is a bounded self-adjoint operator.

Then the operator $\mathcal{A}_R + A_1$ is self-adjoint; the spectrum $\sigma(\mathcal{A}_R + A_1) \subset [-\|A_1\|, +\infty)$, and the set $(\|A_1\|, +\infty) \cap \sigma(\mathcal{A}_R + A_1)$ consists of isolated eigenvalues of finite multiplicity.

The proof follows from Theorems 6.2, 6.3, and Theorem A.7.

Example 6.2. Let us consider the differential-difference operator $\mathcal{A}_R: L_2(0,2) \to L_2(0,2)$ defined in Example 6.1. We introduce the operator $A_1: L_2(0,2) \to L_2(0,2)$ by the formula $(A_1 v)(t) = \sin^2 2\pi t \cdot v(t)$. Evidently, A_1 is a bounded self-adjoint operator, and $\|A_1\| = 1$.

We shall prove that the interval $[0,1]$ belongs to the spectrum $\sigma(\mathcal{A}_R + A_1)$. Let us consider the equation

$$(\mathcal{A}_R + A_1)v - \lambda v = f_0, \tag{6.38}$$

where $f_0(t) = 1$ $(t \in (0,1))$, $f_0(t) = -1$ $(t \in (1,2))$, $\lambda \in (0,1)$. Denote $v_1(t) = v(t)$ $(t \in (0,1))$, $v_2(t) = v(t+1)$ $(t \in (0,1))$. Then from (6.38) it follows that

$$\begin{aligned} -v_1''(t) - v_2''(t) + \sin^2 2\pi t \cdot v_1(t) - \lambda v_1(t) &= 1 \qquad (t \in (0,1)), \\ -v_1''(t) - v_2''(t) + \sin^2 2\pi t \cdot v_2(t) - \lambda v_2(t) &= -1 \qquad (t \in (0,1)). \end{aligned} \tag{6.39}$$

From (6.39) we obtain

$$v_1(t) - v_2(t) = 2/(\sin^2 2\pi t - \lambda) \qquad (t \in (0,1)). \tag{6.40}$$

Since $0 < \lambda < 1$, the function $\sin^2 2\pi t - \lambda$ is equal to zero at some point $0 < t_1 < 1$. Hence, the function $v_1(t) - v_2(t)$, defined by the formula (6.40), does not belong to $L_2(0,1)$. Thus, $\lambda \in \sigma(\mathcal{A}_R + A_1)$.

Boundary Conditions

We shall show that it is meaningless to speak of boundary values for a generalized solution of the boundary value problem (6.1), (6.2). In particular, a generalized solution of the boundary value problem (6.1), (6.2) can have no right limit at the point 0 and left limit at the point d.

Theorem 6.6. *Let conditions* 6.1–6.3 *be fulfilled.*

Then, for each function $\varphi \in L_2(0,\theta)$, *there exist numbers* $\alpha, \beta \neq 0$ *and a function* $v \in \mathcal{N}(\mathcal{A}_R)$ *such that*

$$v(t) = \alpha\varphi(t) \quad (t \in (0,\theta)), \qquad v(t) = \beta\varphi(t-N) \quad (t \in (N,d)). \tag{6.41}$$

Proof. 1. Let us prove that there exists a vector $X = (X_1, \dots, X_{N+1}) \in \mathcal{N}(R_1)$ such that $X_1 \neq 0$, $X_{N+1} \neq 0$.

By virtue of Lemma 6.2, there exists a vector $Y = (Y_1, \dots, Y_{N+1}) \in \mathcal{N}(R_1)$ such that $Y_1 \neq 0$. If $Y_{N+1} \neq 0$, then we set $X = Y$.

Assume $Y_{N+1} = 0$. From the proof of Lemma 6.2 it follows that a vector $\mathcal{Z} = (Y_{N+1}, \dots, Y_1) \in \mathcal{N}(R_1)$. A vector $X = Y + \mathcal{Z} \in \mathcal{N}(R_1)$ and $X_1 = X_{N+1} = Y_1 + Y_{N+1} = Y_1 \neq 0$.

2. Now let us assume that $X \in \mathcal{N}(R_1)$, $X_1 \neq 0$, $X_{N+1} \neq 0$. We denote $V(t) = X\varphi(t)$ $(t \in (0,\theta))$, $\alpha = X_1$, $\beta = X_{N+1}$. Evidently, $V(t) \in \mathcal{N}(R_{Q1})$. Hence, by virtue of (2.7), $v(t) = U_1^{-1}V(t) \in \mathcal{N}(R_Q) \subset \mathcal{N}(\mathcal{A}_R)$ and $v(t) = \alpha\varphi(t)$ $(t \in (0,\theta))$, $v(t) = \beta\varphi(t-N)$ $(t \in (N,d))$. □

Notes

Section 1. Ordinary differential equations with nonlocal boundary conditions were considered by M. Picone [1, 2], A. Sommerfeld [1], Ya. D. Tamarkin [1, 2], W. Feller [1, 2], V. P. Mikhailov [1], A. M. Krall [1, 2], V. A. Il'in and E. I. Moiseev [1, 2], S. Ya. Yakubov and E. Ch. Ibragimov [1] and many others. These problems have interesting applications to mechanics (see A. Sommerfeld [1]) and to the theory of diffusion processes (see W. Feller [1, 2]). A bibliography of publications devoted to this question can be found in the paper of A. M. Krall [1]. The results of Section 1 are applied in Sections 3, 4. The problems, considered in Section 1, are very special cases of nonlocal elliptic boundary value problems studied by A. L. Skubachevskiĭ [4].

Section 2. In Section 2, most of the lemmas follow from the results of A. G. Kamenskiĭ [1]. Theorem 2.1 was proved by A. L. Skubachevskiĭ [1].

Section 3. Generalized solutions of the boundary value problems for ordinary differential-difference equations were considered for the first time by G. A. Kamenskiĭ, A. D. Myshkis [1]. The spectral theory of symmetric differential-difference operators was built by A. G. Kamenskiĭ [1]. In this paper he proved Theorems 3.1, 3.2 in the symmetric case and Theorem 3.6. In Section 3, we used a different method based on reducing a differential-difference equation to a differential equation with nonlocal boundary conditions. Unlike A. G. Kamenskiĭ [1], we used the additional restriction $\det R_2 \neq 0$ if $\theta = 1$. However, our method enables us to consider the asymmetric case. Theorems 3.1, 3.3–3.5 are particular cases of the appropriate results for the elliptic differential-difference operators obtained by A. L. Skubachevskiĭ [1, 10]. Example 3.10, demonstrating the disturbance of smoothness of generalized solutions on a dense set, was constructed in the paper of A. L. Skubachevskiĭ [17]. Theorems 3.7, 3.8 are published here for the first time.

The generalized solutions of functional differential equations were studied by J. Wiener [1] and K. Cooke and J. Wiener [1].

Section 4. The sufficient conditions of existence of smooth solutions were obtained by G. A. Kamenskiĭ, A. D. Myshkis and A. L. Skubachevskiĭ [2] in the case $k = 0$. The necessary and sufficient conditions are adapted from A. L. Skubachevskiĭ [18]. The paper of S. D. Shteingol'd [1] contains Example 4.3. Some generalizations of problem (4.1), (4.2) for variable shifts of argument were studied by G. A. Kamenskiĭ, A. D. Myshkis and A. L. Skubachevskiĭ [3]. Sufficient conditions of existence of smooth solutions for systems of differential-difference equations can be found in the paper of A. Baumstein and A. L. Skubachevskiĭ [1].

Section 5. A variational problem (5.5), (5.2), (5.3) for a control system with delay was considered by N. N. Krasovskiĭ [1] in the case $a = 0$. Variational problems with shifts of argument in dominant terms and their connection with boundary value problems for functional differential equations were studied by G. A. Kamenskiĭ [1, 2]. The applications of the boundary value problems for differential-difference equations to control systems with delay, contained in Section 5, are adapted from A. L. Skubachevskiĭ [19].

For the other problems of the optimal control theory with delay, see L. S. Pontryagin, V. G. Boltyanskiĭ, R. V. Gamkrelidze, and E. F. Mishchenko [1], A. Halanay [1], R. Gabasov, and F. M. Kirillova [1], G. L. Kharatishvili and T. A. Tadumadze [1], F. Colonius [1], G. A. Kent [1], H. T. Banks and G. A. Kent [1], and H. T. Banks and M. Q. Jacobs [1].

Section 6. Theorems 6.1–6.4 and Example 6.2 are due to A. G. Kamenskiĭ [1]. In Section 6, we used different methods, similar to the methods of Chapter IV. Lemmas 6.2, 6.4 were proved by A. L. Skubachevskiĭ [3]. Theorem 6.6 is published here for the first time.

Chapter II

The First Boundary Value Problem for Strongly Elliptic Differential-Difference Equations

In this chapter we study solvability, spectrum and smoothness of generalized solutions of the first boundary value problem for strongly elliptic differential-difference equations. We also apply these results to the investigation of elliptic differential equations with nonlocal conditions.

In Section 7, which is devoted to auxiliary geometric results, we study a decomposition of a domain Q consisting of subdomains Q_r. The definition of this decomposition is closely associated with a form of some difference operator. Subdomains Q_r play an important role in the investigations of necessary and sufficient conditions of strong ellipticity and smoothness of generalized solutions.

Section 8 deals with the properties of the difference operators in the spaces $L_2(\mathbb{R}^n)$, $L_2(Q)$ and in the Sobolev spaces.

In Section 9, we obtain necessary and sufficient conditions of coerciveness in terms of polynomial matrices. Generally speaking, these conditions are not the same. Therefore, we define strong ellipticity using the Gårding inequality. The sufficient condition for strong ellipticity, using the symbol of differential-difference operator, is much rougher. In contrast to strongly elliptic differential equation, the symbol of a strongly elliptic differential-difference equation is quasi-polynomial and can change its sign.

In Section 10, we consider a strongly elliptic differential-difference operator. We prove the Fredholm property and also discreteness and semiboundedness of the spectrum of this operator.

Section 11 deals with the smoothness of generalized solutions in subdomains Q_r. Unlike elliptic differential equations, the smoothness of the generalized solutions of elliptic differential-difference equations can be violated in the domain Q even for an infinitely differentiable right-hand side of the equation.

In Section 12, we establish necessary and sufficient conditions for the conservation of smoothness of solutions on the boundary of adjacent subdomains.

Section 13 deals with applications of the results of Section 10 to elliptic differential equations with nonlocal conditions. It contains theorems concerning the Fredholm property and discreteness of the spectrum of an elliptic operator with nonlocal conditions on the shifts of the boundary.

7 Some Geometrical Constructions

Decomposition of Q

In Chapter II, we assume that $Q \subset \mathbb{R}^n$ is a bounded domain with boundary $\partial Q \in C^\infty$ or a cylinder $(0,d) \times G$, where $G \subset \mathbb{R}^{n-1}$ is a bounded domain (with boundary $\partial G \in C^\infty$ if $n \geq 3$).

Let $\mathcal{M} \subset \mathbb{R}^n$ be a finite set of vectors with integer coordinates. We denote by M the additive group generated by the set $\mathcal{M}$.

Denote by Q_r the open connected components of the set $Q \setminus (\bigcup_{h \in M}(\partial Q + h))$.

Definition 7.1. A set Q_r is called a *subdomain*. A set $\mathcal{R}$ of all subdomains Q_r $(r = 1, 2, \dots)$ is called a *decomposition* of the domain Q.

It is easy to see that a set $\mathcal{R}$ is finite or countable.

Lemma 7.1. $\bigcup\limits_r \partial Q_r = \Big(\bigcup\limits_{h \in M} (\partial Q + h) \Big) \cap \overline{Q}$.

The proof is obvious.

Lemma 7.2. 1) $\bigcup\limits_r Q_r = \overline{Q}$.

2) *For every* Q_{r_1} *and* $h \in M$, *there is either* Q_{r_2} *such that* $Q_{r_2} = Q_{r_1} + h$, *or* $Q_{r_1} + h \subset \mathbb{R}^n \setminus \overline{Q}$.

Proof. 1. The first part of Lemma 7.2 follows from Lemma 7.1. Let us now prove that for every Q_{r_1} and $h \in M$, either $Q_{r_1} + h \subset Q$ or $Q_{r_1} + h \subset \mathbb{R}^n \setminus \overline{Q}$. Assume to the contrary that there exist Q_{r_1} and $h \in M$ such that $(Q_{r_1} + h) \cap (\mathbb{R}^n \setminus \overline{Q}) \neq \emptyset$ and $(Q_{r_1} + h) \cap Q \neq \emptyset$. Then, since the set $Q_{r_1} + h$ is connected, there is $z \in \partial Q \cap (Q_{r_1} + h)$. Hence, $z - h \in (\partial Q - h) \cap Q_{r_1}$. This contradicts the definition of the subdomain Q_{r_1}.

2. Let us prove that, if $Q_{r_1} + h \subset Q$, there exists $Q_{r_2} = Q_{r_1} + h$. Assume to the contrary that for some Q_{r_2} we have $Q_{r_2} \cap (Q_{r_1} + h) \neq \emptyset$ and $Q_{r_2} \Delta (Q_{r_1} + h) \neq \emptyset$. Let $Q_{r_2} \setminus (Q_{r_1} + h) \neq \emptyset$. Then, since the set Q_{r_2} is connected, there is a point $y \in Q_{r_2} \cap (\partial Q_{r_1} + h)$. This contradicts the definition of the set Q_{r_2}. □

We can divide the decomposition $\mathcal{R}$ into disjoint classes in the following way: subdomains $Q_{r_1}, Q_{r_2} \in \mathcal{R}$ belong to the same class if there exists an $h \in M$ such that $Q_{r_2} = Q_{r_1} + h$. We denote the subdomains Q_r by Q_{sl}, where s is the number of a class and l is the number of a subdomain in the sth class.

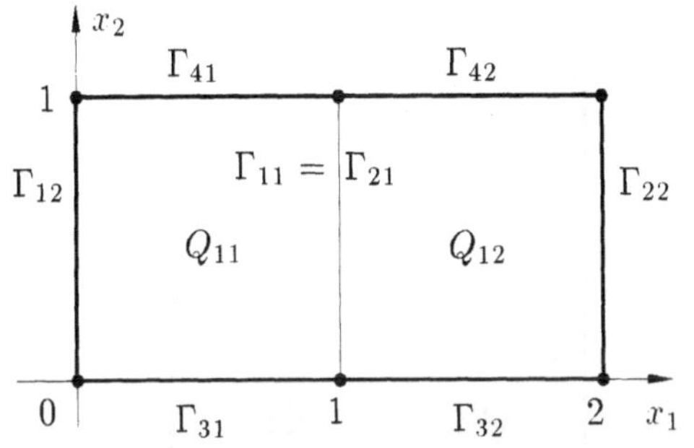

Fig. II.1

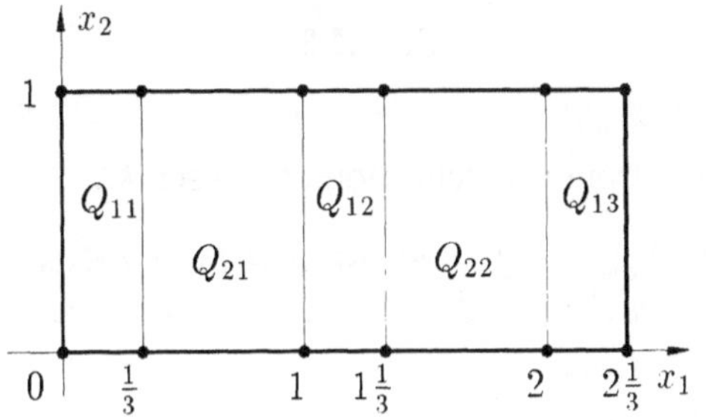

Fig. II.2

Evidently, each class consists of a finite number $N = N(s)$ of subdomains Q_{sl} and $N(s) \leq ([\operatorname{diam} Q] + 1)^n$. A set of classes can be countable.

Example 7.1. Let $Q = (0,2) \times (0,1) \subset \mathbb{R}^2$, $\mathcal{M} = \{(1,0)\}$. Then the decomposition $\mathcal{R}$ consists of one class of subdomains: $Q_{11} = (0,1) \times (0,1)$, $Q_{21} = (1,2) \times (0,1)$ (see Fig. II.1).

Example 7.2. Let $Q = (0, 2\frac{1}{3}) \times (0,1) \subset \mathbb{R}^2$, $\mathcal{M} = \{(1,0)\}$. Then the decomposition $\mathcal{R}$ consists of two classes: $Q_{1l} = (l-1, l-2/3) \times (0,1)$ $(l = 1,2,3)$, and $Q_{2l} = (l - 2/3, l) \times (0,1)$ $(l = 1,2)$ (see Fig. II.2).

Example 7.3. Let $Q \subset \mathbb{R}^2$ be a bounded domain with boundary $\partial Q \in C^\infty$, which, inside the strip $\{x : 0 < x_2 < 2\}$, coincides with two lines $\{x : x_1 = \frac{1}{3}\exp(-1/x_2)\sin(1/x_2)\}$ and $\{x : x_1 = 2\}$, and let $\mathcal{M} = \{(1,0)\}$. Then the decomposition $\mathcal{R}$ consists of countable set of classes.

The Set $\mathcal{K}$

We introduce the set $\mathcal{K}$ by the formula

$$\mathcal{K} = \bigcup_{h_1, h_2 \in M} \{\overline{Q} \cap (\partial Q + h_1) \cap \overline{[(\partial Q + h_2) \setminus (\partial Q + h_1)]}\}. \tag{7.1}$$

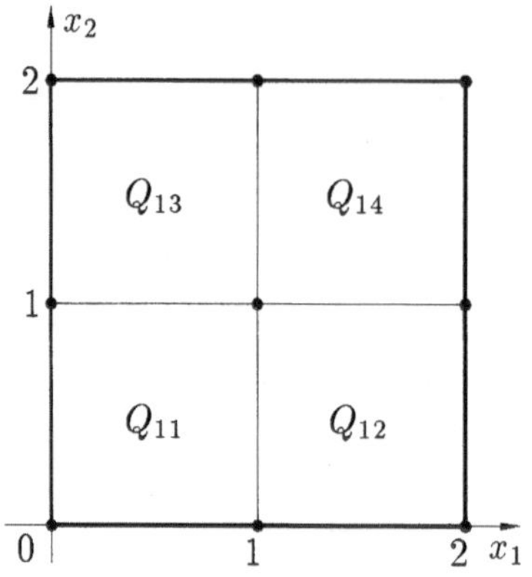

Fig. II.3

Lemma 7.3 follows from the definition of the set $\mathcal{K}$.

Lemma 7.3. *Let $x^0 \in \partial Q_{sl} \cap \partial Q$. Suppose that there exists a sequence of points $x^n \to x^0$ as $n \to \infty$ such that $x^n \in \overline{Q}_{s_n l_n}$, $(s_n, l_n) \neq (s, l)$.*
Then $x^0 \in \mathcal{K}$.

Corollary 7.1. *Let $x^0 \in \partial Q \cap \partial Q_{s_1 l_1} \cap \partial Q_{s_2 l_2}$, $(s_1, l_1) \neq (s_2, l_2)$.*
Then $x^0 \in \mathcal{K}$.

Lemma 7.4 also follows from the definition of the set $\mathcal{K}$.

Lemma 7.4. *Let $x^0 \in Q \cap \partial Q_{pl} \cap \partial Q_{qk}$, $(p, l) \neq (q, k)$. Suppose that there exists a sequence of points $x^n \to x^0$ as $n \to \infty$, and $x^n \in \overline{Q}_{s_n l_n}$, $(s_n, l_n) \neq (p, l), (q, k)$.*
Then $x^0 \in \mathcal{K}$.

Corollary 7.2. *Let $x^0 \in \bigcap_i \partial Q_{s_i l_i}$, where $(s_i, l_i) \neq (s_j, l_j)$ for $i \neq j$ $(i, j = 1, 2, 3)$.*
Then $x^0 \in \mathcal{K}$.

Example 7.4. Let $Q = (0, 2) \times (0, 1) \subset \mathbb{R}^2$, $\mathcal{M} = \{(1, 0)\}$. Then the set $\mathcal{K}$ consists of six points (i, j), where $i = 0, 1, 2$, $j = 0, 1$ (see Fig. II.1).

Example 7.5. Let $Q = (0, 2) \times (0, 2) \subset \mathbb{R}^2$, $\mathcal{M} = \{(1, 0)\} \cup \{(0, 1)\}$. Then the set $\mathcal{K}$ consists of nine points (i, j), where $i, j = 0, 1, 2$ (see Fig. II.3).

A set $\mathcal{K}$ can have a very complicated form even in the case when $\partial Q \in C^\infty$. In particular, in the following example we construct a set $\mathcal{K}$ such that $\mu_{n-1}(\mathcal{K} \cap \partial Q) \neq 0$, where $\mu_{n-1}(\cdot)$ is an $(n-1)$-dimensional Lebesgue measure.

Example 7.6. Denote by A_{11} an open interval with the center at the point $a_{11} = 1/2$ such that its length equals $1/4$. Let us reject A_{11} from the closed interval $[0, 1]$. From each of the remaining two intervals with centers at points a_{2j}, we reject an open interval A_{2j} with its center at a_{2j} having length $1/4^2$. Then from each of the four remaining intervals with centers a_{3j}, we reject an open interval A_{3j} with the same center having length $1/4^3$, and so on. Evidently, $\mu_1(\bigcup_{i=1}^{\infty}\bigcup_{j=1}^{2^{i-1}} A_{ij}) = 1/2$. A set $A = [0,1] \setminus \bigcup_{i,j} A_{ij}$ is closed, nowhere dense and $\mu_1(A) = 1/2$.

We consider a domain $Q \subset \mathbb{R}^2$ with boundary $\partial Q \in C^\infty$ such that $\Gamma_1 \cup \Gamma_2 \subset \partial Q$, where $\Gamma_1 = \{(x_1, x_2) : 0 \le x_1 \le 1, x_2 = 0\}$, $\Gamma_2 = \{(x_1, x_2) : 0 \le x_2 \le 1, x_2 = \varphi(x_1)\}$,

$$\varphi(x_1) = \begin{cases} 1 + \exp(-2^{i^2} - [1 - 2^{4i+2}(x_1 - a_{ij})^2]^{-1}) & (x_1 \in A_{ij}), \\ 1 & (x_1 \in A). \end{cases}$$

Let $\mathcal{M} = \{(0,1)\}$. Then $A \times \{0\} \subset \mathcal{K}$, $\mu_1(A \times \{0\}) = 1/2$.

Decomposition of ∂Q

We shall suppose that the following condition holds:

7.1. $\mu_{n-1}(\mathcal{K} \cap \partial Q) = 0$.

Denote by Γ_p the components of the set $\partial Q \setminus \mathcal{K}$, which are open and connected in the topology of ∂Q.

Lemma 7.5. *If $(\Gamma_p + h) \cap \overline{Q} \neq \emptyset$ for some $h \in M$, then either $\Gamma_p + h \subset Q$, or there is $\Gamma_r \subset \partial Q \setminus \mathcal{K}$ such that $\Gamma_p + h = \Gamma_r$.*

Proof. Let $(\Gamma_p + h) \cap \overline{Q} \neq \emptyset$.

1. We first prove that either $\Gamma_p + h \subset Q$, or $\Gamma_p + h \subset \partial Q$. Assume the contrary, that there are Γ_p and $h \in M$ such that either 1) $(\Gamma_p + h) \cap Q \neq \emptyset$ and $(\Gamma_p + h) \cap \partial Q \neq \emptyset$, or 2) $(\Gamma_p + h) \cap \overline{Q} \neq \emptyset$ and $(\Gamma_p + h) \setminus \overline{Q} \neq \emptyset$. Since the set $\Gamma_p + h$ is connected, in the second case we also have $(\Gamma_p + h) \cap \partial Q \neq \emptyset$. Hence, there exists a point $x^0 \in (\Gamma_p + h) \cap \partial Q$ such that $x^0 \in \overline{(\Gamma_p + h) \setminus \partial Q}$. In the opposite case, the set $(\Gamma_p + h) \cap \partial Q$ would be open in the topology of $\Gamma_p + h$, and this contradicts the connectedness of the set $\Gamma_p + h$. Evidently, $x^0 - h \in \overline{[\Gamma_p \setminus (\partial Q - h)]} \cap (\partial Q - h) \subset \mathcal{K}$ and $x^0 - h \in \Gamma_p$. On the other hand, by definition $\Gamma_p \cap \mathcal{K} \neq \emptyset$. We have a contradiction.

2. Let us prove now that, if $\Gamma_p + h \subset \partial Q$, then there is a Γ_r such that $\Gamma_p + h = \Gamma_r$. Since $\mu_{n-1}(\mathcal{K} \cap \partial Q) = 0$ and the set $\Gamma_p + h$ is open in the topology of ∂Q, there exists a Γ_r such that $(\Gamma_p + h) \cap \Gamma_r \neq \emptyset$. Let $\Gamma_r \Delta (\Gamma_p + h) \neq \emptyset$. Without loss of generality, we assume that $\Gamma_r \setminus (\Gamma_p + h) \neq \emptyset$. From this and from the connectedness of the set Γ_r we have $\partial(\Gamma_p + h) \cap \Gamma_r \neq \emptyset$. On the other hand, $\partial(\Gamma_p + h) \subset \mathcal{K}$ and $\mathcal{K} \cap \Gamma_r = \emptyset$. From this contradiction it follows that $\Gamma_r = \Gamma_p + h$. □

By virtue of Lemma 7.5, we can decompose the set $\{\Gamma_p + h : \Gamma_p + h \subset \overline{Q}, p = 1, 2, \ldots, h \in M\}$ into classes in the following manner. The sets $\Gamma_{p_1} + h_1$ and $\Gamma_{p_2} + h_2$ belong to the same class if 1) there exists an $h \in M$ such that $\Gamma_{p_1} + h_1 = \Gamma_{p_2} + h_2 + h$, and 2) in the case $\Gamma_{p_1} + h_1$, $\Gamma_{p_2} + h_2 \subset \partial Q$, the directions of the inner normals to ∂Q at the points $x \in \Gamma_{p_1}+h_1$ and $x-h \in \Gamma_{p_2}+h_2$ coincide. Clearly, a set $\Gamma_p \subset \partial Q$ can be in only one class, and a set $\Gamma_p+h \subset Q$ is in at most two classes. We denote a set $\Gamma_p + h$ by Γ_{rj}, where r is the number of the class and j is the number of an element in a given class $(1 \le j \le J = J(r))$. Without loss of generality, we shall suppose that $\Gamma_{r1}, \dots, \Gamma_{rJ_0} \subset Q$, $\Gamma_{r,J_0+1}, \dots, \Gamma_{rJ} \subset \partial Q$ $(0 \le J_0 = J_0(r) < J(r))$.

The next lemma follows from the definition of the set $\mathcal{K}$ and Corollary 7.1.

Lemma 7.6. *For every $\Gamma_{rj} \subset \partial Q$, there exists a subdomain Q_{sl} such that $\Gamma_{rj} \subset \partial Q_{sl}$. Moreover, $\Gamma_{rj} \cap \partial Q_{s_1 l_1} = \emptyset$ if $(s_1, l_1) \ne (s, l)$.*

Lemma 7.7 follows from Lemma 7.6.

Lemma 7.7. *For every $r = 1, 2, \dots$, there exists a unique $s = s(r)$ such that $N(s) = J(r)$ and after some renumbering $\Gamma_{rl} \subset \partial Q_{sl}$ $(l = 1, \dots, N(s))$.*

Then Lemma 7.8 follows from Lemma 7.7 and Corollary 7.2.

Lemma 7.8. *For every $\Gamma_{rj} \subset Q$, there exist subdomains $Q_{s_1 l_1}$ and $Q_{s_2 l_2}$ such that $Q_{s_1 l_1} \ne Q_{s_2 l_2}$, $\Gamma_{rj} \subset \partial Q_{s_1 l_1} \cap \partial Q_{s_2 l_2}$ and $\Gamma_{rj} \cap \partial Q_{s_3 l_3} = \emptyset$ if $(s_3, l_3) \ne (s_1, l_1), (s_2, l_2)$.*

Example 7.7 (see Examples 7.1, 7.4). Let $Q = (0,2) \times (0,1) \subset \mathbb{R}^2$, $\mathcal{M} = \{(1,0)\}$. Then there are four classes of the sets Γ_{rl}: 1) $\Gamma_{12} = \{0\} \times (0,1)$, $\Gamma_{11} = \{1\} \times (0,1)$; 2) $\Gamma_{21} = \Gamma_{11}$, $\Gamma_{22} = \{2\} \times (0,1)$; 3) $\Gamma_{31} = (0,1) \times \{0\}$, $\Gamma_{32} = (1,2) \times \{0\}$; 4) $\Gamma_{41} = (0,1) \times \{1\}$, $\Gamma_{42} = (1,2) \times \{1\}$ (see Fig. II.1).

Example 7.8. Let $Q = \{x \in \mathbb{R}^2 : |x| < 1\}$, $\mathcal{M} = \{(1,0)\}$. Then a set $\mathcal{K}$ consists of seven points $(0,0)$, $(\pm 1, 0)$, $(\pm 1/2, \pm\sqrt{3}/2)$. We denote by Γ_{12}, Γ_{22}, Γ_{31}, Γ_{42}, Γ_{52}, Γ_{61} the arcs of the circle ∂Q: $\{\pi(-2+j)/3 < \varphi < \pi(-1+j)/3\}$ $(j = 1, \dots, 6)$.

There are six classes of sets Γ_{rl}: 1) $\Gamma_{11} = \Gamma_{12} - (1,0)$, Γ_{12}; 2) $\Gamma_{21} = \Gamma_{22} - (1,0)$, Γ_{22}; 3) Γ_{31}; 4) $\Gamma_{41} = \Gamma_{42} + (1,0)$, Γ_{42}; 5) $\Gamma_{51} = \Gamma_{52} + (1,0)$, Γ_{52}; 6) Γ_{61} (see Fig. II.4).

The decomposition $\mathcal{R}$ consists of three classes of subdomains: 1) Q_{11} bounded by the curves $\overline{\Gamma}_{42}$, $\overline{\Gamma}_{52}$, $\overline{\Gamma}_{11}$, $\overline{\Gamma}_{21}$, and $Q_{12} = Q_{11} + (1,0)$; 2) Q_{21} bounded by the curves $\overline{\Gamma}_{31}$, $\overline{\Gamma}_{21}$, $\overline{\Gamma}_{41}$; 3) Q_{31} bounded by the curves $\overline{\Gamma}_{61}$, $\overline{\Gamma}_{51}$, $\overline{\Gamma}_{11}$.

8 Difference Operators in the Multidimensional Case

Difference Operators in $L_2(\mathbb{R}^n)$

We consider the properties of the *difference operator* R: $L_2(\mathbb{R}^n) \to L_2(\mathbb{R}^n)$ defined by the formula

$$Ru(x) = \sum_{h \in \mathcal{M}} a_h(x) u(x+h), \tag{8.1}$$

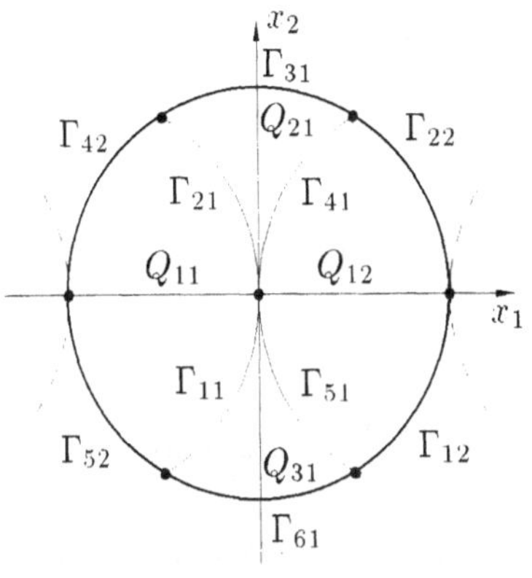

Fig. II.4

where $a_h \in C^\infty(\mathbb{R}^n)$ are complex-valued functions; $\mathcal{M} \subset \mathbb{R}^n$ is a finite set of vectors with integer coordinates; $x = (x_1, \dots, x_n) \in \mathbb{R}^n$.

We introduce the operator $R_Q = P_Q R I_Q \colon L_2(Q) \to L_2(Q)$, where

$I_Q \colon L_2(Q) \to L_2(\mathbb{R}^n)$ is the operator of extension of functions from $L_2(Q)$ by zero in $\mathbb{R}^n \setminus Q$,

$P_Q \colon L_2(\mathbb{R}^n) \to L_2(Q)$ is the operator of restriction of functions from $L_2(\mathbb{R}^n)$ to Q.

Lemma 8.1. *The operators* $I_Q \colon L_2(Q) \to L_2(\mathbb{R}^n)$ *and* $P_Q \colon L_2(\mathbb{R}^n) \to L_2(Q)$ *are bounded, and* $I_Q^* = P_Q$, *i.e.,* $(I_Q u, v)_{L_2(\mathbb{R}^n)} = (u, P_Q v)_{L_2(Q)}$ *for all* $u \in L_2(Q)$, $v \in L_2(\mathbb{R}^n)$.

Lemma 8.2. *The operators* $R \colon L_2(\mathbb{R}^n) \to L_2(\mathbb{R}^n)$ *and* $R_Q \colon L_2(Q) \to L_2(Q)$ *are bounded,*

$$R^* u(x) = \sum_{h \in \mathcal{M}} \overline{a_h(x-h)} u(x-h), \qquad R_Q^* = P_Q R^* I_Q.$$

The proofs are evident.

The next lemma follows from Lemma 8.1.

Lemma 8.3. *If the operator* $R + R^*$ *is positive, then the operator* $R_Q + R_Q^*$ *is positive.*

We denote by $R(\xi) = \sum_{h \in \mathcal{M}} a_h \exp(i(h, \xi))$ the *symbol of the difference operator* R with constant coefficients, where (h, ξ) is the inner product in $\mathbb{R}^n$.

Lemma 8.4. *The operator* $R + R^*$ *with constant coefficients* a_h *is positive if and only if*

$$0 \le \operatorname{Re} R(\xi) \not\equiv 0 \quad \textit{for all } \xi \in \mathbb{R}^n. \tag{8.2}$$

$(l = 1, \dots, k)$, and

$$\begin{aligned} R_1 &= \begin{pmatrix} a_0 & a_1 & \dots & a_k \\ a_{-1} & a_0 & \dots & a_{k-1} \\ \vdots & \vdots & \ddots & \vdots \\ a_{-k} & a_{-k+1} & \dots & a_0 \end{pmatrix}, \\ R_2 &= \begin{pmatrix} a_0 & a_1 & \dots & a_{k-1} \\ a_{-1} & a_0 & \dots & a_{k-2} \\ \vdots & \vdots & \ddots & \vdots \\ a_{-k+1} & a_{-k+2} & \dots & a_0 \end{pmatrix}. \end{aligned} \tag{8.9}$$

Since Q is a bounded domain, by virtue of formula (8.6), a number of different matrices R_s is finite if the coefficients a_h are constants. Let n_1 denote this number and let R_{s_ν} denote all different matrices R_s $(\nu = 1, \dots, n_1)$.

Lemma 8.7. *Let the coefficients a_h be constants.*
Then

$$\sigma(R_Q) = \bigcup_{\nu=1}^{n_1} \sigma(R_{s_\nu}).$$

Proof. By virtue of Lemmas 8.5, 8.6, $\bigcup_\nu \sigma(R_{s_\nu}) \subset \sigma(R_Q)$.
Let $\lambda \notin \bigcup_\nu \sigma(R_{s_\nu})$. Then the linear operator

$$A_\lambda = \sum_s U_s^{-1}(R_{Qs} - \lambda I)^{-1} U_s P_s \colon L_2(Q) \to L_2(Q)$$

is bounded. By virtue of Lemma 8.5,

$$P_s R_Q = R_Q P_s. \tag{8.10}$$

From (8.3), (8.10) it follows that

$$A_\lambda(R_Q - \lambda I) = \sum_s U_s^{-1}(R_{Qs} - \lambda I)^{-1} U_s (R_Q - \lambda I) P_s = I.$$

Analogously, we have $(R_Q - \lambda I)A_\lambda = I$. Hence, if $\lambda \notin \bigcup_\nu \sigma(R_{s_\nu})$, then the operator $R_Q - \lambda I$ has a bounded inverse operator. □

The next lemma follows from Lemma 8.7.

Lemma 8.8. *Let the coefficients a_h be constants.*
Then the operator $R_Q + R_Q^*$ is positive definite if and only if all matrices $R_{s_\nu} + R_{s_\nu}^*$ $(\nu = 1, \dots, n_1)$ are positive definite, where $R_{s_\nu}^*$ are the Hermitian adjoint matrices.

Lemma 8.9. *Let the coefficients a_h be constants. Suppose that the operator $R_Q + R_Q^*$ is positive.*
Then it is positive definite.

Let the operator R have constant coefficients. Then, by virtue of Lemmas 8.3, 8.4, 8.9, the condition (8.2) concerning the symbol of the operator R implies that $R_Q + R_Q^*$ is positive definite. This condition will be used later (see Theorems 9.4, 15.2). However, it is much rougher than the necessary and sufficient condition in Lemma 8.8, since it does not consider the form of the domain Q.

Example 8.3. Let $Q = (0,2) \times (0,1) \subset \mathbb{R}^2$, $Ru(x) = u(x) + a[u(x_1+1, x_2) + u(x_1 - 1, x_2)]$, where $a \in \mathbb{R}$.

Then $R(\xi) = (1 + 2a\cos\xi_1)$. If $|a| \le 1/2$, then, by virtue of Lemmas 8.3, 8.4, 8.9, the operator R_Q is positive definite.

The decomposition $\mathcal{R}$ of the domain Q consists of two subdomains: $Q_{11} = (0,1) \times (0,1)$, $Q_{12} = (1,2) \times (0,1)$. Moreover,

$$R_1 = \begin{pmatrix} 1 & a \\ a & 1 \end{pmatrix}.$$

By virtue of Lemma 8.8, the operator R_Q is positive definite if and only if $|a| < 1$.

Definition 8.1. A function $\varphi \in C(\overline{Q})$ is said to be *M-periodic* in $\overline{Q}$ if $\varphi(x) = \varphi(x+h)$ for all $x, x+h \in \overline{Q}$ and $h \in M$.

Lemma 8.10. *Let a function $\varphi(x)$ be M-periodic in $\overline{Q}$.*

Then $R_Q(\varphi u) = \varphi R_Q u$ for all $u \in L_2(Q)$.

Proof. From (8.3)–(8.5) we have

$$R_Q = \sum_s U_s^{-1} R_s U_s P_s, \tag{8.11}$$

where the operator series converges in $L_2(Q)$. The operators U_s, U_s^{-1}, P_s commute with the operator of multiplication by the function $\varphi(x)$. Hence, Lemma 8.10 follows from (8.11). □

Using Lemmas 8.1, 8.2, and formula (8.11), we obtain the following statements.

Lemma 8.11. *If the operator $R\colon L_2(\mathbb{R}^n) \to L_2(\mathbb{R}^n)$ is self-adjoint, then the operator $R_Q\colon L_2(Q) \to L_2(Q)$ is self-adjoint.*

Lemma 8.12. *Let the coefficients a_h be constants.*

Then the operator $R_Q\colon L_2(Q) \to L_2(Q)$ is self-adjoint if and only if all R_{s_ν} $(\nu = 1, \dots, n_1)$ are the Hermitian matrices.

Difference Operators in Sobolev Spaces

Denote by $\mathring{W}^k(Q)$ the closure of the linear manifold $\dot{C}^\infty(Q)$ in the space $W^k(Q)$. By virtue of Theorem B.10,

$$\mathring{W}^k(Q) = \{u \in W^k(Q) : \mathcal{D}_\nu^{\mu-1} u|_{\partial Q \setminus K} = 0,\ \mu = 1, \dots, k\},$$

where $\mathcal{D}_\nu = -i\partial/\partial\nu$, ν is the inner unit normal to ∂Q at the point $x \in \partial Q \setminus K$, $K = \emptyset$ if $\partial Q \in C^\infty$, $K = (\{0\} \times \partial G) \cup (\{d\} \times \partial G)$ if $Q = (0, d) \times G$.

Lemma 8.13. *The operator R_Q is continuous from $\mathring{W}^k(Q)$ into $W^k(Q)$. If $u \in \mathring{W}^1(Q)$, then*

$$(R_Q u)_{x_i} = R_Q u_{x_i} + R_{iQ} u, \tag{8.12}$$

where $R_i u(x) = \sum_{h \in \mathcal{M}} a_{hx_i}(x) u(x+h)$.

Proof. Evidently, the identity (8.12) is valid for every $u \in \dot{C}^\infty(Q)$. Using this identity k times, from the boundedness of the operators R_Q, $R_{iQ}\colon L_2(Q) \to L_2(Q)$ we have

$$\|R_Q u\|_{W^k(Q)} \le c\|u\|_{W^k(Q)} \tag{8.13}$$

for every $u \in \dot{C}^\infty(Q)$. Since $\dot{C}^\infty(Q)$ is dense in $\mathring{W}^k(Q)$, the boundedness of the operator $R_Q\colon \mathring{W}^k(Q) \to W^k(Q)$ and the correctness of the identity (8.12) for all $u \in \mathring{W}^1(Q)$ follows from (8.13). □

From Lemma 8.13 we obtain

Lemma 8.14. *Let the coefficients a_h be constants.*

Then for all $u \in \mathring{W}^k(Q)$

$$\mathcal{D}^\alpha R_Q u = R_Q \mathcal{D}^\alpha u (|d| \le k). \tag{8.14}$$

Lemma 8.15. *Let the coefficients a_h be constants.*

Then, for all $u \in L_2(Q)$, such that $u \in W^k(Q_{sl})$ ($s = 1, 2, \dots,$ $l = 1, \dots,$ $N(s)$) we have $R_Q u \in W^k(Q_{sl})$ and

$$\|R_Q u\|_{W^k(Q_{sl})} \le c_1 \sum_{j=1}^{N} \|u\|_{W^k(Q_{sj})}. \tag{8.15}$$

Moreover, if $\det R_{s_\nu} \ne 0$ ($\nu = 1, \dots, n_1$), *then $R_Q^{-1} u \in W^k(Q_{sl})$ and*

$$\|R_Q^{-1} u\|_{W^k(Q_{sl})} \le c_2 \sum_{j=1}^{N} \|u\|_{W^k(Q_{sj})}. \tag{8.16}$$

Here constants $c_1, c_2 > 0$ do not depend on s and u.

The proof follows from Lemmas 8.6, 8.7 (cf. Lemma 2.12).

Isomorphism between Sobolev Subspaces Generated by a Difference Operator

In this subsection we prove a theorem on isomorphism for a nondegenerate difference operator. This theorem establishes a connection between the elliptic differential equations with nonlocal boundary conditions of the type (8.17) and the elliptic differential-difference equations with homogeneous boundary conditions of the Dirichlet type. This enables us to apply the results obtained for one of these problems to the study of another (see Sections 13, 23, 25).

Let $\mu_{n-1}(\mathcal{K} \cap \partial Q) = 0$.

In this subsection we also assume that the following condition is fulfilled:

8.1. *For every subdomain* Q_{sl} $(s = 1, 2, \ldots,\ l = 1, \ldots, N(s))$ *and for each* $\varepsilon > 0$, *there exists an open set* $G_{sl} \subset Q_{sl}$ *with boundary* $\partial G_{sl} \in C^1$ *such that* $\mu_n(Q_{sl} \setminus G_{sl}) < \varepsilon$, $\mu_{n-1}(\partial G_{sl} \Delta \partial Q_{sl}) < \varepsilon$.

Denote by $W^k_\gamma(Q)$ $(\gamma = \{\gamma^r_{ij}\})$ the subspace of functions in $W^k(Q)$ satisfying the *nonlocal boundary conditions*

$$\left.\begin{aligned} \mathcal{D}^{\mu-1}_\nu w|_{\Gamma_{rl}} &= \sum_{j=1}^{J_0} \gamma^r_{lj} \mathcal{D}^{\mu-1}_\nu w|_{\Gamma_{rj}} \qquad (r \in B,\ l = J_0 + 1, \ldots, J), \\ \mathcal{D}^{\mu-1}_\nu w|_{\Gamma_{rl}} &= 0 \qquad\qquad\qquad\qquad\quad (r \notin B,\ l = 1, \ldots, J), \end{aligned}\right\} \tag{8.17}$$

where $\mu = 1, \ldots, k$, $J_0 = J_0(r)$, $J = J(r)$, γ^r_{lj} are complex numbers, $B = \{r : J_0 > 0\}$, ν is the inner unit normal vector to ∂Q at the point $x \in \Gamma_{rl}$.

By virtue of Lemma 7.7, for each $r = 1, 2, \ldots$, there exists a unique $s = s(r)$ such that $N(s) = J(r)$ and after some reindexing of subdomains of the s-class $\Gamma_{rl} \subset \partial Q_{sl}$ $(l = 1, \ldots, N)$. We denote by R_{s0} the matrix of order $J_0 \times J_0$ obtained from the matrix R_s $(s = s(r))$ by deleting the last $N - J_0$ rows and columns.

Theorem 8.1. *Let conditions* 7.1, 8.1 *hold. Suppose that the coefficients* a_h *are constants, and the matrices* R_s $(s = 1, 2, \ldots)$, R_{s0} $(s = s(r),\ r \in B)$ *are nonsingular.*

Then there exists a set $\gamma = \{\gamma^r_{lj}\}$ *such that the operator* R_Q *maps* $\mathring{W}^k(Q)$ *onto* $W^k_\gamma(Q)$ *continuously and in a one-to-one manner.*

Proof. 1. First we shall show that $R_Q(\mathring{W}^k(Q)) \subset W^k_\gamma(Q)$ for some γ. For this, by virtue of Lemma 8.13, it is sufficient to prove that a function $R_Q u$ satisfies the conditions (8.17) for $u \in \mathring{W}^k(Q)$.

Denote by e^r_i the ith row of the matrix obtained from the matrix R_s by deleting the last $N - J_0$ columns $(r \in B,\ s = s(r))$. Since $\det R_{s0} \neq 0$, there are numbers γ^r_{lj} such that

$$e^r_l = \sum_{j=1}^{J_0} \gamma^r_{lj} e^r_j \qquad (l = J_0 + 1, \ldots, N = N(s)). \tag{8.18}$$

Since $u \in \mathring{W}^k(Q)$, then $\mathcal{D}_\nu^{\mu-1}(U_s P_s u)_l|_{\Gamma_{r1}} = 0$ $(l = J_0+1,\dots,N,\ \mu = 1,\dots,k)$. From this and from (8.4), (8.10), (8.14), (8.18) we obtain

$$\begin{aligned}\mathcal{D}_\nu^{\mu-1}(R_Q u)|_{\Gamma_{rl}} &= (U_s P_s R_Q \mathcal{D}_\nu^{\mu-1}u)_l|_{\Gamma_{r1}} = (R_s U_s P_s \mathcal{D}_\nu^{\mu-1}u)_l|_{\Gamma_{r1}} \\ &= \sum_{j=1}^{J_0} \gamma_{lj}^r (R_s U_s P_s \mathcal{D}_\nu^{\mu-1}u)_j|_{\Gamma_{r1}} \\ &= \sum_{j=1}^{J_0} \gamma_{lj}^r \mathcal{D}_\nu^{\mu-1}(R_Q u)|_{\Gamma_{rj}}. \qquad (8.19)\end{aligned}$$

2. We now prove that $W_\gamma^k(Q) \subset R_Q(\mathring{W}^k(Q))$. By virtue of Lemma 7.8, for arbitrary fixed r and corresponding $s = s(r)$, there exist $p = p(r)$ and $m = m(r)$ such that $\Gamma_{r1} \subset \partial Q_{pm}$, $Q_{pm} \neq Q_{s1}$. We reindex the subdomains of the pth class so that $\Gamma_{rl} \subset \partial Q_{pl}$ $(l = 1,\dots,J_0)$.

Let $w \in W_\gamma^k(Q)$. Then, by Lemma 8.15, $u = R_Q^{-1}w \in W^k(Q_{sl})$.

a) We shall prove that $u \in \mathring{W}^k(Q)$ if the following relations are valid:

$$\varphi_l^\mu = 0 \qquad (l = J_0+1,\dots,N(s)\,;\mu = 1,\dots,k), \qquad (8.20)$$

$$\varphi_l^\mu = \psi_l^\mu \qquad (l = 1,\dots,J_0,\ \mu = 1,\dots,k), \qquad (8.21)$$

where $\varphi_l^\mu = (U_s P_s \mathcal{D}_\nu^{\mu-1}u)_l|_{\Gamma_{r1}}$ $(l = 1,\dots,N(s))$, $\psi_l^\mu = (U_p P_p \mathcal{D}_\nu^{\mu-1}u)_l|_{\Gamma_{r1}}$ $(l = 1,\dots,N(p))$.

We denote by $u_\alpha(x)$ $(x \in \bigcup_{s,l} Q_{sl})$ a function defined by $u_\alpha(x) = \mathcal{D}^\alpha u(x)$ $(x \in Q_{sl})$, where $|\alpha| \le k$. From Lemma 8.15 it follows that $u_\alpha \in L_2(Q)$. Therefore, by virtue of (8.20), in order to prove that $u \in \mathring{W}^k(Q)$, it is sufficient to show that a function u has a generalized derivative $\mathcal{D}^\alpha u \in L_2(Q)$ and $(\mathcal{D}^\alpha u)(x) = u_\alpha(x)$ $(x \in Q)$, i.e.,

$$\int_Q u_\alpha \bar{v}\,dx = \int_Q u\overline{\mathcal{D}^\alpha v}\,dx \qquad (8.22)$$

for every $v \in \dot{C}^\infty(Q)$.

Since $C^\infty(\overline{Q})$ is dense in $W^k(Q)$, there exists a sequence of functions $w_m \in C^\infty(\overline{Q})$ such that

$$\lim_{m\to\infty} \|w - w_m\|_{W^k(Q)} = 0, \qquad (8.23)$$

$$\lim_{m\to\infty} \sum_{r,l} \|\mathcal{D}^\beta(w - w_m)|_{\Gamma_{rl}}\|_{L_2(\Gamma_{rl})} = 0 \qquad (|\beta| \le k-1). \qquad (8.24)$$

From (8.23) and Lemma 8.15 we have

$$\lim_{m\to\infty} \sum_{s,l} \|u - u_m\|_{W^k(Q_{sl})} = 0, \qquad (8.25)$$

where $u_m = R_Q^{-1}w_m$.

From Lemmas 8.6, 8.14 and formula (8.10) it follows that

$$\begin{aligned} \mathcal{D}^\beta(u-u_m)|_{\Gamma_{rl}} &= (U_sP_s\mathcal{D}^\beta(u-u_m))_l|_{\Gamma_{r1}} \\ &= (R_s^{-1}U_sP_s\mathcal{D}^\beta(w-w_m))_l|_{\Gamma_{r1}}. \end{aligned} \tag{8.26}$$

Since the number of different matrices R_{s_ν} is finite, by virtue of (8.24), (8.26), we obtain

$$\lim_{m\to\infty} \sum_{r,l} \|\mathcal{D}^\beta(u-u_m)|_{\Gamma_{rl}}\|_{L_2(\Gamma_{rl})} = 0. \tag{8.27}$$

A formula for integration by parts for functions $u_m \in C^\infty(\overline{Q}_{sl})$, $v \in C^\infty(\overline{Q}_{sl})$ follows from condition 8.1:

$$\begin{aligned} \int_{Q_{sl}} \mathcal{D}^\alpha u_m \bar{v}\,dx = i\sum_{t=1}^{n}\sum_{j=1}^{\alpha_t} \int_{\partial Q_{sl}\setminus\mathcal{K}} & \mathcal{D}_t^{\alpha_t-j}\mathcal{D}_{t+1}^{\alpha_{t+1}}\dots\mathcal{D}_n^{\alpha_n}u_m \\ & \times \overline{\mathcal{D}_1^{\alpha_1}\dots\mathcal{D}_{t-1}^{\alpha_{t-1}}\mathcal{D}_t^{j-1}v}\cos(x_t,\nu_{sl})\,dx' + \int_{Q_{sl}} u_m\overline{\mathcal{D}^\alpha v}\,dx, \end{aligned}$$

where ν_{sl} is the inner unit normal vector to ∂Q_{sl} at the point $x' \in \partial Q_{sl} \setminus \mathcal{K}$. Passing to the limit as $m \to \infty$ in this formula and summing over s, l, we obtain (8.22). Let us note that, by virtue of (8.21), we have cancelled all integrals over the surfaces $\partial Q_{sl} \setminus \mathcal{K}$.

b) Now we prove the relations (8.20), (8.21). Since $w \in W^k(Q)$, then

$$\begin{aligned} (U_sP_s\mathcal{D}_\nu^{\mu-1}R_Qu)_l|_{\Gamma_{r1}} &= (U_pP_p\mathcal{D}_\nu^{\mu-1}R_Qu)_l|_{\Gamma_{r1}}, \quad \text{i.e.,} \\ (R_sU_sP_s\mathcal{D}_\nu^{\mu-1}u)_l|_{\Gamma_{r1}} &= (R_pU_pP_p\mathcal{D}_\nu^{\mu-1}u)_l|_{\Gamma_{r1}} \qquad (l=1,\dots,J_0). \end{aligned}$$

Thus, the functions φ_l^μ and ψ_l^μ satisfy the conditions

$$\sum_{l=1}^{N(s)} r_{il}^s\varphi_l^\mu = \sum_{l=1}^{N(p)} r_{il}^p\psi_l^\mu \qquad (i=1,\dots,J_0). \tag{8.28}$$

Moreover, the function $w = R_Qu$ satisfies the conditions (8.19), which can be rewritten as

$$\sum_{i=1}^{N(s)} r_{li}^s\varphi_i^\mu = \sum_{j=1}^{J_0}\gamma_{lj}^r\sum_{i=1}^{N(s)} r_{ji}^s\varphi_i^\mu \qquad (l=J_0+1,\dots,N(s)). \tag{8.29}$$

From (8.18) we obtain

$$\sum_{i=1}^{J_0} r_{li}^s\varphi_i^\mu = \sum_{j=1}^{J_0}\gamma_{lj}^r\sum_{i=1}^{J_0} r_{ji}^s\varphi_i^\mu \qquad (l=J_0+1,\dots,N(s)). \tag{8.30}$$

The relations (8.29), (8.30) imply that

$$\sum_{i=J_0+1}^{N(s)} \left(r_{li}^s - \sum_{j=1}^{J_0} \gamma_{lj}^r r_{ji}^s \right) \varphi_i^\mu = 0 \qquad (l = J_0+1, \dots, N(s)). \tag{8.31}$$

By virtue of (8.18), $\det R_s = \det R_{s0} \cdot \det \widetilde{R}_s$, where $\widetilde{R}_s$ is the matrix of order $(N(s) - J_0) \times (N(s) - J_0)$ of the system (8.31). Since $\det R_s \neq 0$, $\det \widetilde{R}_s \neq 0$. Hence, $\varphi_l^\mu = 0$ $(l = J_0+1, \dots, N(s))$. Analogously, we obtain $\psi_l^\mu = 0$ $(l = J_0+1, \dots, N(p))$. Thus, the system of equations (8.28) will have the form

$$\sum_{l=1}^{J_0} (r_{il}^s \varphi_l^\mu - r_{il}^p \psi_l^\mu) = 0 \qquad (i = 1, \dots, J_0). \tag{8.32}$$

By construction, $\Gamma_{rl} \subset \partial Q_{sl} \cap \partial Q_{pl}$ $(l = 1, \dots, J_0)$. Hence, $h_{sl} = h_{pl}$ $(l = 1, \dots, J_0)$. Therefore, by virtue of (8.6), $r_{il}^s = r_{il}^p$ $(i, l = 1, \dots, J_0)$. Since $\det R_{s0} \neq 0$, from the system (8.32) we have $\varphi_l^\mu = \psi_l^\mu$ $(l = 1, \dots, J_0)$. □

Example 8.4. Let $Q = (0,2) \times (0,1) \subset \mathbb{R}^2$, $Ru(x) = u(x) + a_1 u(x_1+1, x_2) + a_{-1} u(x_1 - 1, x_2)$, where $a_1, a_{-1} \in \mathbb{R}$.

Then the decomposition $\mathcal{R}$ of the domain Q consists of two subdomains: $Q_{11} = (0,1) \times (0,1)$, $Q_{12} = (1,2) \times (0,1)$ (see Examples 7.1, 7.4, 7.7). Moreover,

$$R_1 = \begin{pmatrix} 1 & a_1 \\ a_{-1} & 1 \end{pmatrix}.$$

If $a_1 a_{-1} \neq 1$, then, by Theorem 8.1, R_Q maps $\mathring{W}^1(Q)$ onto $W_\gamma^1(Q)$ continuously and in a one-to-one manner. Here $W_\gamma^1(Q)$ is the subspace of functions in $W^1(Q)$ satisfying the conditions

$$w|_{x_2=0} = w|_{x_2=1} = 0, \quad w|_{x_1=0} = a_1 w|_{x_1=1}, \quad w|_{x_1=2} = a_{-1} w|_{x_1=1}.$$

Remark 8.1. Let the coefficients a_h be constants, and let the operator $R_Q + R_Q^*$ be positive definite. Then the conditions of Theorem 8.1 concerning the matrices R_s, R_{s0} are fulfilled. In fact, by Lemma 8.8, the matrices $R_s + R_s^*$ are positive definite. Hence, the matrices $R_{s0} + R_{s0}^*$ are also positive definite. Thus $\det R_s \neq 0$, $\det R_{s0} \neq 0$.

Self-Adjoint Difference Operators

In this subsection we shall assume that the operator $R_Q : L_2(Q) \to L_2(Q)$ is self-adjoint and the coefficients a_h are constants.

Denote by R_Q^R the restriction of R_Q to $\mathcal{R}(R_Q)$. Denote by $P^R : L_2(Q) \to L_2(Q)$ and $P_s^R : L_2^N(Q_{s1}) \to L_2^N(Q_{s1})$ the operators of orthogonal projection onto the subspaces $\mathcal{R}(R_Q)$ and $\mathcal{R}(R_{Qs})$, respectively. We shall further prove that the subspaces $\mathcal{R}(R_Q)$ and $\mathcal{R}(R_{Qs})$ are closed.

Lemma 8.16. $L_2^N(Q_{s1}) = \mathcal{N}(R_{Qs}) \oplus \mathcal{R}(R_{Qs})$.

Proof. By virtue of Lemma 8.12, the operator $R_{Qs}: L_2^N(Q_{s1}) \to L_2^N(Q_{s1})$ is self-adjoint. Therefore, it is sufficient to prove that a linear manifold $\mathcal{R}(R_{Qs})$ is closed in $L_2^N(Q_{s1})$. Since the matrices R_s are Hermitian, we have $\mathbb{C}^N = \mathcal{N}(R_s) \oplus \mathcal{R}(R_s)$. Thus, $\mathcal{R}(R_s)$ is a set of vectors $y \in \mathbb{C}^N$ satisfying the equations

$$\sum_{j=1}^{N} \overline{a_{ij}} y_j = 0 \qquad (i = 1, \dots, m),$$

where $m = \dim \mathcal{N}(R_s)$, and the vectors $(a_{i1}, \dots, a_{iN})$ $(i = 1, \dots, m)$ form the orthogonal basis in $\mathcal{N}(R_s)$. Hence, $\mathcal{R}(R_{Qs})$ is a set of vector-valued functions $Y \in L_2^N(Q_{s1})$ satisfying the equations

$$\sum_{j=1}^{N} \overline{a_{ij}} Y_j = 0 \qquad (i = 1, \dots, m) \tag{8.33}$$

for almost all $x \in Q_{s1}$. From this it follows that $\mathcal{R}(R_{Qs})$ is closed in $L_2^N(Q_{s1})$. □

Lemma 8.17. $L_2(Q) = \mathcal{N}(R_Q) \oplus \mathcal{R}(R_Q)$. *The operator* $R_Q^R: \mathcal{R}(R_Q) \to \mathcal{R}(R_Q)$ *has a bounded inverse, and for every* $u \in L_2(Q)$

$$\|P^R u\|_{L_2(Q)} \le c \|R_Q u\|_{L_2(Q)}, \tag{8.34}$$

where $c > 0$ *is constant.*

Proof. In order to prove the first part of Lemma 8.17, it is sufficient to show that $\mathcal{R}(R_Q)$ is closed in $L_2(Q)$. Let us consider a sequence $u_m \in \mathcal{R}(R_Q)$ such that $u_m \to u_0$ in $L_2(Q)$ as $m \to \infty$. We shall prove that $u_0 \in \mathcal{R}(R_Q)$.

Clearly $U_s P_s u_m \in \mathcal{R}(R_{Qs})$. Hence, by Lemma 8.16, there is a function $V_{ms} \in \mathcal{R}(R_{Qs})$ such that $U_s P_s u_m = R_s V_{ms}$. There exists a matrix $\widehat{R}_s$ such that $\widehat{R}_s R_s y = y$ for every $y \in \mathcal{R}(R_s)$. Thus $V_{ms} = \widehat{R}_s U_s P_s u_m$, i.e., $U_s^{-1} V_{ms} = U_s^{-1} \widehat{R}_s U_s P_s u_m$. Since the number of different matrices R_s is finite, we obtain

$$\|P_s v_m - P_s v_k\|^2_{L_2(\bigcup_l Q_{sl})} \le c_1 \|P_s u_m - P_s u_k\|^2_{L_2(\bigcup_l Q_{sl})}$$

for all m, k, where $c_1 > 0$ does not depend on s, m, k, $v_m = \sum_s U_s^{-1} V_{ms} \in L_2(Q)$. Hence, summing over s, by virtue of (8.3), we have

$$\|v_m - v_k\|^2_{L_2(Q)} \le c_1 \|u_m - u_k\|^2_{L_2(Q)}.$$

Thus there exists a limit $v_0 = \lim_{m \to \infty} v_m$ in the space $L_2(Q)$. Since $R_Q v_m = u_m$, by virtue of Lemma 8.2, $u_0 = R_Q v_0$.

From Banach's inverse operator theorem it follows that the operator R_Q^R: $\mathcal{R}(R_Q) \to \mathcal{R}(R_Q)$ has a bounded inverse and inequality (8.34) holds. □

Remark 8.2. The operator $P_s^R\colon L_2^N(Q_{s1}) \to L_2^N(Q_{s1})$ is the operator of multiplication by some matrix. We denote this matrix also by P_s^R. Clearly the multiplication by the matrix P_s^R in the complex space $\mathbb{C}^N$ is the operator of orthogonal projection in $\mathbb{C}^N$ onto the range of the matrix R_s.

Lemma 8.18.

$$P^R = \sum_s U_s^{-1} P_s^R U_s P_s, \tag{8.35}$$

$$U_s P_s P^R = P_s^R U_s P_s. \tag{8.36}$$

Proof. Evidently $P_s^R = U_s P^R U_s^{-1}$. Therefore, using (8.3), we obtain

$$P^R = \sum_s P^R P_s = \sum_s U_s^{-1} U_s P^R U_s^{-1} U_s P_s = \sum_s U_s^{-1} P_s^R U_s P_s.$$

The equation (8.36) follows from (8.35). □

Example 8.5. Let $Ru(x) = u(x) + u(x_1+1, x_2) + u(x_1-1, x_2)$ and $Q = (0,2)\times(0,1)$ (see Example 8.3). Then

$$R_1 = \begin{pmatrix} 1 & 1 \\ 1 & 1 \end{pmatrix}.$$

Therefore, by virtue of Lemma 8.7, $0 \in \sigma(R_Q)$. Evidently $\mathcal{N}(R_Q) = \{u \in L_2(Q) : -u(x_1,x_2) = u(x_1+1, x_2) \text{ for } x \in Q_{11}\}$, $\mathcal{R}(R_Q) = \{u \in L_2(Q) : u(x_1,x_2) = u(x_1+1,x_2) \text{ for } x \in Q_{11}\}$.

Lemma 8.19. *Let Q'_{sl} be open, connected sets such that $Q'_{sl} \subset Q_{sl}$ and $Q'_{sl} = Q'_{s1} + h_{sl}$ $(s = 1,2,\ldots,\ l = 1,\ldots,N(s))$.*

Then, for all $u \in L_2(Q)$ such that $R_Q u \in W^k(Q'_{sl})$, we have $P^R u \in W^k(Q'_{sl})$ and

$$\|P^R u\|_{W^k(Q'_{sl})} \le c_1 \sum_{j=1}^{N} \|R_Q u\|_{W^k(Q'_{sj})}, \tag{8.37}$$

where $c_1 > 0$ does not depend on s and u, $s = 1,2,\ldots,\ l = 1,\ldots,N(s)$.

Proof. Evidently $(U_s P_s R_Q u)_l \in W^k(Q'_{s1})$ for each fixed s and arbitrary $1 \le l \le N$. By virtue of (8.5), (8.10), and Lemma 8.17,

$$U_s P_s R_Q u = U_s P_s R_Q P^R u = R_s U_s P_s P^R u.$$

From this, by virtue of formula (8.36) and Remark 8.2, we obtain

$$U_s P_s P^R u = \widehat{R}_s (R_s U_s P_s P^R u) = \widehat{R}_s (U_s P_s R_Q u), \tag{8.38}$$

where the matrix $\widehat{R}_s$ was defined in the proof of Lemma 8.17. Hence, $P^R u \in W^k(Q'_{sl})$, and the inequality (8.37) holds. □

9 Necessary and Sufficient Conditions for Strong Ellipticity

Necessary Conditions

We consider the equation

$$\mathcal{A}_R u = \sum_{|\alpha|,|\beta|\le m} \mathcal{D}^\alpha R_{\alpha\beta Q}\mathcal{D}^\beta u(x) = f_0(x) \qquad (x \in Q) \tag{9.1}$$

with the boundary conditions

$$\mathcal{D}_\nu^{\mu-1} u|_{\partial Q\setminus K} = 0 \qquad (x \in \partial Q \setminus K,\ \mu = 1, \dots, m). \tag{9.2}$$

Here

$$R_{\alpha\beta Q} = P_Q R_{\alpha\beta} I_Q, \qquad R_{\alpha\beta} u(x) = \sum_{h\in\mathcal{M}} a_{\alpha\beta h}(x) u(x+h)$$

are difference operators, $a_{\alpha\beta h} \in C^\infty(\mathbb{R}^n)$, $f_0 \in L_2(Q)$ are complex-valued functions, $\mathcal{M} \subset \mathbb{R}^n$ is a finite set of vectors with integer coordinates, $Q \subset \mathbb{R}^n$ is a bounded domain with boundary $\partial Q \in C^\infty$ or a cylinder $(0,d) \times G$, where $G \subset \mathbb{R}^{n-1}$ is a bounded domain (with boundary $\partial G \in C^\infty$ if $n \ge 3$). $K = \emptyset$ if $\partial Q \in C^\infty$ and $K = (\{0\} \times \partial G) \cup (\{d\} \times \partial G)$ if $Q = (0,d) \times G$.

Definition 9.1. The equation (9.1) is called a *differential-difference equation.*

Definition 9.2. The differential-difference equation (9.1) is said to be *strongly elliptic* in $\overline{Q}$ if for all $u \in \dot{C}^\infty(Q)$

$$\mathrm{Re}\left(\sum_{|\alpha|,|\beta|\le m} \mathcal{D}^\alpha R_{\alpha\beta Q}\mathcal{D}^\beta u, u\right)_{L_2(Q)} \ge c_1\|u\|^2_{W^m(Q)} - c_2\|u\|^2_{L_2(Q)}, \tag{9.3}$$

where $c_1 > 0$, $c_2 \ge 0$ do not depend on u.

Definition 9.3. The problem (9.1), (9.2) is called the *first boundary value problem.*

In order to state the necessary conditions for strong ellipticity in an algebraic form, we introduce the matrices $R_{\alpha\beta s}(x)$ $(x \in \overline{Q}_{s1})$ of order $N(s) \times N(s)$ with the elements

$$r_{ij}^{\alpha\beta s} = \begin{cases} a_{\alpha\beta h}(x + h_{si}) & (h = h_{sj} - h_{si} \in \mathcal{M}), \\ 0 & (h_{sj} - h_{si} \notin \mathcal{M}). \end{cases} \tag{9.4}$$

Theorem 9.1. *Let equation* (9.1) *be strongly elliptic in* $\overline{Q}$.

Then, for all $s = 1, 2, \dots,$ $x \in \overline{Q}_{s1}$, *and* $0 \ne \xi \in \mathbb{R}^n$, *the matrices*

$$\sum_{|\alpha|,|\beta|=m} (R_{\alpha\beta s}(x) + R^*_{\alpha\beta s}(x))\xi^{\alpha+\beta}$$

are positive definite.

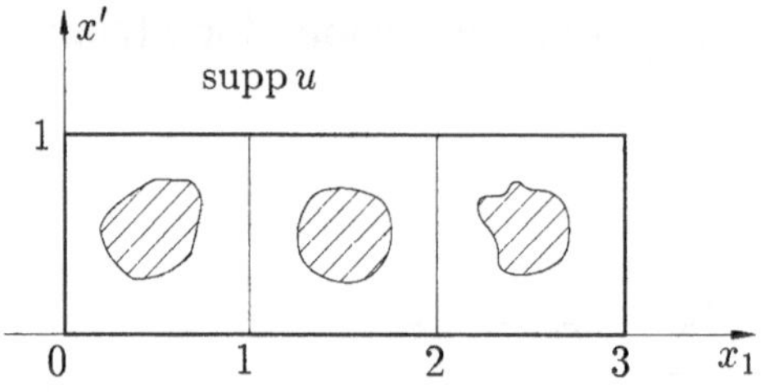

Fig. II.5

Proof. 1. For a fixed s, we put $u \in \dot{C}^\infty(\bigcup_l Q_{sl})$ in the integral inequality (9.3). We introduce the vector-valued function

$$V_s = U_s u \in \dot{C}^{\infty,N}(Q_{s1}) \qquad (N = N(s))$$

(see Fig. II.5), where $\dot{C}^{\infty,N}(Q_{s1}) = \prod_{l=1}^N \dot{C}^\infty(Q_{s1})$. Then by virtue of (8.5), we have

$$\operatorname{Re}\Bigg(\sum_{|\alpha|,|\beta|\le m} \mathcal{D}^\alpha R_{\alpha\beta s}(x)\mathcal{D}^\beta V_s, V_s\Bigg)_{L_2^N(Q_{s1})} \ge c_1\|V_s\|^2_{W^{m,N}(Q_{s1})} - c_2\|V_s\|^2_{L_2^N(Q_{s1})} \quad (9.5)$$

for every $V_s \in \dot{C}^{\infty,N}(Q_{s1})$, where $W^{m,N}(Q_{s1}) = \prod_{l=1}^N W^m(Q_{s1})$.

Using (9.5), (B.20) and the inequality

$$ab \le q^{-m}a^2 + q^m b^2 \qquad (a, b \in \mathbb{R},\ q > 0), \quad (9.6)$$

we obtain

$$\operatorname{Re}\Bigg(\sum_{|\alpha|,|\beta|= m} \mathcal{D}^\alpha R_{\alpha\beta s}(x)\mathcal{D}^\beta V_s, V_s\Bigg)_{L_2^N(Q_{s1})} \ge k_1\|V_s\|^2_{W^{m,N}(Q_{s1})} - k_2\|V_s\|^2_{L_2^N(Q_{s1})}. \quad (9.7)$$

2. Let $x^0 \in \overline{Q}_{s1}$ be an arbitrary point. We choose x^1, r so that $\overline{S_r(x^1)} \subset Q_{s1} \cap S_\delta(x^0)$, where $\delta > 0$ will be defined later. Assume that $\operatorname{supp} V_s \subset S_r(x^1)$. Then from (9.7) it follows that

$$b_1 + b_2 \ge k_1\|V_s\|^2_{W^{m,N}(S_r(x^1))} - k_2\|V_s\|^2_{L_2^N(S_r(x^1))}, \quad (9.8)$$

where

$$b_1 = \operatorname{Re}\Bigg(\sum_{|\alpha|,|\beta|= m} R_{\alpha\beta s}(x^0)\mathcal{D}^\beta V_s, \mathcal{D}^\alpha V_s\Bigg)_{L_2^N(S_r(x^1))},$$

$$b_2 = \operatorname{Re}\Bigg(\sum_{|\alpha|,|\beta|= m} (R_{\alpha\beta s}(x) - R_{\alpha\beta s}(x^0))\mathcal{D}^\beta V_s, \mathcal{D}^\alpha V_s\Bigg)_{L_2^N(S_r(x^1))}.$$

Since the functions $a_{\alpha\beta h}(x)$ are uniformly continuous on the compact $\overline{Q}$, we have

$$|b_2| < \varepsilon(\delta)\|V_s\|_{W^{m,N}(S_r(x^1))},$$

where $\varepsilon(\delta) \to 0$ as $\delta \to 0$. Let us choose $\delta > 0$ such that $\varepsilon(\delta) < k_1/2$. Hence, from (9.8) we obtain

$$\operatorname{Re}\Big(\sum_{|\alpha|,|\beta|=m} R_{\alpha\beta s}(x^0)\mathcal{D}^\beta V_s, \mathcal{D}^\alpha V_s\Big)_{L_2^N(S_r(x^1))} \geq \frac{k_1}{2}\|V_s\|^2_{W^{m,N}(S_r(x^1))} - k_2\|V_s\|^2_{L_2^N(S_r(x^1))}.$$

3. Now we shall obtain the appropriate estimate for $W_s \in \dot{C}^{\infty,N}(S_R(0))$, where $t = R/r > 1$. Let us change variables $y = t(x - x^1)$. Denote $W_s(y) = V_s(x(y))$. Then from the last inequality we obtain

$$\begin{aligned}
&\operatorname{Re}\Big(\sum_{|\alpha|,|\beta|=m} R_{\alpha\beta s}(x^0)\mathcal{D}_y^\beta W_s(y), \mathcal{D}_y^\alpha W_s(y)\Big)_{L_2^N(S_R(0))}\\
&= t^{n-2m}\operatorname{Re}\Big(\sum_{|\alpha|,|\beta|=m} R_{\alpha\beta s}(x^0)\mathcal{D}_x^\beta V_s(x), \mathcal{D}_x^\alpha V_s(x)\Big)_{L_2^N(S_r(x^1))}\\
&\geq \frac{k_1}{2}t^{n-2m}\sum_{|\alpha|=m}\|\mathcal{D}_x^\alpha V_s(x)\|^2_{L_2^N(S_r(x^1))} - k_2 t^{n-2m}\|V_s(x)\|^2_{L_2^N(S_r(x^1))}\\
&\geq \frac{k_1}{2}\sum_{|\alpha|=m}\|\mathcal{D}_y^\alpha W_s(y)\|^2_{L_2^N(S_R(0))} - k_2\|W_s(y)\|^2_{L_2^N(S_R(0))}.
\end{aligned} \tag{9.9}$$

4. Suppose that $W_s = v_s Y$, where $v_s \in \dot{C}^\infty(S_R(0))$, $Y \in \mathbb{C}^N$. Let a function v_s be extended by zero in $\mathbb{R}^n \setminus S_R(0)$. Then, using the Fourier transform, from (9.9) we have

$$\begin{aligned}
&\frac{1}{2}\int_{\mathbb{R}^n}\Big(\sum_{|\alpha|,|\beta|=m}(R_{\alpha\beta s}(x^0) + R^*_{\alpha\beta s}(x^0))\xi^{\alpha+\beta}Y, Y\Big)|\hat{v}_s(\xi)|^2 d\xi\\
&\geq k_3\int_{\mathbb{R}^n}|\xi|^{2m}|Y|^2|\hat{v}_s(\xi)|^2 d\xi - k_2\int_{\mathbb{R}^n}|Y|^2|\hat{v}_s(\xi)|^2 d\xi,
\end{aligned} \tag{9.10}$$

where $k_3 = k_3(n, m) > 0$ does not depend on v_s, $(\cdot,\cdot)$ is the inner product in $\mathbb{C}^N$.

Since a set $\dot{C}^\infty(\mathbb{R}^n)$ is dense in $L_2(\mathbb{R}^n)$, from (9.10) it follows that

$$\Big(\sum_{|\alpha|,|\beta|=m}(R_{\alpha\beta s}(x^0) + R^*_{\alpha\beta s}(x^0))\xi^{\alpha+\beta}Y, Y\Big) \geq 2k_3|\xi|^{2m}|Y|^2 - 2k_2|Y|^2.$$

If $|\xi|^{2m} > 2k_2/k_3$, then

$$\left(\sum_{|\alpha|,|\beta|=m} (R_{\alpha\beta s}(x^0) + R^*_{\alpha\beta s}(x^0))\xi^{\alpha+\beta}Y, Y \right) \geq k_3|\xi|^{2m}|Y|^2. \tag{9.11}$$

Since inequality (9.11) is homogeneous with respect to ξ, this inequality is valid for all $\xi \in \mathbb{R}^n$. □

From the proof of Theorem 9.1 it follows that the strongly elliptic differential-difference equation (9.1) in $\bigcup_l Q_{sl}$ is equivalent to a strongly elliptic system of differential equations in Q_{s1}. However, this approach does not allow us to obtain sufficient conditions for strong ellipticity of the equation (9.1). In fact, a formal transform U_s of the problem (9.1), (9.2) will give a finite or countable number of systems of partial differential equations in subdomains Q_{s1} with angular points. Furthermore, the unknown vector-valued functions $U_s u$ must satisfy some conjunction conditions at the pieces of ∂Q_{s1}. These difficulties do not allow us to apply classical results on coerciveness of systems of differential equations (see S. Agmon [1], D. G. Figueiredo [1]).

Sufficient Conditions

In order to formulate the sufficient conditions for strong ellipticity, we shall introduce some auxiliary notation. Let $x \in \overline{Q}_{s1}$ be an arbitrary point. Consider all points $x^i \in \overline{Q}$ such that $x^i - x \in M$. Since the domain Q is bounded, the set $\{x^i\}$ consists of a finite number of points $I = I(s,x)$ $(I \geq N(s))$. We shall number the points x^i so that $x^i = x + h_{si}$ for $i = 1, \dots, N = N(s)$, $x^1 = x$, where h_{si} satisfies the condition $Q_{si} = Q_{s1} + h_{si}$. We introduce the $I \times I$-matrices $A_{\alpha\beta s}(x)$ with elements $a_{ij}^{\alpha\beta s}(x)$ by the formula

$$a_{ij}^{\alpha\beta s}(x) = \begin{cases} a_{\alpha\beta h}(x^i) & (h = x^j - x^i \in \mathcal{M}), \\ 0 & (x^j - x^i \notin \mathcal{M}). \end{cases} \tag{9.12}$$

Remark 9.1. If $I = N$, then the matrix $R_{\alpha\beta s}(x)$ is equal to the matrix $A_{\alpha\beta s}(x)$. If $N < I$, then the matrix $R_{\alpha\beta s}(x)$ is obtained from the matrix $A_{\alpha\beta s}(x)$ by deleting the last $I - N$ rows and columns.

We shall construct a special partition of unity. Suppose that $\{S_{\delta/2}(x)\}$ cover the set $\overline{Q}$, where $S_{\delta/2}(x)$ are the open balls with radius $\delta/2$ and centers at the points $x \in \overline{Q}$. For each $x \in \overline{Q}$, we shall take $\delta = \delta(x)$ so that $2\delta(x) < \min\{1/2, r, a\}$. Here, since Q is a bounded domain, $r = r(x) = \inf \rho(x+h, Q) > 0$ $(h : x+h \notin \overline{Q})$. The number a does not depend on x and will be chosen later. Since $\overline{Q}$ is compact, there is a finite subcovering of $\overline{Q}$ by the balls $S_{\delta/2}(y^j)$ $(y^j \in \overline{Q}$, $j = 1, \dots, J)$. Denote $G = \bigcup_j S_{\delta/2}(y^j)$.

Lemma 9.1. *There exist non-negative functions, M-periodic in $\overline{G}$, $\varphi \in \dot{C}^\infty(\mathbb{R}^n)$ $(j = 1, \dots, J)$ such that:*

1) $\sum_j \varphi_j^2(x) \le 1$ *for* $x \in \mathbb{R}^n$;
2) $\sum_j \varphi_j^2 = 1$ *for* $x \in \overline{G}$;
3) $\varphi_j(x) = 0$ *for* $x \notin \Omega_j$, *where* $\Omega_j = \bigcup_h (S_\delta(y^j) + h)$ $(h \in M : (S_\delta(y^j) + h) \cap \overline{G} \neq \emptyset)$.

Proof. We define functions $\eta_j \in \dot{C}^\infty(\mathbb{R}^n)$ so that $\eta_j(x) = 1$ $(x \in S_{\delta/2}(y^j))$, $\operatorname{supp} \eta_j \subset S_\delta(y^j)$, $\eta_j(x) \ge 0$. Let $\psi_j = \sum_h \eta_j(x+h)$ $(h \in M : (S_\delta(y^j)+h) \cap \overline{G} \neq \emptyset)$. By definition, the functions $\psi_j \in \dot{C}^\infty(\mathbb{R}^n)$ are M-periodic in $\overline{G}$.

Let $\Lambda = \{x \in \mathbb{R}^n : \sum_j \psi_j^2(x) > 0\}$. Clearly, $\overline{Q} \subset G$ and $\overline{G} \subset \Lambda$. Define a function $g \in \dot{C}^\infty(\mathbb{R}^n)$ such that $0 \le g(x) \le 1$, $g(x) = 1$ $(x \in \overline{G})$, $g(x) = 0$ $(x \notin \Lambda)$. We denote

$$\varphi_j(x) = \left\{1 - g(x) + \sum_j \psi_j^2(x)\right\}^{-1/2} \psi_j(x).$$

Evidently, the functions $\varphi_j(x) \in \dot{C}^\infty(\mathbb{R}^n)$ are M-periodic in $\overline{G}$ and satisfy conditions 1)–3). □

Theorem 9.2. *Let the matrices*

$$\sum_{|\alpha|,|\beta|=m} (A_{\alpha\beta s}(x) + A_{\alpha\beta s}^*(x))\xi^{\alpha+\beta}$$

be positive definite for all $s = 1, 2, \ldots$, $x \in \overline{Q}_{s1}$, *and* $0 \neq \xi \in \mathbb{R}^n$.

Then the equation (9.1) *is strongly elliptic in* $\overline{Q}$.

Proof. Suppose that $u \in \dot{C}^\infty(Q)$. For a given point $y^j \in \overline{Q}$, there exists a subdomain Q_{sl} such that $y^j \in \overline{Q}_{sl}$. Denote $z^j = y^j - h_{sl}$. Then $z^j \in \overline{Q}_{s1}$. We introduce the vector-valued functions $W^j \in \dot{C}^{\infty,I}(S_\delta(z^j))$ with coordinates

$$W_i^j(x) = (\varphi_j u)(x + z^{ji} - z^j) \qquad (x \in S_\delta(z^j)), \tag{9.13}$$

where $i = 1, \ldots, I = I(s, z^j)$, and the points z^{ji} are put in accordance with the point z^j in the manner described at the beginning of this subsection. Let us note that in $S_\delta(z^j)$ the matrices $A_{\alpha\beta s}(x)$ may have different order at different points x (see Example 9.1). Therefore, we also introduce the $I(s, z^j) \times I(s, z^j)$-matrices $A^j_{\alpha\beta s}(x)$ $(x \in S_\delta(z^j))$ with elements $a_{ik}^{\alpha\beta s j}(x)$ defined by the formula

$$a_{ik}^{\alpha\beta sj}(x) = \begin{cases} a_{\alpha\beta h}(x + z^{ji} - z^j) & (h = z^{jk} - z^{ji} \in \mathcal{M}), \\ 0 & (z^{jk} - z^{ji} \notin \mathcal{M}). \end{cases} \tag{9.14}$$

Obviously, $A^j_{\alpha\beta s}(z^j) = A_{\alpha\beta s}(z^j)$.

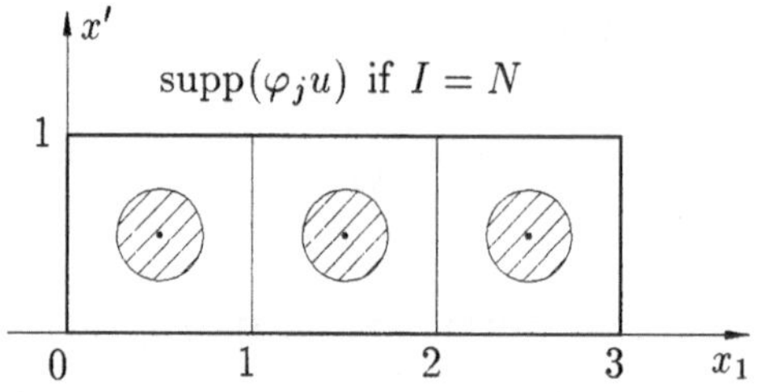

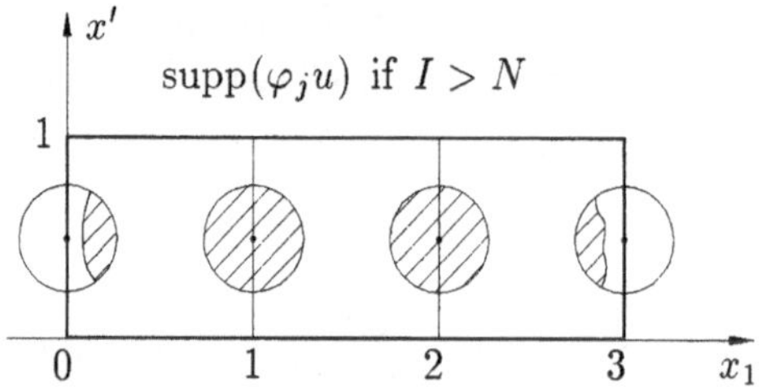

Fig. II.6

Integrating by parts, using the Leibniz formula and Lemmas 9.1, 8.10 and 8.2, we obtain

$$\begin{aligned}
&\operatorname{Re}\Big(\sum_{|\alpha|,|\beta|\le m}\mathcal{D}^\alpha R_{\alpha\beta Q}\mathcal{D}^\beta u,u\Big)_{L_2(Q)}\\
&\quad=\operatorname{Re}\sum_{|\alpha|,|\beta|\le m}\sum_j(\varphi_j R_{\alpha\beta Q}\mathcal{D}^\beta u,\varphi_j\mathcal{D}^\alpha u)_{L_2(Q)}\\
&\quad\ge b-k_1(a)\|u\|_{W^m(Q)}\|u\|_{W^{m-1}(Q)},
\end{aligned}\tag{9.15}$$

where $b=\operatorname{Re}\sum_{|\alpha|,|\beta|=m}\sum_j(R_{\alpha\beta Q}\mathcal{D}^\beta(\varphi_j u),\mathcal{D}^\alpha(\varphi_j u))_{L_2(Q)}$ (see Fig. II.6).

Since $\delta<r$, for every z^j+h ($h\in M$) satisfying the condition $(S_\delta(z^j)+h)\cap Q\ne\emptyset$, there exists $z^{ji}=z^j+h$. From this, by virtue of (9.13), (9.14), we have

$$b=b_1+b_2,$$

where

$$b_1=\operatorname{Re}\sum_{|\alpha|,|\beta|=m}\sum_j((A^j_{\alpha\beta s}(x)-A^j_{\alpha\beta s}(z^j))\mathcal{D}^\beta W^j,\mathcal{D}^\alpha W^j)_{L_2^I(S_\delta(z^j))},$$

$$b_2=\operatorname{Re}\sum_{|\alpha|,|\beta|=m}\sum_j(A^j_{\alpha\beta s}(z^j)\mathcal{D}^\beta W^j,\mathcal{D}^\alpha W^j)_{L_2^I(S_\delta(z^j))}.$$

Since $I(s,x)\le([\operatorname{diam}Q]+1)^n$ for all $s=1,2,\dots$, $x\in\overline{Q}_{s1}$ and the functions $a_{\alpha\beta h}(x)$ are uniformly continuous on the compact $\overline{Q}^1=\{x\in\mathbb{R}^n:\rho(x,Q)\le 1\}$, then

$$|b_1| \le \varepsilon(a) \sum_j \|W^j\|^2_{W^{m,I}(S_\delta(z^j))}, \tag{9.16}$$

where $\varepsilon(a) \to 0$ as $a \to 0$. Using the Leibniz formula, from (9.16) we obtain

$$|b_1| \le \varepsilon(a)\|u\|^2_{W^m(Q)} + k_2(a)\|u\|_{W^m(Q)}\|u\|_{W^{m-1}(Q)}. \tag{9.17}$$

According to the assumption of Theorem 9.2, we have

$$\mathrm{Re}\Bigg(\sum_{|\alpha|,|\beta|=m} A^j_{\alpha\beta s}(z^j)\xi^{\alpha+\beta}Y, Y\Bigg) \ge k_3\Bigg(\sum_{i=1}^n \xi_i^2\Bigg)^m (Y, Y)$$

for all $0 \ne \xi \in \mathbb{R}^n$, $Y \in \mathbb{C}^I$, $j = 1, \dots, J$, where $(\cdot, \cdot)$ is the inner product in $\mathbb{C}^I$. Let the vector-valued function $W^j(x)$ be extended by zero in $\mathbb{R}^n \setminus S_\delta(z^j)$. Using the Fourier transform and Theorem B.11 concerning the equivalent norms in $\mathring{W}^m(Q)$, we obtain

$$\begin{aligned} b_2 &= \sum_j \mathrm{Re}\Bigg(\sum_{|\alpha|,|\beta|=m} A^j_{\alpha\beta s}(z^j)\xi^{\alpha+\beta}\widehat{W}^j, \widehat{W}^j\Bigg)_{L_2^I(\mathbb{R}^n)} \\ &\ge k_3 \sum_j \Bigg(\Bigg(\sum_{i=1}^n \xi_i^2\Bigg)^m \widehat{W}^j, \widehat{W}^j\Bigg)_{L_2^I(\mathbb{R}^n)} \\ &\ge k_4 \sum_j \sum_{|\alpha|=m} \|\xi^\alpha \widehat{W}^j\|^2_{L_2^I(\mathbb{R}^n)} \ge k_5 \sum_j \|\varphi_j u\|^2_{W^m(Q)} \\ &\ge k_5\|u\|^2_{W^m(Q)} - k_6(a)\|u\|_{W^m(Q)}\|u\|_{W^{m-1}(Q)}. \end{aligned} \tag{9.18}$$

From inequalities (9.15), (9.17), (9.18), (B.20), and (9.6) it follows that

$$\begin{aligned} &\mathrm{Re}\Bigg(\sum_{|\alpha|,|\beta|\le m} \mathcal{D}^\alpha R_{\alpha\beta Q}\mathcal{D}^\beta u, u\Bigg)_{L_2(Q)} \\ &\qquad \ge (k_5 - \varepsilon(a) - q^{-1}k_7(a))\|u\|^2_{W^m(Q)} - q^{2m-1}k_8(a)\|u\|^2_{L_2(Q)} \end{aligned} \tag{9.19}$$

for every $q > 0$, where constants $k_7(a), k_8(a) > 0$ do not depend on q. Let a be such that $4\varepsilon(a) < k_5$. Then, choosing $q > 0$ so that $4q^{-1}k_7(a) < k_5$, we obtain the inequality (9.3). □

Comparison of Necessary and Sufficient Conditions

From Examples 9.1, 9.2 it follows that, generally the necessary and sufficient conditions for strong ellipticity are not the same.

Example 9.1. Consider the equation (9.1) in the domain $Q = \{x \in \mathbb{R}^2 : |x| < 1\}$, where $R_{\alpha\beta}u(x) = a_{\alpha\beta 0}u(x) + a_{\alpha\beta 1}(u(x_1+1, x_2) + u(x_1-1, x_2))$, $a_{\alpha\beta 0}, a_{\alpha\beta 1} \in \mathbb{R}$. The decomposition $\mathcal{R}$ of the domain Q consists of three classes of subdomains:

1) Q_{11} and Q_{12}, 2) Q_{21} and 3) Q_{31} (see Example 7.8).
Let $s=1$. Then

$$R_{\alpha\beta 1}(x) = \begin{pmatrix} a_{\alpha\beta 0} & a_{\alpha\beta 1} \\ a_{\alpha\beta 1} & a_{\alpha\beta 0} \end{pmatrix} \quad \text{for } x \in \overline{Q}_{11},$$

$$A_{\alpha\beta 1}(x) = R_{\alpha\beta 1}(x) \quad \text{for } x \in \overline{Q}_{11},\ x \notin (-1,0),(0,0).$$

If $x=(-1,0)$, then

$$A_{\alpha\beta 1}(x) = \begin{pmatrix} a_{\alpha\beta 0} & a_{\alpha\beta 1} & 0 \\ a_{\alpha\beta 1} & a_{\alpha\beta 0} & a_{\alpha\beta 1} \\ 0 & a_{\alpha\beta 1} & a_{\alpha\beta 0} \end{pmatrix}.$$

Example 9.2. We consider the equation (9.1) in the domain $Q=(0,d)\times G$, where $d=k+\theta$, $k>0$ is a natural number, $0<\theta\le 1$, $G\subset\mathbb{R}^{n-1}$ is a bounded domain (with boundary $\partial G\in C^\infty$ if $n\ge 3$),

$$R_{\alpha\beta}u(x) = \sum_{i=-k}^{k} a_{\alpha\beta i}(x)u(x_1+i,x_2,\dots,x_n), \qquad a_{\alpha\beta i}\in C^\infty(\mathbb{R}^n).$$

If $\theta=1$, then the decomposition $\mathcal{R}$ of the domain Q consists of one class of subdomains $Q_{1l}=(l-1,l)\times G$ $(l=1,\dots,k+1)$. If $\theta<1$, then the decomposition $\mathcal{R}$ of the domain Q consists of two classes of subdomains $Q_{1l}=(l-1,l-1+\theta)\times G$ $(l=1,\dots,k+1)$ and $Q_{2l}=(l-1+\theta,l)\times G$ $(l=1,\dots,k)$ (see Example 8.2).

Denote $h_{sl}=(l-1,0,\dots,0)$ $(l=1,\dots,k+1$ if $s=1$, $l=1,\dots,k$ if $s=2)$. By virtue of (9.4), the elements of the matrix $R_{\alpha\beta s}(x)$ $(x\in\overline{Q}_{s1})$ are defined by the formula $r_{ij}^{\alpha\beta s}(x)=a_{\alpha,\beta,j-i}(x_1+i-1,x_2,\dots,x_n)$. Let $0<\theta<1$. Then, by virtue of (9.12), the matrices $A_{\alpha\beta 1}(x)$ of order $(k+1)\times(k+1)$ coincide with the matrices $R_{\alpha\beta 1}(x)$ for $0\le x_1\le\theta$, and the matrices $A_{\alpha\beta 2}(x)$ of order $k\times k$ coincide with the matrices $R_{\alpha\beta 2}(x)$ for $\theta<x_1<1$. Moreover, the matrices $A_{\alpha\beta 2}(x)$ of order $(k+1)\times(k+1)$ coincide with the matrices $R_{\alpha\beta 1}(x)$ if $x_1=\theta$, and with the matrices $R_{\alpha\beta 1}(x)$ (after reindexing of some rows and columns) if $x_1=1$. Thus, if d is a noninteger, then the necessary and sufficient conditions of strong ellipticity coincide. Clearly, if d is an integer, then these conditions do not coincide.

Thus, in Example 9.2, the set of numbers $d\ge 0$, for which the necessary and sufficient conditions for strong ellipticity are not the same, is nowhere dense in a metric space $\{d\in\mathbb{R}: d\ge 0\}$. We shall further show that in the general case a similar result is also valid in some sense.

Denote $Q^\sigma=\{x\in\mathbb{R}^n:\rho(x,Q)<\sigma\}$, where $\rho(x,Q)=\inf_{y\in Q}|x-y|$, $\sigma>0$ is such that $\sigma<\min_j\delta(y^j)/2$ and $\overline{Q^\sigma}\subset G$ (see the previous subsection). Let Ω be an arbitrary domain with boundary $\partial\Omega\in C^\infty$, and let $\overline{Q}\subset\Omega$, $\overline{\Omega}\subset Q^\sigma$ (see Fig. II.7). We consider the formula (9.4), in which we use a decomposition of the domain Ω into subdomains Ω_{sl}. This formula defines the matrix $R^\Omega_{\alpha\beta s}(x)$ $(x\in\overline{\Omega}_{s1})$.

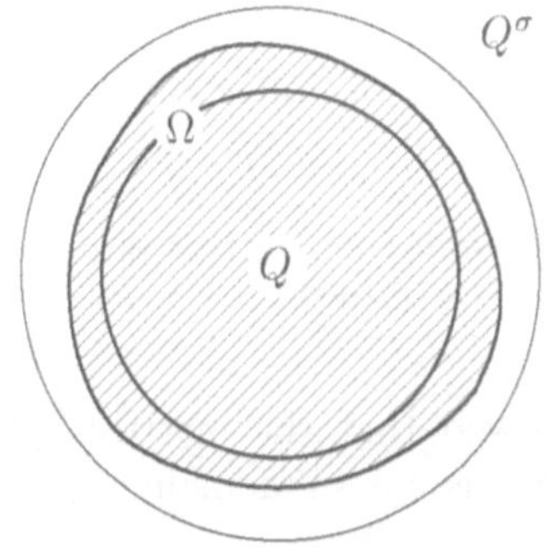

Fig. II.7

Theorem 9.3. *Let Ω be a domain with boundary $\partial\Omega \in C^\infty$ such that $\overline{Q} \subset \Omega$, $\overline{\Omega} \subset Q^\sigma$.*

Then the equation

$$\sum_{|\alpha|,|\beta|\le m} \mathcal{D}^\alpha R_{\alpha\beta\Omega}\mathcal{D}^\beta u = f_0(x) \qquad (x \in \Omega) \tag{9.20}$$

is strongly elliptic in $\overline{\Omega}$ if and only if the matrices

$$\sum_{|\alpha|=|\beta|=m} (R^{\Omega}_{\alpha\beta s}(x) + R^{\Omega *}_{\alpha\beta s}(x))\xi^{\alpha+\beta} \tag{9.21}$$

are positive definite for all $s = 1, 2, \ldots$, $x \in \overline{\Omega}_{s1}$, and $0 \ne \xi \in \mathbb{R}^n$.

Proof. 1. The necessity follows from Theorem 9.1. Now we prove the sufficiency. Let the matrices (9.21) be positive definite for all $s = 1, 2, \ldots$, $x \in \overline{\Omega}_{s1}$, and $0 \ne \xi \in \mathbb{R}^n$. By virtue of Lemma 9.1, repeating the proof of Theorem 9.2, we obtain the inequality (9.15) for Ω instead of Q.

Let $(S_\delta(y^j) + h) \cap \overline{\Omega} \ne \emptyset$ $(h \in M)$. Then, since $\sigma < \min_j \delta(y^j)/2$, $\overline{\Omega} \subset Q^\sigma$, we have

$$\rho(y^j + h, Q) \le \rho(y^j + h, y) + \rho(y, Q) \le \delta(y^j) + \sigma < 2\delta(y^j)$$

for $y \in (S_\delta(y^j) + h) \cap \overline{\Omega}$. From this, by virtue of condition $2\delta(y^j) < r(y^j)$ (see the previous subsection), we obtain $y^j + h \subset \overline{Q}$. Hence, there exists $z^{ji} = y^j + h$. Thus, the dimensions of vector-valued functions W^j and matrices $A^j_{\alpha\beta s}(z^j)$ used in the proof of inequalities (9.16)–(9.18) will not change. Assume that the matrices

$$\sum_{|\alpha|,|\beta|=m} (A_{\alpha\beta s}(z^j) + A^*_{\alpha\beta s}(z^j))\xi^{\alpha+\beta} \tag{9.22}$$

are positive definite for all $s = 1, 2, \ldots,$ $j = 1, \ldots, J$, and $0 \neq \xi \in \mathbb{R}^n$. Then, repeating the proof of Theorem 9.2, for all $u \in \dot{C}^\infty(\Omega)$ we obtain the inequality

$$\operatorname{Re}\left(\sum_{|\alpha|,|\beta|\le m} \mathcal{D}^\alpha R_{\alpha\beta\Omega}\mathcal{D}^\beta u, u\right)_{L_2(\Omega)} \ge c_1\|u\|^2_{W^m(\Omega)} - c_2\|u\|^2_{L_2(\Omega)}. \qquad (9.23)$$

2. It remains to be proved that the matrices (9.22) are positive definite. For a fixed $z^j \in \overline{Q}_{s1}$, there exists a subdomain Ω_{pk} such that $z^j \in \overline{\Omega}_{pk}$. Then $x^j = z^j - h_{pk} \in \overline{\Omega}_{p1}$, where $\Omega_{pk} = \Omega_{p1} + h_{pk}$. From part 1 of the proof it follows that if $x^j + h \in \overline{\Omega}$, then $x^j + h \subset \overline{Q}$. Therefore, there exists $1 \le i \le I(s, z^j)$ such that $z^{ji} = x^j + h$. Moreover, a number $N(p)$, corresponding to the domain Ω, is equal to a number $I(s, z^j)$, corresponding to the domain Q. From this and from the formulas (9.4) for the domain Ω and (9.12) for the domain Q it follows that the matrices $R^\Omega_{\alpha\beta p}(x^j)$ are equal to $A_{\alpha\beta s}(z^j)$ (possibly after reindexing of some rows and columns). □

Some Special Cases

We now consider some cases of differential-difference equations, for which the necessary and sufficient conditions for strong ellipticity coincide for every domain.

Example 9.3. We consider the differential-difference equation

$$AR_Qu = f_0(x) \qquad (x \in Q), \qquad (9.24)$$

where

$$R_Q = P_Q R I_Q,$$

$$Ru(x) = \sum_{h\in\mathcal{M}} a_h(u(x+h) + u(x-h)), \qquad a_h \in \mathbb{R},$$

$$A = -\sum_{i,j=1}^n \frac{\partial}{\partial x_i} a_{ij} \frac{\partial}{\partial x_j}, \qquad a_{ij} = a_{ji} \in C^\infty(\mathbb{R}^n)$$

are real-valued M-periodic functions in $\overline{Q}$. Let the necessary condition for strong ellipticity of the equation (9.24) in $\overline{Q}$ be fulfilled, i.e., the matrices $(\sum_{i,j=1}^n a_{ij}(x)\xi_i\xi_j)R_s$ are positive definite for all $s = 1, 2, \ldots,$ $x \in \overline{Q}_{s1}$, and $0 \neq \xi \in \mathbb{R}^n$. This condition holds if and only if the quadratic form $\sum_{i,j} a_{ij}(x)\xi_i\xi_j$ for all $x \in \overline{Q}_{s1}$ and the matrix R_s $(s = 1, 2, \ldots)$ are both positive definite or negative definite. Suppose, for example, that they are positive definite. Since the number of different matrices R_s is finite, integrating by parts, using M-periodicity of coefficients

$a_{ij}(x)$ in $\overline{Q}$ and Theorem B.11 concerning the equivalent norms, we have

$$\begin{aligned}(AR_Q u, u)_{L_2(Q)} &= \sum_{i,j}(a_{ij}(R_Q u)_{x_j}, u_{x_i})_{L_2(Q)}\\ &= \sum_s \sum_{i,j}(a_{ij}R_s(U_sP_su)_{x_j}, (U_sP_su)_{x_i})_{L_2^N(Q_{s1})}\\ &= \sum_s \sum_{i,j}\left(a_{ij}\left(\sqrt{R_s}U_sP_su\right)_{x_j}, \left(\sqrt{R_s}U_sP_su\right)_{x_i}\right)_{L_2^N(Q_{s1})}\\ &\geq k_1 \sum_s \sum_i\left(\left(\sqrt{R_s}U_sP_su\right)_{x_i}, \left(\sqrt{R_s}U_sP_su\right)_{x_i}\right)_{L_2^N(Q_{s1})}\\ &\geq k_2 \sum_i \|u_{x_i}\|^2_{L_2(Q)} \geq k_3\|u\|^2_{W^1(Q)} \qquad (u \in \dot{C}^\infty(Q)).\end{aligned}$$

Example 9.4. Consider the differential-difference equation

$$\sum_{i=1}^m \mathcal{D}_i^m R_{iQ}\mathcal{D}_i^m u(x) = f_0(x) \qquad (x \in Q), \tag{9.25}$$

where $R_{iQ} = P_Q R_i I_Q$, $R_i u(x) = \sum_{h\in\mathcal{M}} a_{ih}(u(x+h) + u(x-h))$, $a_{ih} \in \mathbb{R}$. The necessary condition for strong ellipticity of the equation (9.25) in $\overline{Q}$ has the form: the matrices $\sum_{i=1}^n R_{is}\xi_i^{2m}$ are positive definite for all $s = 1, 2, \ldots$, $0 \neq \xi \in \mathbb{R}^n$. Hence, the matrices R_{is} $(i = 1, \ldots, n,\ s = 1, 2, \ldots)$ are positive definite. Integrating by parts, using Lemma 8.8 and Theorem B.11, we obtain

$$\begin{aligned}\left(\sum_{i=1}^n \mathcal{D}_i^m R_{iQ}\mathcal{D}_i^m u, u\right)_{L_2(Q)} &= \sum_{i=1}^n (R_{iQ}\mathcal{D}_i^m u, \mathcal{D}_i^m u)_{L_2(Q)}\\ &\geq k_1 \sum_{i=1}^n \|\mathcal{D}_i^m u\|^2_{L_2(Q)} \geq k_2\|u\|^2_{W^m(Q)}\end{aligned}$$

for all $u \in \dot{C}^\infty(Q)$. Thus the equation (9.25) is strongly elliptic.

Now we can formulate the following unsolved problem.

Problem 9.1. *Is there a necessary and sufficient condition for strong ellipticity in an algebraic form for an arbitrary differential-difference equation of order* $2m$ (9.1) *and an arbitrary domain* Q?

Symbol of a Differential-Difference Operator

Now let us consider the sufficient conditions for strong ellipticity using a *symbol of a differential-difference operator*

$$\mathcal{A}_R(x,\xi) = \sum_{|\alpha|,|\beta|=m} \sum_{h\in\mathcal{M}} a_{\alpha\beta h}(x)\exp(i(h,\xi))\xi^{\alpha+\beta}.$$

Theorem 9.4. *Let the coefficients* $a_{\alpha\beta h}(x) \in C^\infty(\mathbb{R}^n)$ *be* M*-periodic in* $\mathbb{R}^n$. *Suppose there exist a finite set of vectors with integer coordinates* $\mathcal{M}_1 \subset \mathbb{R}^n$ *and numbers* $a_{ph} \in \mathbb{R}$ $(p = 1, \ldots, n,\ h \in \mathcal{M}_1)$ *such that*

$$\mathcal{A}_R(x,\xi) \geq 2\sum_{p=1}^{n}\sum_{h\in\mathcal{M}_1} a_{ph}\cos(h,\xi)\xi_p^{2m} \qquad (x \in \mathbb{R}^n,\ \xi \in \mathbb{R}^n), \tag{9.26}$$

$$0 \leq \sum_{h\in\mathcal{M}_1} a_{ph}\cos(h,\xi) \not\equiv 0 \qquad (\xi \in \mathbb{R}^n,\ p = 1, \ldots, n). \tag{9.27}$$

Then the equation (9.1) *is strongly elliptic in* $\overline{Q}$.

Proof. Let $u \in \dot{C}^\infty(Q)$. Consider an arbitrary point $y^j \in \overline{Q}_{sl}$. Denote $z^j = y^j - h_{sl} \in \overline{Q}_{s1}$ (see the subsection on Sufficient Conditions). We introduce the operators $R^j_{\alpha\beta}$ by the formulas

$$R^j_{\alpha\beta}u(x) = \sum_{h\in\mathcal{M}} a_{\alpha\beta h}(z^j)u(x+h).$$

From the Leibniz formula and Lemmas 9.1, 8.10, and 8.2 we obtain

$$\operatorname{Re}\Bigg(\sum_{|\alpha|,|\beta|\leq m} \mathcal{D}^\alpha R_{\alpha\beta Q}\mathcal{D}^\beta u, u\Bigg)_{L_2(Q)} \\ \geq b_1 + b_2 - k_1(a)\|u\|_{W^m(Q)}\|u\|_{W^{m-1}(Q)}, \tag{9.28}$$

where

$$b_1 = \operatorname{Re}\sum_{|\alpha|,|\beta|=m}\sum_j((R_{\alpha\beta Q} - R^j_{\alpha\beta Q})\mathcal{D}^\beta(\varphi_j u), \mathcal{D}^\alpha(\varphi_j u))_{L_2(Q)},$$

$$b_2 = \operatorname{Re}\sum_{|\alpha|,|\beta|=m}\sum_j(\mathcal{D}^{\alpha+\beta}R^j_{\alpha\beta Q}(\varphi_j u), \varphi_j u)_{L_2(Q)}.$$

Using the uniform continuity of the functions $a_{\alpha\beta h}(x)$ on the compact $\overline{Q}$ and the Leibniz formula, we have

$$|b_1| \leq \varepsilon(a)\|u\|^2_{W^m(Q)} + k_2(a)\|u\|_{W^m(Q)}\|u\|_{W^{m-1}(Q)}, \tag{9.29}$$

where $\varepsilon(a) \to 0$ as $a \to 0$.

Denote

$$R_p u(x) = \sum_{h\in\mathcal{M}_1} a_{ph}(u(x+h) + u(x-h)).$$

Then, by virtue of Lemma 8.1, the Plancherel theorem, the conditions (9.26), (9.27), Lemmas 8.4, 8.3, 8.9 and the Leibniz formula, we obtain

$$b_2 = \operatorname{Re}\sum_j\sum_{|\alpha|,|\beta|=m}(\mathcal{D}^{\alpha+\beta}R^j_{\alpha\beta}I_Q(\varphi_j u), I_Q(\varphi_j u))_{L_2(\mathbb{R}^n)}$$

$$
\begin{aligned}
&= \sum_j \int_{\mathbb{R}^n} \operatorname{Re}\Big\{ \sum_{|\alpha|,|\beta|=m} \sum_{h\in\mathcal{M}} a_{\alpha\beta h}(z^j) \exp(i(h,\xi))\xi^{\alpha+\beta} \Big\} |I_Q\widehat{(\varphi_j u)}|^2 d\xi \\
&\geq \sum_j \int_{\mathbb{R}^n} 2\sum_{p=1}^{n} \sum_{h\in\mathcal{M}_1} a_{ph} \cos(h,\xi)\xi_p^{2m} |I_Q\widehat{(\varphi_j u)}|^2 \, d\xi \\
&= \sum_j \sum_{p=1}^{n} (R_{pQ}\mathcal{D}_p^m(\varphi_j u), \mathcal{D}_p^m(\varphi_j u))_{L_2(Q)} \geq k_3 \sum_j \sum_{p=1}^{n} \|\mathcal{D}_p^m(\varphi_j u)\|^2_{L_2(Q)} \\
&\geq k_3 \sum_{p=1}^{n} \|\mathcal{D}_p^m u\|^2_{L_2(Q)} - k_4(a)\|u\|_{W^m(Q)}\|u\|_{W^{m-1}(Q)}. \qquad (9.30)
\end{aligned}
$$

From inequalities (9.28)–(9.30), (B.20), (9.6), and Theorem B.11 the inequality (9.3) follows. □

Example 9.5. We consider the differential-difference equation

$$
-\sum_{i,j=1}^{2} \frac{\partial}{\partial x_i} R_{ijQ} \frac{\partial u(x)}{\partial x_j} = f_0(x) \qquad (x \in Q), \qquad (9.31)
$$

where

$$
\begin{aligned}
R_{11}u(x) &= 2u(x) + \frac{1}{2}(u(x_1+1,x_2) + u(x_1-1,x_2)), \\
R_{12}u(x) = R_{21}u(x) &= \frac{1}{2}(u(x_1,x_2+1) + u(x_1,x_2-1)), \\
R_{22}u(x) &= \frac{3}{2}u(x) + \frac{1}{2}(u(x_1+1,x_2) + u(x_1-1,x_2)) \\
&\quad + \frac{1}{4}(u(x_1,x_2+2) + u(x_1,x_2-2)).
\end{aligned}
$$

Let us verify fulfillment of the conditions (9.26), (9.27) for the equation (9.31). In fact,

$$
\begin{aligned}
&\xi_1^2(2+\cos\xi_1) + 2\xi_1\xi_2\cos\xi_2 + \xi_2^2\left(\frac{3}{2} + \cos\xi_1 + \frac{1}{2}\cos 2\xi_2\right) \\
&\quad = (\xi_1^2+\xi_2^2)(1+\cos\xi_1) + \xi_1^2 + 2\xi_1\xi_2\cos\xi_2 + \xi_2^2\cos^2\xi_2^2 \\
&\quad \geq (\xi_1^2+\xi_2^2)(1+\cos\xi_1).
\end{aligned}
$$

Thus the equation (9.31) is strongly elliptic.

Remark 9.2. Sufficient conditions for strong ellipticity of the type (9.26), (9.27) are fulfilled in some problems of elasticity theory (see Section 15).

We show that if the conditions of Theorem 9.4 hold, then the conditions of Theorem 9.2 are also fulfilled.

Theorem 9.5. *Let the conditions of Theorem 9.4 be fulfilled.*

Then the matrices

$$\sum_{|\alpha|,|\beta|=m} (A_{\alpha\beta s}(x) + A^*_{\alpha\beta s}(x))\xi^{\alpha+\beta}$$

are positive definite for all $s = 1, 2, \dots,$ $x \in \overline{Q}_{s1}$, *and* $0 \neq \xi \in \mathbb{R}^n$.

Proof. The conclusion of Theorem 9.4 is valid for an arbitrary bounded domain. Therefore the differential-difference equation (9.20) is strongly elliptic in $\overline{\Omega}$, where Ω is a domain with boundary $\partial\Omega \in C^\infty$ such that $\overline{Q} \subset \Omega$, $\overline{\Omega} \subset Q^\sigma$, $0 < \sigma < \min_j \delta(y^j)/2$ and $\overline{Q^\sigma} \subset G$. Hence, by virtue of Theorem 9.3, the matrices

$$\sum_{|\alpha|=|\beta|=m} (R^\Omega_{\alpha\beta s}(x) + R^{\Omega *}_{\alpha\beta s}(x))\xi^{\alpha+\beta}$$

are positive definite for all $s = 1, 2, \dots,$ $x \in \overline{\Omega}_{s1}$ and $0 \neq \xi \in \mathbb{R}^n$. For any s and $z^0 \in \overline{Q}_{s1}$, there is Ω_{pk} such that $z^0 \in \overline{\Omega_{pk}}$. Then $x^0 = z^0 - h_{pk} \in \overline{\Omega_{p1}}$, where $\Omega_{pk} = \Omega_{p1} + h_{pk}$. Without loss of generality, we can assume that $\sigma < \inf \rho(z^0 + h, Q)$ $(h : z^0 + h \notin \overline{Q})$. Then, as in the proof of Theorem 9.3, we can reindex rows and columns of the matrix $A_{\alpha\beta s}(z^0)$ so that $A_{\alpha\beta s}(z^0) = R^\Omega_{\alpha\beta s}(x^0)$. Therefore the matrices

$$\sum_{|\alpha|,|\beta|=m} (A_{\alpha\beta s}(x) + A^*_{\alpha\beta s}(x))\xi^{\alpha+\beta}$$

are positive definite. □

It follows from Theorem 9.4 that a differential-difference equation can be strongly elliptic even if its symbol vanishes for $\xi \neq 0$. Furthermore, there exist strongly elliptic differential-difference equations such that their symbol can change sign.

Example 9.6. Let us consider the equation

$$-\Delta R_Q u(x) = f_0(x) \qquad (x \in Q = (0,2) \times (0,1)), \tag{9.32}$$

where $Ru(x) = u(x) + a[u(x_1+1, x_2) + u(x_1-1, x_2)]$. A symbol of the differential-difference operator $-\Delta R_Q$ has the form $(\xi_1^2 + \xi_2^2)(1 + 2a\cos\xi_1)$. From Examples 8.3, 9.3, it follows that the equation (9.32) is strongly elliptic if and only if $|a| < 1$. We remark that in the case $1/2 < |a| < 1$ the symbol changes its sign.

10 Solvability and Spectrum

Definitions

We define the unbounded operator $\mathcal{A}_R : L_2(Q) \to L_2(Q)$ with domain $\mathcal{D}(\mathcal{A}_R) = \{u \in \mathring{W}^m(Q) : \mathcal{A}_R u \in L_2(Q)\}$ acting in the space of distributions $\mathcal{D}'(Q)$ by the formula

$$\mathcal{A}_R u = \sum_{|\alpha|,|\beta|\le m} \mathcal{D}^\alpha R_{\alpha\beta Q} \mathcal{D}^\beta u.$$

The operator $\mathcal{A}_R$ is called a *differential-difference operator.*

In this section we assume that the differential-difference equations are strongly elliptic. We say that the operator $\mathcal{A}_R$ is *strongly elliptic* if it corresponds to the strongly elliptic differential-difference equation (9.1).

Definition 10.1. A function u is called a *generalized solution* of the boundary value problem (9.1), (9.2) if $u \in \mathcal{D}(\mathcal{A}_R)$ and

$$\mathcal{A}_R u = f_0. \tag{10.1}$$

We can formulate an equivalent definition of a generalized solution.

Definition 10.2. Let $f_0 \in L_2(Q)$. A function u is called a *generalized solution* of the boundary value problem (9.1), (9.2) if $u \in \mathring{W}^m(Q)$ and

$$\sum_{|\alpha|,|\beta|\le m} (R_{\alpha\beta Q}\mathcal{D}^\beta u, \mathcal{D}^\alpha v)_{L_2(Q)} = (f_0, v)_{L_2(Q)} \tag{10.2}$$

for all $v \in \mathring{W}^m(Q)$.

We introduce the unbounded operator $\mathcal{A}_R^+ : L_2(Q) \to L_2(Q)$ with domain $\mathcal{D}(\mathcal{A}_R^+) = \{u \in \mathring{W}^m(Q) : \mathcal{A}_R^+ u \in L_2(Q)\}$ acting in the space of distributions $\mathcal{D}'(Q)$ by the formula

$$\mathcal{A}_R^+ u = \sum_{|\alpha|,|\beta|\le m} \mathcal{D}^\alpha R^*_{\beta\alpha Q} \mathcal{D}^\beta u.$$

Remark 10.1. The operators $\mathcal{A}_R$ and $\mathcal{A}_R^+$ are formally adjoint, i.e.,

$$(\mathcal{A}_R u, v)_{L_2(Q)} = (u, \mathcal{A}_R^+ v)_{L_2(Q)}$$

for all $u, v \in \dot{C}^\infty(Q)$.

Remark 10.2. According to the identity

$$\operatorname{Re}(\mathcal{A}_R u, u)_{L_2(Q)} = \operatorname{Re}(\mathcal{A}_R^+ u, u)_{L_2(Q)} \qquad (u \in \dot{C}^\infty(Q))$$

the operator $\mathcal{A}_R^+$ is strongly elliptic.

Spectrum

We introduce the sesquilinear form $b_R[u, v]$ with domain $\mathcal{D}(b_R) = \mathring{W}^m(Q)$ by the formula

$$b_R[u, v] = \sum_{|\alpha|,|\beta|\le m} (R_{\alpha\beta Q}\mathcal{D}^\beta u, \mathcal{D}^\alpha v)_{L_2(Q)}. \tag{10.3}$$

Theorem 10.1. *Let the operator $\mathcal{A}_R$ be strongly elliptic.*

Then we have:

(a) $\mathcal{A}_R\colon L_2(Q) \to L_2(Q)$ *is the m-sectorial operator associated with the form* b_R.

(b) *The operator* $\mathcal{A}_R\colon L_2(Q) \to L_2(Q)$ *is Fredholm, and* $\operatorname{ind} \mathcal{A}_R = 0$.

(c) *The spectrum* $\sigma(\mathcal{A}_R)$ *is discrete, and* $\sigma(\mathcal{A}_R) \subset \{\lambda \in \mathbb{C} : \operatorname{Re} \lambda > -c_2\}$, *where* $c_2 \geq 0$ *is a constant in* (9.3).

(d) *If* $\lambda \notin \sigma(\mathcal{A}_R)$, *then the resolvent* $R(\lambda, \mathcal{A}_R)\colon L_2(Q) \to L_2(Q)$ *is a compact operator.*

(e) $\mathcal{A}_R^* = \mathcal{A}_R^+$, $(\mathcal{A}_R^+)^* = \mathcal{A}_R$.

Proof. Since the operator $\mathcal{A}_R$ is strongly elliptic, Lemma 8.2 implies that b_R is a bounded sectorial form on $\mathring{W}^m(Q)$ with vertex $\gamma = -c_2$. We denote by B_{b_R} the m-sectorial operator associated with b_R. Since Definitions 10.1, 10.2 are equivalent, by virtue of Theorem A.10, $\mathcal{A}_R = B_{b_R}$, $\mathcal{A}_R^+ = B_{b_R^*}$. Now the statements of Theorem 10.1 follow from Theorems A.14, A.15 and the compactness of the imbedding operator $\mathring{W}^m(Q)$ into $L_2(Q)$. □

Remark 10.3. Let $A_R\colon L_2(Q) \to L_2(Q)$ be the unbounded operator given by

$$\left.\begin{aligned} &\mathcal{D}(A_R) = \dot{C}^\infty(Q),\\ &A_R u = \sum_{|\alpha|,|\beta|\leq m} \mathcal{D}^\alpha R_{\alpha\beta Q} \mathcal{D}^\beta u \quad (u \in \mathcal{D}(A_R)). \end{aligned}\right\} \tag{10.4}$$

Then $\mathcal{A}_R$ is the Friedrichs extension of A_R.

Example 10.1. Consider the equation

$$-\Delta R_Q u(x) = f_0(x) \qquad (x \in Q = (0,2) \times (0,1)) \tag{10.5}$$

with the boundary conditions

$$u|_{\partial Q} = 0, \tag{10.6}$$

where $Ru(x) = 2u(x) + u(x_1+1, x_2) + u(x_1-1, x_2)$. From Examples 8.3, 9.3, we obtain

$$(-\Delta R_Q u, u)_{L_2(Q)} \geq c_1 \|u\|^2_{W^1(Q)} \qquad (u \in \dot{C}^\infty(Q)).$$

Hence, by Theorem 10.1, the boundary value problem (10.5), (10.6) has a unique generalized solution $u \in \mathring{W}^1(Q)$ for every $f_0 \in L_2(Q)$.

Theorem 10.2. *Let the strongly elliptic operator $\mathcal{A}_R$ be symmetric, i.e.,*

$$(\mathcal{A}_R u, v)_{L_2(Q)} = (u, \mathcal{A}_R v)_{L_2(Q)} \tag{10.7}$$

for all $u, v \in \dot{C}^\infty(Q)$.

Then the operator $\mathcal{A}_R\colon L_2(Q) \to L_2(Q)$ *is self-adjoint, the spectrum* $\sigma(\mathcal{A}_R)$ *consists of real isolated eigenvalues* $\lambda_s > -c_2$ *of finite multiplicity. The eigenfunctions* v_s *of the operator* $\mathcal{A}_R$ *form an orthonormal basis in* $L_2(Q)$. *Moreover, the functions* $v_s/\sqrt{\lambda_s+c_2}$ *form an orthonormal basis in* $\mathring{W}^m(Q)$ *with inner product given by*

$$(u,v)'_{\mathring{W}^m(Q)} = b_R[u,v] + c_2(u,v)_{L_2(Q)}.$$

A proof follows from Theorems 10.1 and A.16.

11 Smoothness of Generalized Solutions in Subdomains

Interior Smoothness of Solutions in Subdomains

Theorem 11.1. *Let the equation* (9.1) *be strongly elliptic in* $\overline{Q}$. *Suppose that* u *is a generalized solution of the boundary value problem* (9.1), (9.2), *while* $f_0 \in L_2(Q) \cap W^k_{\mathrm{loc}}(Q_{sl})$ $(s=1,2,\dots,\ l=1,\dots,N(s))$.

Then $u \in W^{k+2m}_{\mathrm{loc}}(Q_{sl})$ *for all* s,l.

Proof. Let s be a fixed number. In the integral identity (10.2) we assume that $v \in \dot{C}^\infty(\bigcup_l Q_{sl})$, $v(x)=0$ for $x \notin \bigcup_l Q_{sl}$. By virtue of (8.4), (8.6), it follows from (10.2) that

$$\sum_{|\alpha|,|\beta|\le m} \int_{Q_{s1}} (R_{\alpha\beta s}\mathcal{D}^\beta(U_sP_su), \mathcal{D}^\alpha(U_sv))\,dx = \int_{Q_{s1}} (U_sP_sf_0, U_sv)\,dx,$$

where $(\cdot,\cdot)$ is the inner product in $\mathbb{C}^N$, $N=N(s)$. Hence, the vector-valued function $U_sP_su \in W^{m,N}(Q_{s1})$ is the generalized solution of N partial differential equations

$$\sum_{|\alpha|,|\beta|\le m} \mathcal{D}^\alpha R_{\alpha\beta s}(x)\mathcal{D}^\beta(U_sP_su)(x) = (U_sP_sf_0)(x) \qquad (x\in Q_{s1}). \tag{11.1}$$

By Theorem 9.1, the matrices $\sum_{|\alpha|,|\beta|=m}(R_{\alpha\beta s}(x)+R^*_{\alpha\beta s}(x))\xi^{\alpha+\beta}$ are positive definite for $x \in \overline{Q}_{s1}$ and $0 \ne \xi \in \mathbb{R}^n$, i.e., the system of equations (11.1) is strongly elliptic. According to the theorem on the interior smoothness of generalized solutions of strongly elliptic systems, $(U_sP_su)_l \in W^{k+2m}_{\mathrm{loc}}(Q_{sl})$ (see O. V. Guseva [1], C. B. Morrey [1]). Thus, $u \in W^{k+2m}_{\mathrm{loc}}(Q_{sl})$. □

Smoothness of Solutions near Boundaries of Subdomains for a Second Order Equation

We now study the smoothness of generalized solutions near the set $\partial Q_{sl} \setminus \mathcal{K}$. The proof is based on the approximation of differential operators by difference operators (see Theorem B.16). However, elliptic differential-difference equations are nonlocal, in contrast to elliptic differential equations. Therefore, considering the smoothness of the solution in the neighborhood of the point $y \in \partial Q_{sl} \setminus \mathcal{K}$, we must consider simultaneously the respective neighborhoods of all points $y+h \in \overline{Q}$, where $h \in M$

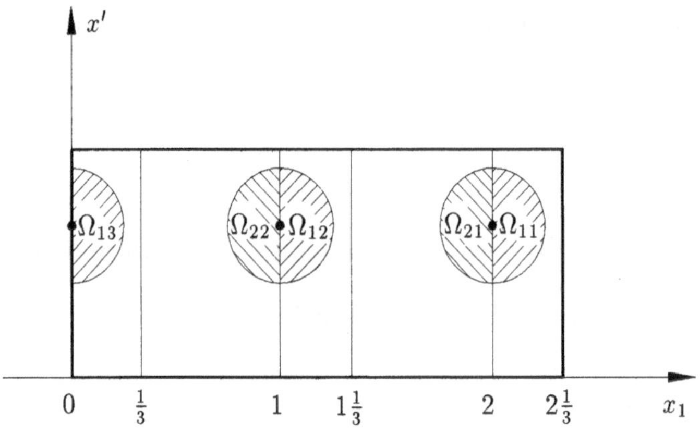

Fig. II.8

(see Fig. II.8). Changing the variables, we can straighten the boundaries in the neighborhoods of the points $y+h$. The domain Q_{sl} transforms locally into the half-space $\{x_n > 0\}$. We construct a function $v(x)$ from (10.2) using a smooth function $\xi(x)$ with support in the corresponding neighborhoods and the difference operator δ^r_{-t} $(1 \le r \le n-1)$. Then we obtain a priori estimates of the difference relations. From this, passing to a limit as $t \to 0$, we have the appropriate estimates for the generalized derivatives.

The proof of the smoothness of generalized solutions near the boundaries of subdomains Q_{sl} in the general case is very complicated. Therefore, in order to demonstrate the main ideas of the proof, we first study the simplest case.

Example 11.1. We consider the boundary value problem

$$-\sum_{i,j=1}^{n} \frac{\partial}{\partial x_i} R_{ijQ} \frac{\partial u(x)}{\partial x_j} = f_0(x) \qquad (x \in Q), \tag{11.2}$$

$$u|_{\partial Q \setminus K} = 0 \qquad (x \in \partial Q \setminus K), \tag{11.3}$$

where $R_{ijQ} = P_Q R_{ij} I_Q$, $R_{ij} u(x) = \sum_{h \in \mathcal{M}} a_{ijh} u(x+h)$, $a_{ijh} \in \mathbb{R}$, $f_0 \in L_2(Q)$.

Theorem 11.2. *Let $Q \subset \mathbb{R}^n$ be a bounded domain with boundary $\partial Q \in C^\infty$ satisfying the condition 7.1, or a cylinder $(0,d) \times G$, where $G \subset \mathbb{R}^{n-1}$ is a bounded domain (with boundary $\partial G \in C^\infty$ if $n \ge 3$). Let the equation (11.2) be strongly elliptic in $\overline{Q}$. Suppose that u is a generalized solution of the boundary value problem (11.2), (11.3) and $f_0 \in L_2(Q)$.*

Then, $u \in W^2(Q_{sl} \setminus \overline{\mathcal{K}^\varepsilon})$ for every $\varepsilon > 0$ $(s = 1, 2, \dots,\ l = 1, \dots, N(s))$, where $\mathcal{K}^\varepsilon = \{x \in \mathbb{R}^n : \rho(x, \mathcal{K}) < \varepsilon\}$.

Proof. 1. By virtue of Theorem 11.1, it suffices to show that for an arbitrary given point $y \in \partial Q_{pi} \setminus \mathcal{K}$ there exists a ball $S_\delta(y)$ such that $u \in W^2(Q_{pi} \cap S_\delta(y))$.

Let $h_{pl} \in M$ denote a vector satisfying the condition $Q_{pl} = Q_{p1} + h_{pl}$ $(l = 1, \dots, N = N(p))$, $h_{p1} = 0$. We introduce the points $y^1, \dots, y^N$ so that $y^i = y$, $y^l = y^i - h_{pi} + h_{pl}$ $(l = 1, \dots, N = N(p))$. By definition of the sets Γ_{rl} (see Section 7), there exists a unique r such that $J(r) \geq N(p)$ and, after some renumbering of the sets Q_{pl}, Γ_{rl}, we have $y^l \in \Gamma_{rl} \subset \partial Q_{pl} \setminus \mathcal{K}$ $(l = 1, \dots, N(p))$, $\Gamma_{rl} \subset Q$ $(l = 1, \dots, J_0 = J_0(r))$, $\Gamma_{rl} \subset \partial Q$ $(l = 1 + J_0, \dots, J(r))$.

By virtue of Lemma 7.8, there exists a unique subdomain $Q_{qj} \neq Q_{p1}$ such that $\Gamma_{r1} \subset \partial Q_{qj}$. Let us reindex the subdomains of the qth class so that $\Gamma_{rl} \subset \partial Q_{ql}$ $(l = 1, \dots, J_0)$. We introduce the points $z^1, \dots, z^N \in \overline{Q}$ by the formula $z^l = y^1 + h_{ql}$ $(l = 1, \dots, N,\ N = N(q) \geq J_0)$, where $h_{ql} \in M$, $Q_{ql} = Q_{q1} + h_{ql}$ $(l = 1, \dots, N(q))$, $h_{q1} = 0$. According to the construction, $z^l = y^l \in Q$ $(l = 1, \dots, J_0)$, $z^l \in \partial Q$ $(l = J_0 + 1, \dots, N(q))$, $z^l \in \partial Q_{ql} \setminus \mathcal{K}$ $(l = 1, \dots, N(q))$. Clearly $(\bigcup_{l > J_0} y^l) \cap (\bigcup_{l > J_0} z^l) = \emptyset$.

We consider the balls $S_{4\delta}(x^{sl})$ $(l = 1, \dots, N(s),\ s = p, q)$, where $x^{pl} = y^l$, $x^{ql} = z^l$. By virtue of the condition $K \subset \mathcal{K}$ and Lemmas 7.3, 7.4, we can choose $\delta > 0$ so small that $4\delta < \min_{l,s} \min\{\rho(x^{sl}, \mathcal{K}), 1/2\}$, the sets $\partial Q_{sl} \cap S_{4\delta}(x^{sl})$ are connected and belong to the class C^∞ $(l = 1, \dots, N(s),\ s = p, q)$, $S_{4\delta}(x^{sl}) \subset \Gamma_{rl} \cup Q_{pl} \cup Q_{ql}$ $(l = 1, \dots, J_0)$, $S_{4\delta}(x^{sl}) \cap Q = S_{4\delta}(x^{sl}) \cap Q_{sl}$ $(l = J_0 + 1, \dots, N(s))$, $s = p, q$.

2. By definition, a function u satisfies the integral identity

$$\sum_{i,j} \int_Q R_{ijQ} u_{x_j} \overline{v}_{x_i}\, dx = \int_Q f_0 \overline{v}\, dx \tag{11.4}$$

for all $v \in \mathring{W}^1(Q)$.

Suppose that $v = \xi v_0$, where $v_0 \in \mathring{W}^1(Q)$,

$$\xi(x) = \sum_{1 \leq l \leq N(p)} \eta(x - h_{pl}) + \sum_{J_0 < l \leq N(q)} \eta(x - h_{ql}),$$

$$\eta \in \dot{C}^\infty(\mathbb{R}^n), \quad 0 \leq \eta(x) \leq 1 \quad (x \in S_{2\delta}(y^1)),$$

$$\eta(x) = 1 \quad (x \in S_\delta(y^1)), \quad \eta(x) = 0 \quad (x \notin S_{2\delta}(y^1)).$$

Then, using the Leibniz formula, we obtain

$$\begin{aligned} &\sum_{i,j} \sum_{s,l} \int_{\Omega_{sl}} (R_{ijQ} u_{x_j}) \xi \overline{v}_{0x_i}\, dx \\ &\quad = \sum_{s,l} \int_{\Omega_{sl}} \Big\{ f_0 \xi \overline{v}_0 - \sum_{i,j} (R_{ijQ} u_{x_j}) \xi_{x_i} \overline{v}_0 \Big\} dx, \end{aligned} \tag{11.5}$$

where $\Omega_{sl} = Q_{sl} \cap S_{4\delta}(x^{sl})$ (see Fig. II.8). In formula (11.5) and further we sum over $i, j = 1, \dots, n$, $l = 1, \dots, N(s)$, $s = p, q$.

We introduce the operator $U_s \colon L_2(\bigcup_l Q_{sl}) \to L_2^N(Q_{s1})$ by formula (8.4), where $N = N(s)$, $s = p, q$. Since the operators of multiplication by the functions ξ and ξ_{x_i} commute with the operator U_s, from (11.5) it follows that

$$\sum_{i,j}\sum_{s}\int(\eta R_{ijs}W_{sx_j},V^0_{sx_i})\,dx = \sum_{s}\int\Big\{(\eta F_s,V^0_s)-\sum_{i,j}(\eta_{x_i}R_{ijs}W_{sx_j},V^0_s)\Big\}dx, \tag{11.6}$$

where $W_s=(U_sP_su)(x)$, $F_s=(U_sP_sf_0)(x)$, $V^0_s=(U_sP_sv_0)(x)$ for $x\in\Omega_s$. In formula (11.6) and further we integrate over the set $\Omega_s=\Omega_{s1}$. Using the Leibniz formula again, we have

$$\sum_{s}\sum_{i,j}\int(R_{ijs}(\eta W_s)_{x_j},V^0_{sx_i})\,dx = \sum_{s}\int\Big\{(\eta F_s,V^0_s)-\sum_{i,j}(\eta_{x_i}R_{ijs}W_{sx_j},V^0_s) + \sum_{i,j}(\eta_{x_j}R_{ijs}W_s,V^0_{sx_i})\Big\}dx. \tag{11.7}$$

Without loss of generality, we assume that $y^1=0$ and the equation for the surface $\Gamma_{p1}\cap S_{4\delta}(0)$ has the form $x_n=0$. Otherwise, we can obtain the plane boundary, introducing the new variables (see Lemma 20 of N. Dunford and J. Schwartz [2, Chapter XIV, Section 6]). Let $\mathring{W}^1_\delta$ denote the space of vector-valued functions $V=(V_p,V_q)\in W^{1,N(p)}(\Omega_p)\times W^{1,N(q)}(\Omega_q)$ such that $\operatorname{supp}V_s\subset\overline{\Omega_s\cap S_{2\delta}(0)}$, $SV\in\mathring{W}^1(Q)$, where $(SV)(x)=V_{sl}(x-h_{sl})$ for $x\in\Omega_{sl}$, $(SV)(x)=0$ for $x\notin\bigcup_{s,l}\Omega_{sl}$ ($l=1,\dots,N(s)$, $s=p,q$). Suppose that $v_0=SV^0$, $V^0=\delta^r_{-t}V^1=(\delta^r_{-t}V^1_p,\delta^r_{-t}V^1_q)$, where $V^1\in\mathring{W}^1_{3\delta/2}$, $1\le r\le n-1$, $0<t<\delta$. The operator $\delta^r_{\pm t}$ is defined by the formulas:

$$\mathcal{D}(\delta^r_{\pm t})=\{W\in L_2^{N(s)}(\Omega_s):\operatorname{supp}W\subset\overline{\Omega_s\cap S_{3\delta}(0)}\},$$
$$\delta^r_{\pm t}W=(W^r_{\pm t}-W)/(\pm t),$$

where $W^r_{\pm t}=W(x_1,\dots,x_r\pm t,\dots,x_n)$. By the construction, $v_0\in\mathring{W}^1(Q)$ and $V^0_s=\delta^r_{-t}V^1_s$. Since the operators $-\delta^r_{-t}$ and δ^r_t are formally adjoint, from (11.7) we have

$$\sum_{s}\sum_{i,j}\int(R_{ijs}\delta^r_t(\eta W_s)_{x_j},V^1_{sx_i})=I, \tag{11.8}$$

where

$$I=\sum_{s}\int\Big\{-(\eta F_s,\delta^r_{-t}V^1_s)+\sum_{i,j}(\eta_{x_i}R_{ijs}W_{sx_j},\delta^r_{-t}V^1_s) + \sum_{i,j}(R_{ijs}\delta^r_t(\eta_{x_j}W_s),V^1_{sx_i})\Big\}dx.$$

3. We set $V_s^1 = \delta_{t_1}^r(\eta W_s)$ $(0 < t_1 < \delta)$. Clearly $V^1 = (V_p^1, V_q^1) \in \mathring{W}^1_{3\delta/2}$. By virtue of the Schwarz inequality and Theorem B.16, we obtain

$$|I| \leq k_1\Big(\sum_s \|V_s^1\|_{1,N}^2\Big)^{1/2} \sum_s (\|W_s\|_{1,N} + \|F_s\|_{0,N}), \tag{11.9}$$

where $\|V_s^1\|_{k,N} = \big\{\sum_{l=1}^{N(s)} \|V_{sl}^1\|^2_{W^k(\Omega_s)}\big\}^{1/2}$, $k = 0, 1$.

We now estimate the left part of (11.8). We set $t_1 = t$, i.e., $V_s^1 = \delta_t^r(\eta W_s)$. By virtue of definition of strong ellipticity, formulas (8.4), (8.6) and Theorem B.16, we have

$$\begin{aligned}
&\operatorname{Re}\sum_s \sum_{i,j} \int (R_{ijs} V^1_{sx_j}, V^1_{sx_i})\, dx \\
&= \operatorname{Re}\sum_{i,j} (R_{ijQ} v_{1x_j}, v_{1x_i})_{L_2(Q)} \geq c_1\|v_1\|^2_{W^1(Q)} - c_2\|v_1\|^2_{L_2(Q)} \\
&= \sum_s (c_1\|V_s^1\|^2_{1,N} - c_2\|V_s^1\|^2_{0,N}) \\
&\geq c_1 \sum_s \|V_s^1\|^2_{1,N} - k_2\Big(\sum_s \|V_s^1\|^2_{1,N}\Big)^{1/2} \sum_s \|W_s\|_{1,N},
\end{aligned} \tag{11.10}$$

where $v_1 = SV^1$.

Thus from (11.8)–(11.10) it follows that

$$\Big\{\sum_s \|V_s^1\|^2_{1,N}\Big\}^{1/2} \leq k_3 \sum_s (\|W_s\|_{1,N} + \|F_s\|_{0,N}).$$

From this inequality and from Theorem B.16 we obtain $(\eta W_s)_{x_i x_r} \in L_2^N(\Omega_s)$ for all $i = 1, \ldots, n$, $r = 1, \ldots, n-1$. Hence, $W_{px_ix_r} \in L_2^N(Q_{p1} \cap S_\delta(0))$.

4. Now we shall show that $W_{px_nx_n} \in L_2^N(Q_{p1} \cap S_\delta(0))$. Suppose that in (11.7) $V_p^0 = U_p P_p v_0 \in \dot{C}^{\infty,N}(Q_{p1} \cap S_\delta(0))$ is an arbitrary vector-valued function, $V_q^0 = 0$. Then we obtain that the vector-valued function W_p is a generalized solution of the differential equations

$$-\sum_{i,j} R_{ijp} W_{px_ix_j} = F_p, \tag{11.11}$$

where $F_p \in L_2^N(Q_{p1} \cap S_\delta(0))$. By virtue of Theorem 9.1, the matrix $R_{nnp} + R^*_{nnp}$ is positive definite. Hence, there exists the inverse matrix R^{-1}_{nnp}. Therefore $W_{px_nx_n} = R^{-1}_{nnp}(F_p - \sum_{i+j<2n} R_{ijp} W_{px_ix_j})$. We have proved above that $W_{px_ix_j} \in L_2^N(Q_{p1} \cap S_\delta(0))$ $(i + j < 2n)$. Thus $W_{px_nx_n} \in L_2^N(Q_{p1} \cap S_\delta(0))$.

Hence, $u \in W^2(Q_{pi} \cap S_\delta(y))$. □

The General Case

Theorem 11.3. *Let* $Q \subset \mathbb{R}^n$ *be a bounded domain with boundary* $\partial Q \in C^\infty$ *satisfying the condition 7.1, or a cylinder* $(0,d) \times G$, *where* $G \subset \mathbb{R}^{n-1}$ *is a bounded domain (with boundary* $\partial G \in C^\infty$ *if* $n \geq 3$). *Let the equation (9.1) be strongly elliptic in* $\overline{Q}$. *Suppose that* u *is a generalized solution of the boundary value problem (9.1), (9.2) and* $f_0 \in L_2(Q) \cap W^k(Q_{sl})$ $(s = 1, 2, \dots,\ l = 1, \dots, N(s))$.

Then $u \in W^{k+2m}(Q_{sl} \setminus \overline{\mathcal{K}^\varepsilon})$ *for each* $\varepsilon > 0$ $(s = 1, 2, \dots,\ l = 1, \dots, N(s))$.

Proof. 1. By Theorem 11.1, it suffices to show that for an arbitrary given point $y \in \partial Q_{pi} \setminus \mathcal{K}$ there exists a ball $S_a(y)$ such that $u \in W^{k+2m}(Q_{pi} \cap S_a(y))$.

We introduce the points $y^1, \dots, y^N$ by the formulas $y^i = y$, $y^l = y^i - h_{pi} + h_{pl}$ $(l = 1, \dots, N = N(p))$, where $Q_{pl} = Q_{p1} + h_{pl}$. By definition of the sets Γ_{rl}, there is a unique r such that $J(r) \geq N(p)$ and, after some reindexing of the sets Q_{pl}, Γ_{rl}, we have $y^l \in \Gamma_{rl} \subset \partial Q_{pl} \setminus \mathcal{K}$ $(l = 1, \dots, N(p))$, $\Gamma_{rl} \subset Q$ $(l = 1, \dots, J_0 = J_0(r))$, $\Gamma_{rl} \subset \partial Q$ $(l = J_0 + 1, \dots, J(r))$.

By virtue of Lemma 7.8, there exists a unique subdomain $Q_{qj} \neq Q_{p1}$ such that $\Gamma_{r1} \subset \partial Q_{qj}$. Let us reindex the subdomains of the qth class so that $\Gamma_{rl} \subset \partial Q_{ql}$ $(l = 1, \dots, J_0)$. We introduce the points $z^1, \dots, z^N \in \overline{Q}$ so that $z^l = y^1 + h_{ql}$ $(l = 1, \dots, N,\ N = N(q) \geq J_0)$. According to the construction, $z^l = y^l \in Q$ $(l = 1, \dots, J_0)$, $z^l \in \partial Q$ $(l = J_0 + 1, \dots, N(q))$, $z^l \in \partial Q_{ql} \setminus \mathcal{K}$ $(l = 1, \dots, N(q))$. Clearly

$$\Big(\bigcup_{l>J_0} y^l\Big) \cap \Big(\bigcup_{l>J_0} z^l\Big) = \emptyset.$$

We consider the balls $S_{4\delta}(x^{sl})$ $(l = 1, \dots, N(s),\ s = p, q)$, where $x^{pl} = y^l$, $x^{ql} = z^l$. By virtue of Lemmas 7.3, 7.4, we can choose $\delta > 0$ so small that $4\delta < \min_{l,s} \min\{\rho(x^{sl}, \mathcal{K}), 1/2\}$, the sets $\partial Q_{sl} \cap S_{4\delta}(x^{sl})$ are connected and belong to the class C^∞ $(l = 1, \dots, N(s),\ s = p, q)$, $S_{4\delta}(x^{sl}) \subset \Gamma_{rl} \cup Q_{pl} \cup Q_{ql}$ $(l = 1, \dots, J_0)$, $S_{4\delta}(x^{sl}) \cap Q = S_{4\delta}(x^{sl}) \cap Q_{sl}$ $(l = J_0 + 1, \dots, N(s))$, $s = p, q$.

2. We set $v = \xi^\delta v_0$ in (10.2), where $v_0 \in \mathring{W}^m(Q)$,

$$\xi^\delta(x) = \sum_{1 \leq l \leq N(p)} \eta^\delta(x - h_{pl}) + \sum_{J_0 < l \leq N(q)} \eta^\delta(x - h_{ql}),$$

$\eta^\delta \in \dot{C}^\infty(\mathbb{R}^n)$, $0 \leq \eta^\delta(x) \leq 1$ $(x \in S_{2\delta}(y^1))$, $\eta^\delta(x) = 1$ $(x \in S_\delta(y^1))$, $\eta^\delta(x) = 0$ $(x \notin S_{2\delta}(y^1))$.

Then, using the Leibniz formula, we obtain

$$\sum_{\alpha,\beta} \sum_{s,l} \int_{\Omega_{sl}^\delta} (R_{\alpha\beta Q} \mathcal{D}^\beta u) \xi^\delta \overline{\mathcal{D}^\alpha v_0}\, dx$$

$$= \sum_{s,l} \int_{\Omega_{sl}^\delta} \Big\{ f_0 \xi^\delta \overline{v_0} + \sum_{\alpha,\beta,\nu} (R_{\alpha\beta Q} \mathcal{D}^\beta u) a_{\alpha\nu}^\delta \overline{\mathcal{D}^\nu v_0} \Big\} dx. \quad (11.12)$$

Here $\nu = (\nu_1, \dots, \nu_n)$, $\Omega_{sl}^\delta = Q_{sl} \cap S_{4\delta}(x^{sl})$, $a_{\alpha\nu}^\delta = c_{\alpha\nu}\mathcal{D}^{\alpha-\nu}\xi^\delta(x)$, $c_{\alpha\nu}$ are constants. In formula (11.12) and further we sum over the indices α, β, ν, l, s such that $|\alpha|, |\beta| \le m$, $|\nu| \le m-1$, $l = 1, \dots, N(s)$, $s = p, q$.

We introduce the operator $U_s \colon L_2(\bigcup_l Q_{sl}) \to L_2^N(Q_{s1})$ by formula (8.4), where $N = N(s)$, $s = p, q$. Since the operator of multiplication by the functions ξ^δ and $a_{\alpha\nu}^\delta$ commutes with the operator U_s, it follows from (11.12) that

$$\sum_{\alpha,\beta}\sum_s \int (\eta^\delta R_{\alpha\beta s}\mathcal{D}^\beta W_s, \mathcal{D}^\alpha V_s^0)\,dx$$
$$= \sum_s \int \Big\{ (\eta^\delta F_s, V_s^0) + \sum_{\alpha,\beta,\nu} (a_{\alpha\nu}^\delta R_{\alpha\beta s}\mathcal{D}^\beta W_s, \mathcal{D}^\nu V_s^0) \Big\} dx, \qquad (11.13)$$

where $W_s = (U_s P_s u)(x)$, $F_s = (U_s P_s f_0)(x)$, $V_s^0 = (U_s P_s v_0)(x)$ for $x \in \Omega_s^\delta$. In formula (11.13) and further we integrate over the set $\Omega_s^\delta = \Omega_{s1}^\delta$. Using the Leibniz formula again, we have

$$\sum_s \sum_{\alpha,\beta} \int (R_{\alpha\beta s}\mathcal{D}^\beta(\eta^\delta W_s), \mathcal{D}^\alpha V_s^0)\,dx$$
$$= \sum_s \int \Big\{ (\eta^\delta F_s, V_s^0) + \sum_{\alpha,\beta,\nu} (a_{\alpha\nu}^\delta R_{\alpha\beta s}\mathcal{D}^\beta W_s, \mathcal{D}^\nu V_s^0)$$
$$+ \sum_{\alpha,\beta,\mu} (b_{\beta\mu}^\delta R_{\alpha\beta s}\mathcal{D}^\mu W_s, \mathcal{D}^\alpha V_s^0) \Big\} dx, \qquad (11.14)$$

where $b_{\beta\mu}^\delta = d_{\beta\mu}\mathcal{D}^{\beta-\mu}\eta^\delta(x)$, $d_{\beta\mu}$ are constants. In formula (11.14) and further we sum over μ such that $|\mu| \le m-1$.

Without loss of generality, we assume that $y^1 = 0$ and that the equation for the surface $\Gamma_{p1} \cap S_{4\delta}(0)$ has the form $x_n = 0$. Otherwise, we can obtain a plane boundary, introducing the new variables. We denote by $\mathring{W}_\delta^m$ the space of vector-valued functions $V = (V_p, V_q) \in W^{m,N(p)}(\Omega_p^\delta) \times W^{m,N(q)}(\Omega_q^\delta)$ such that $\operatorname{supp} V_s \subset \overline{\Omega_s^\delta} \cap \overline{S_{2\delta}(0)}$, $SV \in \mathring{W}^m(Q)$, where $(SV)(x) = V_{sl}(x - h_{sl})$ for $x \in \Omega_{sl}^\delta$, $(SV)(x) = 0$ for $x \notin \bigcup_{s,l}\Omega_{sl}^\delta$. Suppose that $v_0 = SV^0$, $V^0 = -i\delta_{-t}^r V^1 = (-i\delta_{-t}^r V_p^1, -i\delta_{-t}^r V_q^1)$, where $V^1 \in \mathring{W}_{3\delta/2}^m$, $1 \le r \le n-1$, $0 < t < \delta$. According to the construction, $v_0 \in \mathring{W}^m(Q)$ and in the equation (11.14) $V_s^0 = -i\delta_{-t}^r V_s^1$. Since the operators $-i\delta_t^r$ and $-i\delta_{-t}^r$ are formally adjoint, we have

$$\sum_s \sum_{\alpha,\beta} \int (-i\delta_t^r[R_{\alpha\beta s}\mathcal{D}^\beta(\eta^\delta W_s)], \mathcal{D}^\alpha V_s^1)\,dx$$
$$= \sum_s \int \Big\{ (\eta^\delta F_s, -i\delta_{-t}^r V_s^1) + \sum_{\beta,\nu} (A_{\nu\beta s}^\delta \mathcal{D}^\beta W_s, -i\delta_{-t}^r \mathcal{D}^\nu V_s^1)$$
$$+ \sum_{\alpha,\mu} (-i\delta_t^r(B_{\alpha\mu s}^\delta \mathcal{D}^\mu W_s), \mathcal{D}^\alpha V_s^1) \Big\} dx, \qquad (11.15)$$

where $A^{\delta}_{\nu\beta s}$, $B^{\delta}_{\alpha\mu s}$ are matrices of order $N(s) \times N(s)$ with infinitely differentiable elements, obtained by reducing similar terms in (11.14). Since $\operatorname{supp}\eta^{\delta} \subset \overline{S_{2\delta}(0)}$, all the elements of the matrices $A^{\delta}_{\nu\beta s}(x)$, $B^{\delta}_{\alpha\mu s}(x)$ vanish for $x \notin S_{2\delta}(0)$.

3. Suppose that for some j $(1 \le j \le m+k)$ there exists $\delta_j > 0$ such that $W_s \in W^{m+j-1,N}(\Omega_s^{\delta_j})$ and W_s satisfies the equation

$$\sum_s \sum_{\alpha,\beta} \int (-i\delta_t^r [R_{\alpha\beta s} \mathcal{D}^{\beta+\gamma}(\eta^{\delta_j} W_s)], \mathcal{D}^{\alpha} V_s^j)\, dx = I \tag{11.16}$$

for all $V^j \in \mathring{W}^m_{3\delta_j/2}$, $0 < t < \varepsilon_1 < \delta_j$, $\gamma \in \{|\gamma| = j-1 : \gamma_n = 0\}$, $r = 1, \ldots, n-1$. Here $\lambda_n = 0$,

$$\begin{aligned} I = \sum_s \int \Big\{ & \sum_{|\lambda| \le j-1} \sum_{\beta,\nu} (A^{\delta_j}_{\nu\beta\lambda s} \mathcal{D}^{\beta+\lambda} W_s, -i\delta^r_{-t} \mathcal{D}^{\nu} V_s^j) \\ & + \sum_{\alpha} \Big(\sum_{|\omega| \le m-1} \sum_{|\lambda| \le j-1} + \sum_{|\omega| \le m} \sum_{|\lambda| \le j-2} \Big) (-i\delta_t^r (B^{\delta_j}_{\alpha\omega\lambda s} \mathcal{D}^{\omega+\lambda} W_s), \mathcal{D}^{\alpha} V_s^j) \\ & + \sum_{|\lambda| \le j-1} a^{\delta_j}_{\lambda s} \Big\} dx; \end{aligned}$$

we integrate over the set $\Omega_s^{\delta_j}$; $A^{\delta_j}_{\nu\beta\lambda s}$, $B^{\delta_j}_{\alpha\omega\lambda s}$ are the matrices with infinitely differentiable elements vanishing for $x \notin S_{2\delta_j}(0)$; $a^{\delta_j}_{\lambda s} = (b^{\delta_j}_{\lambda s} F_s, -i\delta^r_{-t} \mathcal{D}^{\lambda} V_s^j)$ if $|\lambda| \le m-1$; $a^{\delta_j}_{\lambda s} = (-i\delta_t^r (b^{\delta_j}_{\lambda s} \mathcal{D}^{\lambda^2} F_s), \mathcal{D}^{\lambda^1} V_s^j)$ if $|\lambda| > m-1$; $|\lambda^1| = m$, $\lambda^1 + \lambda^2 = \lambda$, $b^{\delta_j}_{\lambda s} \in \dot{C}^{\infty}(S_{2\delta_j}(0))$.

We shall prove that in this case there exists $0 < \delta_{j+1} < \delta_j/4$ such that $W_s \in W^{m+j,N}(\Omega_s^{\delta_{j+1}})$.

Let us note that for $j = 1$, equation (11.16) coincides with equation (11.15) if we set $\delta = \delta_1$.

a) First we prove that $\mathcal{D}^{\alpha+\sigma} W_s \in L_2^N(\Omega_s^{\delta_j/4})$ for all α and σ such that $|\alpha| \le m$, $|\sigma| \le j$, $\sigma_n = 0$. Let $V_s^j = -i\delta^r_{t_1} \mathcal{D}^{\gamma}(\eta^{\delta_j} W_s)$ $(0 < t_1 < \varepsilon_1)$. Clearly $V^j = (V_p^j, V_q^j) \in \mathring{W}^m_{3\delta_j/2}$. Using the identity

$$\delta_t^r(aw) = a_t^r \delta_t^r w + w \delta_t^r a,$$

we obtain from (11.16) that

$$I_1 + I_2 = I_3 + I, \tag{11.17}$$

where

$$I_1 = \sum_s \sum_{\alpha,\beta} \int (R_{\alpha\beta s}\mathcal{D}^\beta(-i\delta_t^r)\mathcal{D}^\gamma(\eta^{\delta_j} W_s), \mathcal{D}^\alpha V_s^j)\, dx,$$

$$I_2 = \sum_s \sum_{\alpha,\beta} \int ([(R_{\alpha\beta s})_t^r - R_{\alpha\beta s}]\mathcal{D}^\beta(-i\delta_t^r)\mathcal{D}^\gamma(\eta^{\delta_j} W_s), \mathcal{D}^\alpha V_s^j)\, dx,$$

$$I_3 = -\sum_s \sum_{\alpha,\beta} \int ((-i\delta_t^r R_{\alpha\beta s})\mathcal{D}^{\beta+\gamma}(\eta^{\delta_j} W_s), \mathcal{D}^\alpha V_s^j)\, dx.$$

Using the Schwarz inequality and Theorem B.16, we obtain

$$|I| + |I_3| \le k_1 \Big(\sum_s \|V_s^j\|_{m,N}^2 \Big)^{1/2} \sum_s (\|W_s\|_{m+j-1,N} + \|F_s\|_{k,N}), \tag{11.18}$$

where $\|F_s\|_{k,N} = \{\sum_{l=1}^{N(s)} \|F_{sl}\|_{W^k(\Omega_s^\delta)}^2\}^{1/2}$, $\delta = \delta_j$.

We estimate the left-hand side of (11.17). We set $t_1 = t$, i.e., $V_s^j = -i\delta_t^r \mathcal{D}^\gamma$ $(\eta^{\delta_j} W_s)$. By Definition 9.2 and formulas (8.4), (8.6), we have

$$\begin{aligned}
\operatorname{Re} I_1 &= \operatorname{Re} \sum_{\alpha,\beta} (R_{\alpha\beta Q}\mathcal{D}^\beta v_j, \mathcal{D}^\alpha v_j)_{L_2(Q)} \\
&\ge c_1 \|v_j\|_{W^m(Q)}^2 - c_2 \|v_j\|_{L_2(Q)}^2 = \sum_s (c_1 \|V_s^j\|_{m,N}^2 - c_2 \|V_s^j\|_{0,N}^2) \\
&\ge c_1 \sum_s \|V_s^j\|_{m,N}^2 \\
&\qquad - k_2 \Big(\sum_s \|V_s^j\|_{m,N}^2 \Big)^{1/2} \sum_s \|W_s\|_{m+j-1,N},
\end{aligned} \tag{11.19}$$

where $v_j = SV^j$.

Since the elements of the matrix $R_{\alpha\beta s}(x)$ are uniformly continuous in $\overline{\Omega_s^{\delta_j}}$, it follows that

$$|I_2| \le k_3(\varepsilon_1) \sum_s \|V_s^j\|_{m,N}^2, \tag{11.20}$$

where $k_3(\varepsilon_1) \to 0$ as $\varepsilon_1 \to 0$. Choosing $\varepsilon_1 > 0$ so that $2k_3(\varepsilon_1) < c_1$, from (11.17)–(11.20) we obtain

$$\Big\{ \sum_s \|V_s^j\|_{m,N}^2 \Big\}^{1/2} \le k_4 \sum_s (\|W_s\|_{m+j-1,N} + \|F_s\|_{k,N}).$$

Since $1 \le r \le n-1$ is arbitrary, by virtue of Theorem B.16, $\mathcal{D}^{\alpha+\sigma}(\eta^{\delta_j} W_s) \in L_2^N(\Omega_s^{\delta_j})$ for all $|\alpha| \le m$ and $|\sigma| \le j$, $\sigma_n = 0$. Thus, $\mathcal{D}^{\alpha+\sigma} W_s \in L_2^N(\Omega_s^{\delta_j/4})$.

b) We now prove that there exists $\delta_{j+1} > 0$ such that $\mathcal{D}^{\alpha+\gamma}W_s \in L_2^N(\Omega_s^{\delta_{j+1}})$ for all $|\alpha| \le m$, $|\gamma| \le j$. For this it suffices to show that for each $g = 0, \dots, j-1$ from

$$\mathcal{D}_n^{\kappa}\mathcal{D}^{\alpha+\sigma}W_s \in L_2^N(\Omega_s^{\delta_{jg}}) \qquad (\kappa \le g,\ |\sigma| \le j-\kappa,\ \sigma_n = 0,\ |\alpha| \le m) \tag{11.21}$$

it follows that

$$\mathcal{D}_n^{g+1}\mathcal{D}^{\alpha+\zeta}W_s \in L_2^N(\Omega_s^{\delta_{j,g+1}}) \qquad (|\zeta| \le j-g-1,\ \zeta_n = 0,\ |\alpha| \le m), \tag{11.22}$$

where $0 < \delta_{j,g+1} < \delta_{jg}/4$, $\delta_{j0} = \delta_j/4$.

For $g = 0$, the relations (11.21) were proved above.

Now let the relations (11.21) be fulfilled for some g. Suppose that $\delta = \delta_{jg}$ in (11.14) and $V_s^0 \in \dot{C}^{\infty,N}(\Omega_s^{\delta/4})$ is an arbitrary vector-valued function. Then a vector-valued function W_s is a generalized solution of the system of differential equations

$$\sum_{\alpha,\beta} \mathcal{D}^{\alpha}R_{\alpha\beta s}\mathcal{D}^{\beta}W_s = F_s \qquad (x \in \Omega_s^{\delta/4}). \tag{11.23}$$

Applying the operator $\mathcal{D}_n^g\mathcal{D}^{\zeta}$ ($|\zeta| \le j-g-1$, $\zeta_n = 0$), we obtain

$$\widehat{A}_s\mathcal{D}_n^{2m+g}\mathcal{D}^{\zeta}W_s = -\sum_{\alpha,\beta} R_{\alpha\beta s}\mathcal{D}_n^g\mathcal{D}^{\alpha+\beta+\zeta}W_s + \sum_{\theta,\kappa,\lambda} C_{\theta\kappa\lambda s}\mathcal{D}_n^{\kappa}\mathcal{D}^{\theta+\lambda}W_s + \mathcal{D}_n^g\mathcal{D}^{\zeta}F_s, \tag{11.24}$$

where $\widehat{A}_s = R_{\alpha\beta s}(x)$ for $\alpha = \beta = (0, \dots, 0, m)$; $C_{\theta\kappa\lambda s}$ are matrices with infinitely differentiable elements; we sum over α, β, θ, κ, λ such that $|\alpha|, |\beta| \le m$, $\alpha_n + \beta_n < 2m$ and $|\theta| \le 2m$, $\kappa \le g$, $|\lambda| \le \zeta$, $|\theta| + \kappa + |\lambda| \le 2m + g + |\zeta| - 1$, $\lambda_n = 0$.

Since $F_s \in W^{k,N}(\Omega_s^{\delta/4})$, from (11.21) it follows that the right part of (11.24) belongs to $W^{1-m,N}(\Omega_s^{\delta/4})$. Here $W^{-k}(\Omega_s^{\delta/4})$ is the space dual to $\mathring{W}^k(\Omega_s^{\delta/4})$ ($k \ge 0$) (see Appendix B) and $W^{-k,N}(\Omega_s^{\delta/4}) = \prod_l W^{-k}(\Omega_s^{\delta/4})$. By virtue of Theorem 9.1, the matrix $\widehat{A}_s$ has an inverse matrix $\widehat{A}_s^{-1}$ with infinitely differentiable elements in $\overline{\Omega_s^{\delta/4}}$. Thus $\mathcal{D}_n^{2m+g}\mathcal{D}^{\zeta}W^s \in W^{1-m,N}(\Omega_s^{\delta/4})$. Hence, from (11.21) we have $\mathcal{D}^{\theta}(\mathcal{D}_n^{g+1}\mathcal{D}^{\zeta}W_s) \in W^{1-m,N}(\Omega_s^{\delta/4})$ ($|\theta| \le 2m-1$, $|\zeta| \le j-g-1$, $\zeta_n = 0$), where $\delta = \delta_{jg}$. Thus, by virtue of Lemma 16 of N. Dunford and J. Schwartz [2, Chapter XIV, Section 6], we obtain $\mathcal{D}_n^{g+1}\mathcal{D}^{\zeta}W_s \in W^{m,N}(\Omega_s^{\delta_{j,g+1}})$ for some $0 < \delta_{j,g+1} < \delta_{jg}/4$, i.e., $\mathcal{D}_n^{g+1}\mathcal{D}^{\alpha+\zeta}W_s \in L_2^N(\Omega_s^{\delta_{j,g+1}})$ ($|\zeta| \le j-g-1$, $|\alpha| \le m$).

We have proved that $\mathcal{D}^{\alpha+\gamma}W_s \in L_2^N(\Omega_s^{\delta_{j+1}})$ for all $|\gamma| \le j$, $|\alpha| \le m$, where $\delta_{j+1} = \delta_{jj}$. Hence, $W_s \in W^{m+j,N}(\Omega_s^{\delta_{j+1}})$.

4. It remains to show that, if $W_s \in W^{m+j-1,N}(\Omega_s^{\delta_j})$ satisfies the equation (11.16) for all $V^j \in \mathring{W}^m_{3\delta_j/2}$, then the function W_s satisfies the equation

$$\sum_s \sum_{\alpha,\beta} \int (-i\delta_t^r[R_{\alpha\beta s}\mathcal{D}^{\beta+\gamma}(\eta^{\delta_{j+1}}W_s)], \mathcal{D}^\alpha V_s^{j+1})\,dx$$

$$= \sum_s \int \Big\{ \sum_{|\lambda|\le j} \sum_{\beta,\nu} (A^{\delta_{j+1}}_{\nu\beta\lambda s}\mathcal{D}^{\beta+\lambda}W_s, -i\delta^r_{-t}\mathcal{D}^\nu V_s^{j+1})$$

$$+ \sum_\alpha \Big(\sum_{|\omega|\le m-1} \sum_{|\lambda|\le j} + \sum_{|\omega|\le m} \sum_{|\lambda|\le j-1} \Big) (-i\delta_t^r(B^{\delta_{j+1}}_{\alpha\omega\lambda s}\mathcal{D}^{\omega+\lambda}W_s), \mathcal{D}^\alpha V_s^{j+1})$$

$$+ \sum_{|\lambda|\le j} a^{\delta_{j+1}}_{\lambda s} \Big\} dx \tag{11.25}$$

for all $V^{j+1} \in \mathring{W}^m_{3\delta_{j+1}/2}$, $0 < t < \varepsilon_1 < \delta_{j+1}$, $\gamma \in \{|\gamma| = j : \gamma_n = 0\}$, $r = 1, \dots, n-1$, where $\lambda_n = 0$, $A^{\delta_{j+1}}_{\nu\beta\lambda s}(x)$, $B^{\delta_{j+1}}_{\alpha\omega\lambda s}(x)$ are the matrices with infinitely differentiable elements vanishing for $x \notin S_{2\delta_{j+1}}(0)$, we integrate over the set $\Omega_s^{\delta_{j+1}}$.

We proved above that $\mathcal{D}_r\mathcal{D}^{\beta+\gamma}(\eta^{\delta_j}W_s) \in L_2^N(\Omega_s^{\delta_j})$ for $|\beta| \le m$, $|\gamma| \le j-1$, $\gamma_n = 0$. Hence, by virtue of Theorem B.16, we can pass to the limit in (11.16) as $t \to 0$. Then integrating by parts and using the Leibniz formula, we obtain

$$\sum_s \sum_{\alpha,\beta} \int (R_{\alpha\beta s}\mathcal{D}_r\mathcal{D}^{\beta+\gamma}(\eta^{\delta_j}W_s), \mathcal{D}^\alpha V_s^j)\,dx$$

$$= \sum_s \int \Big\{ \sum_{|\lambda|\le j} \sum_{\beta,\nu} (\widehat{A}^{\delta_j}_{\nu\beta\lambda s}\mathcal{D}^{\beta+\lambda}W_s, \mathcal{D}^\nu V_s^j)$$

$$+ \sum_\alpha \Big(\sum_{|\omega|\le m-1} \sum_{|\lambda|\le j} + \sum_{|\omega|\le m} \sum_{|\lambda|\le j-1} \Big) (\widehat{B}^{\delta_j}_{\alpha\omega\lambda s}\mathcal{D}^{\omega+\lambda}W_s, \mathcal{D}^\alpha V_s^j)$$

$$+ \sum_{|\lambda|\le j-1} \hat{a}^{\delta_j}_{\lambda s} \Big\} dx, \tag{11.26}$$

where $|\gamma| = j-1$, $\gamma_n = \lambda_n = 0$, $\widehat{A}^{\delta_j}_{\nu\beta\lambda s}(x)$, $\widehat{B}^{\delta_j}_{\alpha\omega\lambda s}(x)$ are the matrices with infinitely differentiable elements in $\overline{\Omega_s^{\delta_j}}$; $\hat{a}^{\delta_j}_{\lambda s} = (b^{\delta_j}_{\lambda s}F_s, \mathcal{D}_r\mathcal{D}^\lambda V_s^j)$ if $|\lambda| \le m-1$; $\hat{a}^{\delta_j}_{\lambda s} = (\mathcal{D}_r(b^{\delta_j}_{\lambda s}\mathcal{D}^{\lambda^2}F_s), \mathcal{D}^{\lambda^1}V_s^j)$ if $|\lambda| > m-1$; we integrate over $\Omega_s^{\delta_j}$.

We set $V_s^j = \eta^{\delta_{j+1}}(-i\delta^r_{-t})V_s^{j+1}$ $(r = 1, \dots, n-1)$, where $V^{j+1} \in \mathring{W}^m_{3\delta_{j+1}/2}$, $0 < t < \varepsilon < \delta_{j+1}$.

According to the construction, $V_s^j \in \mathring{W}_{\delta_j}^m$. Since the operators $-i\delta_{-t}^r$ and $-i\delta_t^r$ are formally adjoint, using the Leibniz formula, we obtain from (11.26)

$$\sum_s \sum_{\alpha,\beta} \int \big(- i\delta_t^r [R_{\alpha\beta s}\mathcal{D}_r \mathcal{D}^{\beta+\gamma}(\eta^{\delta_{j+1}}\eta^{\delta_j} W_s)], \mathcal{D}^\alpha V_s^{j+1}\big) dx$$

$$= \sum_s \int \Big\{ \sum_{|\lambda|\le j} \sum_{\beta,\nu} (A_{\nu\beta\lambda s}^{\delta_{j+1}} \mathcal{D}^{\beta+\lambda} W_s, -i\delta_{-t}^r \mathcal{D}^\nu V_s^{j+1})$$

$$+ \sum_\alpha \Big(\sum_{|\omega|\le m-1} \sum_{|\lambda|\le j} + \sum_{|\omega|\le m} \sum_{|\lambda|\le j-1} \Big) (-i\delta_t^r (B_{\alpha\omega\lambda s}^{\delta_{j+1}} \mathcal{D}^{\omega+\lambda} W_s), \mathcal{D}^\alpha V_s^{j+1})$$

$$+ \sum_{|\lambda|\le j} a_{\lambda s}^{\delta_{j+1}} \Big\} dx. \qquad (11.27)$$

Since $\eta^{\delta_j}(x) = 1$ for $x \in S_{\delta_j}(0)$, $\delta_{j+1} < \delta_j/4$ and $\operatorname{supp} \eta^{\delta_{j+1}} \subset \overline{S_{2\delta_{j+1}}(0)}$, $\eta^{\delta_{j+1}}(x)\eta^{\delta_j}(x) = \eta^{\delta_{j+1}}(x)$ for $x \in \Omega_s^{\delta_{j+1}}$. Therefore, from (11.27) we obtain (11.25).

Thus $W_s \in W^{k+2m,N}(\Omega_s^{\delta_{m+k+1}})$. Hence, $u \in W^{k+2m}(Q_{pi} \cap S_a(y))$, where $a = 4\delta_{m+k+1}$. □

Remark 11.1. In the first part of the proof of Theorem 11.3 it is possible that $y^1 \in \partial Q_{p1} \cap \partial Q_{pj}$ $(j \neq 1)$. In other words, the point y^1 can belong to the boundaries of two different subdomains of the same class (see Example 9.2 for $\theta = 1$). Then, we must reindex the subdomains Q_{qj} so that $y^l \in \partial Q_{pl} \cap \partial Q_{ql}$ $(l = 1, \dots, J_0)$. Evidently, in that case the index q does not correspond to a new class of subdomains, but to the same class, which has been renumbered.

Remark 11.2. If $J_0(r) = 0$ in the first part of the proof, then we must not introduce the subdomains Q_{ql}, and the proof is simplified.

Remark 11.3. In Section 23 we shall consider the case of a cylindrical domain $Q = (0,d) \times G$ and a difference operator with shifts along the x_1 axis. In this case Theorem 11.3 is valid for $\varepsilon = 0$. But from the following example it is easy to see that in the general case this theorem is not true for $\varepsilon = 0$.

Example 11.2. We consider the boundary value problem

$$-\Delta R_Q u = f_0(x) \qquad (x \in Q), \qquad (11.28)$$

$$u|_{\partial Q} = 0 \qquad (x \in \partial Q). \qquad (11.29)$$

Here the domain $Q \subset \mathbb{R}^n$ has boundary $\partial Q \in C^\infty$, which outside the disks $S_{1/8}((i4/3, j4/3))$ $(i,j = 0,1)$ coincides with the boundary of the square $(0,4/3) \times (0,4/3)$,

$$Ru(x) = u(x) + au(x_1+1, x_2+1) + au(x_1-1, x_2-1), \quad 0 < a < 1.$$

Clearly the equation (11.28) is strongly elliptic.

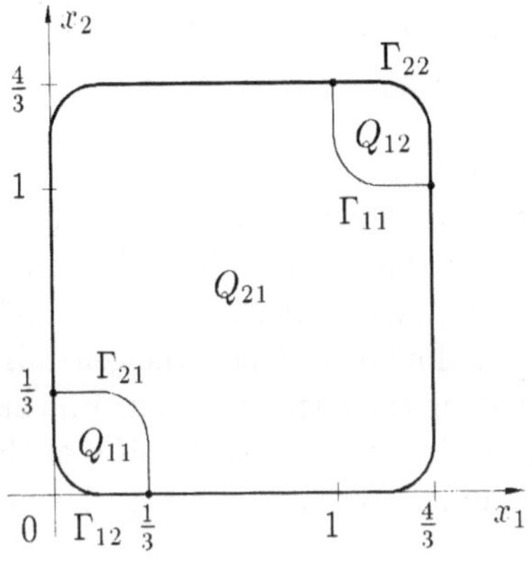

Fig. II.9

We denote $\Gamma_{12} = \{x \in \partial Q : x_1 < 1/3,\ x_2 < 1/3\}$, $\Gamma_{11} = \Gamma_{12} + (1,1)$, $\Gamma_{22} = \{x \in \partial Q : 1 < x_1, 1 < x_2\}$, $\Gamma_{21} = \Gamma_{22} - (1,1)$. The decomposition $\mathcal{R}$ consists of two classes: 1) Q_{11}, bounded by the curves $\overline{\Gamma_{12}}$ and $\overline{\Gamma_{21}}$; Q_{12}, bounded by the curves $\overline{\Gamma_{11}}$ and $\overline{\Gamma_{22}}$, and 2) $Q_{21} = Q \setminus (\overline{Q_{11}} \cup \overline{Q_{12}})$. The set $\mathcal{K} \subset \partial Q$ and consists of four points: $g^1 = (1/3, 0)$, $g^2 = (4/3, 1)$, $g^3 = (0, 1/3)$, $g^4 = (1, 4/3)$ (see Fig. II.9).

We introduce the function $u(x)$ by the formula

$$u(x) = \begin{cases} \dfrac{u_1(x_1 - 1/3, x_2) - a u_2(x_1 - 1/3, x_2)}{1 - a^2} & (x \in Q_{11}), \\ \dfrac{-a u_1(x_1 - 4/3, x_2 - 1) + u_2(x_1 - 4/3, x_2 - 1)}{1 - a^2} & (x \in Q_{12}), \\ u_1(x_1 - 1/3, x_2) + u_2(x_1 - 4/3, x_2 - 1) & (x \in Q_{21}), \end{cases} \tag{11.30}$$

where the functions u_1 and u_2 have the form

$$u_1(r, \varphi) = \xi(r) r^\lambda \sin \lambda\varphi, \qquad u_2(r, \varphi) = \xi(r) r^\lambda \sin \lambda(3\pi/2 - \varphi),$$

$\xi(r) \in \dot{C}^\infty(\mathbb{R})$, $0 \le \xi(r) \le 1$, $\xi(r) = 1$ for $r \le 1/8$, $\xi(r) = 0$ for $r \ge 1/6$, $\lambda = (2/\pi) \arccos(a/2)$, r, φ are polar coordinates.

Hence,

$$R_Q u(x) = u_1(x_1 - 1/3, x_2) + u_2(x_1 - 4/3, x_2 - 1).$$

Since $0 < \lambda < 1$, it is easy to see that $u \in \mathring{W}^1(Q)$, $-\Delta R_Q u \in L_2(Q)$, while $u \notin W^2(Q_{s1} \cap S_\delta(g^1))$ for every $\delta > 0$ $(s = 1, 2)$.

12 Smoothness of Solutions on a Boundary of Neighboring Subdomains

In this section we shall show that a generalized solution may not have corresponding smoothness on a boundary of neighboring subdomains (see Example 12.1). This is connected with the fact that differential-difference operators are nonlocal. A normal derivative of the solution of order $k > m-1$ can have a discontinuity on a boundary of neighboring subdomains. Hence, the generalized derivatives of order $2m$ may contain addenda having the form of a δ-function and its derivatives. However, the difference operators contain different shifts of the solution $u(x)$. Thus the discontinuities will be annihilated, and $(\mathcal{A}_R u)(x)$ will belong to the Sobolev space $W^k(Q)$.

Necessary and Sufficient Conditions of Smoothness

Let the differential-difference operator $\mathcal{A}_R$ be strongly elliptic, and let a domain Q satisfy the condition 7.1. Suppose that $u(x)$ is a generalized solution of the boundary value problem (9.1), (9.2), where $f_0 \in W^k(Q)$. We fix $s = p$ and consider a point $y^1 \in Q \cap (\partial Q_{p1} \setminus \mathcal{K})$. Let $y^l = y^1 + h_{pl} \in \partial Q_{pl} \setminus \mathcal{K}$ $(l = 1, \dots, N(p))$. As in the proof of Theorem 10.3, we assume that $y^l \in Q$ $(l = 1, \dots, J_0)$, $y^l \in \partial Q$ $(l = J_0 + 1, \dots, N(p))$. We establish the conditions when for a given $1 \le l_0 \le J_0$ there exists $a > 0$ such that $u \in W^{k+2m}(S_a(y^{l_0}))$ for all $f_0 \in W^k(Q)$, i.e., the solution has a corresponding smoothness in some neighborhood of the point y^{l_0}.

By virtue of Lemma 7.8, there exists a unique subdomain $Q_{qj} \ne Q_{p1}$ such that $y^1 \in \partial Q_{qj}$. We introduce the points $z^1, \dots, z^N \in \overline{Q}$ such that $z^l = z^j - h_{qj} + h_{ql} \in \partial Q_{ql} \setminus \mathcal{K}$ $(l = 1, \dots, N(q))$, $z^j = y^1$. Without loss of generality, we can assume that $y^l = z^l$ $(l = 1, \dots, J_0)$, $z^l \in \partial Q$ $(l = J_0 + 1, \dots, N(q))$. By virtue of Lemmas 7.3, 7.4, we can choose $a > 0$ so small that the sets $\partial Q_{sl} \cap S_a(x^{sl})$ are connected and belong to the class C^∞ $(l = 1, \dots, N(s),\ s = p, q)$, while $a < \min_{l,s} \min\{\rho(x^{sl}, \mathcal{K}), 1/2\}$, $S_a(x^{sl}) \subset Q$, $S_a(x^{sl}) \cap Q_{s_1 l_1} = \emptyset$ $(l = 1, \dots, J_0,$ $(s_1, l_1) \ne (s, l),\ s = p, q)$, $S_a(x^{sl}) \cap Q = S_a(x^{sl}) \cap Q_{sl}$ $(l = J_0 + 1, \dots, N(s),$ $s = p, q)$, $x^{pl} = y^l$, $x^{ql} = z^l$.

For simplicity we assume that $y^1 = 0$,

$$Q_{p1} \cap S_a(0) = \{x \in \mathbb{R}^n : x_n < 0\} \cap S_a(0),$$
$$\partial Q_{p1} \cap S_a(0) = \{x \in \mathbb{R}^n : |x| < a, x_n = 0\}.$$

Using the identity (8.12) m times, we can write the operator

$$\mathcal{A}_R = \sum_{|\alpha|, |\beta| \le m} \mathcal{D}^\alpha R_{\alpha\beta Q} \mathcal{D}^\beta \tag{12.1}$$

in the form

$$\mathcal{A}_R = \sum_{|\alpha| \le 2m} \mathcal{D}^\alpha R_{\alpha Q}, \tag{12.2}$$

where

$$R_\alpha u(x) = \sum_{h\in\mathcal{M}} a_{\alpha h}(x)u(x+h), \qquad a_{\alpha h} \in C^\infty(\mathbb{R}^n).$$

Hence, a function $u(x)$ satisfies the integral identity

$$\int_{S_a(y^l)} \Big(\sum_{|\alpha|\le 2m} \mathcal{D}^\alpha R_{\alpha Q} u\Big)\overline{\varphi}\,dx = \int_{S_a(y^l)} f_0\overline{\varphi}\,dx$$

$$(\varphi \in \dot{C}^\infty(S_a(y^l)),\ l = 1,\dots,J_0).$$

Then, for all $w \in \dot{C}^\infty(S_a(0))$,

$$I = \int_{S_a(0)} \Big(\sum_{|\alpha|\le 2m} \mathcal{D}^\alpha v_{\alpha l}\Big)\overline{w}\,dx = \int_{S_a(0)} f_l\overline{w}\,dx, \tag{12.3}$$

where $f_l(x) = (U_sP_sf_0)_l(x)$, $v_{\alpha l}(x) = (R_{\alpha s}U_sP_su)_l(x)$ for $x \in \omega_s = Q_{s1}\cap S_a(0)$ $(s = p, q,\ l = 1,\dots,J_0)$. By Theorem 11.3, $v_{\alpha l} \in W^{k+2m}(\omega_s)$. Therefore, we can integrate by parts m times over ω_s $(s = 1,2)$. Each time, we integrate by parts the term $\int_{\omega_s} \mathcal{D}^{\alpha-\lambda} v_{\alpha l}\overline{\mathcal{D}^\lambda w}\,dx$ with respect to x_n if $\alpha_n - \lambda_n \ge 1$, $|\lambda| \le m-1$, and with respect to x_i $(i \ne n)$ if $\alpha_n - \lambda_n = 0$, $|\lambda| \le m-1$, $1 \le |\alpha-\lambda|$. Then we obtain

$$I = i\sum_s\sum_{j=1}^m \sum_{\alpha:|\alpha|\le 2m, j\le\alpha_n} (-1)^{\mu(s)} \int_{\gamma_s} (\mathcal{D}_n^{\alpha_n-j}\mathcal{D}_{x'}^{\alpha'} v_{\alpha l})|_{\gamma_s}\overline{(\mathcal{D}_n^{j-1}w)|_{\gamma_s}}\,dx'$$

$$+\sum_s\sum_{|\alpha|\le 2m}\int_{\omega_s} \mathcal{D}^{\alpha-\beta} v_{\alpha l}\overline{\mathcal{D}^\beta w}\,dx. \tag{12.4}$$

Here $(\mathcal{D}_n^{\alpha_n-j}\mathcal{D}_{x'}^{\alpha'} v_{\alpha l})|_{\gamma_s}$ is the trace of the function $\mathcal{D}_n^{\alpha_n-j}\mathcal{D}_{x'}^{\alpha'} v_{\alpha l}$, which is defined on ω_s, $\gamma = \gamma_s = \{x \in \partial Q_{s1} : |x| < a\}$, $x' = (x_1,\dots,x_{n-1})$, $\mathcal{D}_{x'}^{\alpha'} = \mathcal{D}_1^{\alpha_1}\cdots\mathcal{D}_{n-1}^{\alpha_{n-1}}$, $\mu(p) = 1$, $\mu(q) = 2$, a multi-index $\beta = \beta(\alpha)$ is such that $\beta_i \le \alpha_i$, $|\beta| \le m$, $|\alpha-\beta| \le m$, $\beta_n = \min(\alpha_n, m)$. On the other hand, using the definition of a generalized derivative in the space of distributions $\mathcal{D}'(S_a(0))$ m times, from (12.3) we have

$$\int_{S_a(0)} \Big(\sum_{|\alpha|\le 2m} \mathcal{D}^{\alpha-\beta} v_{\alpha l}\Big)\overline{\mathcal{D}^\beta w}\,dx = \int_{S_a(0)} f_l\overline{w}\,dx. \tag{12.5}$$

Since a function $w(x)$ is arbitrary, from formulas (12.3)–(12.5) it follows that

$$\sum_s\sum_\alpha (-1)^{\mu(s)+1}(\mathcal{D}_n^{\alpha_n-j}\mathcal{D}_{x'}^{\alpha'} v_{\alpha l})|_{\gamma_s} = 0. \tag{12.6}$$

In (12.6) we sum over α such that $|\alpha| \le 2m$, $j \le \alpha_n$ $(j = 1,\dots,m)$ and over $s = p, q$. Since $f_l \in W^k(S_a(0))$, formulas (12.6) are valid for $j = -k+1,\dots,m$.

Since $h_{pl} = h_{ql}$ $(l = 1, \ldots, J_0)$, by formulas (9.4), the $J_0 \times J_0$-matrices, obtained from the matrices $R_{\alpha s}(x)$ $(s = p, q)$ by eliminating the last $N(s) - J_0$ rows and columns, are equal. We denote this $J_0 \times J_0$-matrix by $R_{\alpha p0}(x)$ (see Section 8). Let $R_{\alpha s1}(x)$ be the matrices of order $J_0 \times (N(s) - J_0)$, obtained from the matrix $R_{\alpha s}(x)$ by deleting the first J_0 columns and the last $N(s) - J_0$ rows. We introduce the vector-valued functions $V^s(x) = ((U_s P_s u)_1(x), \ldots, (U_s P_s u)_{J_0}(x))$ and vector-valued functions $W^s(x) = ((U_s P_s u)_{J_0+1}(x), \ldots, (U_s P_s u)_N(x))$. Then (12.6) may be written in the form

$$\sum_s \sum_\alpha (-1)^{\mu(s)+1} \mathcal{D}_n^{\alpha_n - j} \mathcal{D}_{x'}^{\alpha'} (R_{\alpha p0} V^s)|_\gamma = \sum_s \sum_\alpha (-1)^{\mu(s)} \mathcal{D}_n^{\alpha_n - j} \mathcal{D}_{x'}^{\alpha'} (R_{\alpha s1} W^s)|_\gamma \quad (j = -k+1, \ldots, m). \tag{12.7}$$

Since $u \in \mathring{W}^m(Q)$, we have

$$\mathcal{D}_n^{\alpha_n} V^p|_\gamma = \mathcal{D}_n^{\alpha_n} V^q|_\gamma, \quad \mathcal{D}_n^{\alpha_n} W^s|_\gamma = 0 \qquad (s = p, q) \tag{12.8}$$

for $0 \le \alpha_n \le m - 1$, $s = 1, 2$. Using formulas (12.8) and the Leibniz formula, we obtain from (12.7)

$$\sum_s (-1)^{\mu(s)+1} \mathcal{D}_n^{m+j} (B_p V^s)|_\gamma = \sum_s \sum_\alpha (-1)^{\mu(s)} \mathcal{D}_n^{\alpha_n - m + j} \mathcal{D}_{x'}^{\alpha'} (R_{\alpha p0} V^s)|_\gamma + \sum_s \sum_\alpha (-1)^{\mu(s)} \mathcal{D}_n^{\alpha_n - m + j} \mathcal{D}_{x'}^{\alpha'} (R_{\alpha s1} W^s)|_\gamma$$
$$(j = 0, \ldots, m + k - 1), \tag{12.9}$$

where $B_p(x) = R_{\alpha p0}(x)$ for $\alpha = (0, \ldots, 0, 2m)$. In the first group of terms on the right-hand side, we sum over $|\alpha| \le 2m$, $2m - j \le \alpha_n \le 2m - 1$; in the second group of terms we sum over $|\alpha| \le 2m$, $2m - j \le \alpha_n$.

Using the Leibniz formula and formulas (12.8), we obtain from (12.9) for $j = 0$,

$$\sum_s (-1)^{\mu(s)+1} B_p \mathcal{D}_n^m V^s|_\gamma = \sum_s B_{0\alpha s} \mathcal{D}_n^m W^s|_\gamma, \tag{12.10}$$

where $\alpha = (0, \ldots, 0, 2m)$, $B_{0\alpha s}(x) = (-1)^{\mu(s)} R_{\alpha s1}(x)$. By Theorem 9.1, the matrix $R_{\alpha s}(x) + R_{\alpha s}^*(x)$ is positive definite for $\alpha = (0, \ldots, 0, 2m)$, $x \in \overline{\omega_s}$. Thus, the matrix $B_p(x) + B_p^*(x)$ is also positive definite. Hence, there exists the inverse matrix $B_p^{-1}(x)$ with infinitely differentiable elements in $\overline{\omega_s}$. Therefore, from (12.10) we obtain

$$\sum_s (-1)^{\mu(s)+1} \mathcal{D}_n^m V^s|_\gamma = \sum_s B_p^{-1} B_{0\alpha s} \mathcal{D}_n^m W^s|_\gamma. \tag{12.11}$$

Using the equality (12.9) for $j=1$, the Leibniz formula and formulas (12.8), (12.11), we have

$$\sum_s (-1)^{\mu(s)+1} B_p \mathcal{D}_n^{m+1} V^s|_\gamma = \sum_s \sum_\alpha B_{1\alpha s} \mathcal{D}_n^{\alpha_n - m+1} \mathcal{D}_{x'}^{\alpha'} W^s|_\gamma, \tag{12.12}$$

where $B_{1\alpha s}$ are the $J_0 \times (N(s) - J_0)$-matrices with infinitely differentiable elements in $\overline{\omega_s}$. In (12.12) we sum over $|\alpha| \le 2m$, $2m-1 \le \alpha_n$.

In the same way, since $\det B_p(x) \ne 0$, we obtain by induction

$$B_p(x) Y^j(x) = F^j(x) \qquad (x \in \gamma), \tag{12.13}$$

where

$$Y^j = \mathcal{D}_n^{m+j}(V^p(x) - V^q(x))|_\gamma, \tag{12.14}$$

$$F^j = \sum_s \sum_\alpha B_{j\alpha s}(x) \mathcal{D}_n^{\alpha_n - m+j} \mathcal{D}_{x'}^{\alpha'} W^s(x)|_\gamma, \tag{12.15}$$

$B_{j\alpha s}(x)$ are the $J_0 \times (N(s) - J_0)$-matrices with infinitely differentiable elements in $\overline{\omega_s}$. Here, we sum over α such that $|\alpha| \le 2m$, $2m - j \le \alpha_n$ $(j = 0, \dots, m+k-1)$. We note that $\alpha_i \ge 0$ $(i = 1, \dots, n-1)$ and α_n can be negative.

Denote by $B_{pl}(x)$ the $J_0 \times (J_0 - 1)$-matrix obtained from $B_p(x)$ by deleting the lth column, where $l = l_0$.

Theorem 12.1. *Let $Q \subset \mathbb{R}^n$ be a bounded domain with boundary $\partial Q \in C^\infty$ satisfying the condition* 7.1 *or a cylinder $(0,d) \times G$, where $G \subset \mathbb{R}^{n-1}$ is a bounded domain (with boundary $\partial G \in C^\infty$ if $n \ge 3$). Let the equation* (9.1) *be strongly elliptic in $\overline{Q}$.*

Then, for a given l $(1 \le l \le J_0)$, a generalized solution $u(x)$ of the boundary value problem (9.1), (9.2) *belongs to $W^{k+2m}(S_a(y^l))$ for all $f_0 \in W^k(Q)$ iff for every $x \in \gamma$ each column of the matrices $B_{j\alpha s}(x)$ is a linear combination of the columns of $B_{pl}(x)$ $(j = 0, \dots, m+k-1,\ |\alpha| \le 2m,\ 2m - j \le \alpha_n,\ s = p, q)$.*

Proof. 1. Sufficiency. By Theorem 11.3, the solution $u(x)$ of the boundary value problem (9.1), (9.2) belongs to $W^{k+2m}(S_a(y^l))$ if and only if

$$Y_l^j(x) = 0 \qquad (x \in \gamma) \tag{12.16}$$

for all $j = 0, 1, \dots, m+k-1$. By what has been proved above, the solution of the boundary value problem (9.1), (9.2) satisfies the equations (12.13). Fix j. Since $\det B_p(x) \ne 0$ for all $x \in \gamma$, there exists a unique solution $Y^j(x)$ of the system (12.13). Suppose that every column of the matrices $B_{j\alpha s}(x)$ is a linear combination of the columns of $B_{pl}(x)$ for all $x \in \gamma$. Then, the matrix of the system (12.13), (12.16) and the extended matrix have the same rank J_0. Hence, a solution $Y^j(x)$ of the system (12.13) also satisfies the equation (12.16). Since j is arbitrary, $u \in W^{k+2m}(S_a(y^l))$.

2. Necessity. Suppose that for $x = g \in \gamma$, $j = c$, $\alpha = \beta$, $s = p$ the rth column $B^r_{j\alpha s}(x)$ of the matrix $B_{j\alpha s}(x)$ is not a linear combination of the columns of $B_{pl}(x)$. We shall prove that there exists $u \in \mathcal{D}(\mathcal{A}_R)$ such that $\mathcal{A}_R u \in W^k(Q)$, while $u \notin W^{k+2m}(S_a(y^l))$.

By virtue of the continuity of the matrices $B_{pl}(x)$, $B_{c\beta p}(x)$, there exists $\varepsilon > 0$ such that the column $B^r_{c\beta p}(x)$ is not a linear combination of the columns of matrix $B_{pl}(x)$ for all $x \in S_{3\varepsilon}(g) \cap \gamma \subset S_a(0)$. Denote

$$\begin{aligned} W^p_r(x) &= i^{|\beta|-m+c}(x_1-g_1)^{\beta_1}\cdots(x_{n-1}-g_{n-1})^{\beta_{n-1}}x_n^{\beta_n-m+c}\xi(x) \\ &\qquad (x \in \omega_p), \\ W^p_t(x) &= 0 \qquad (t \neq r,\ x \in \omega_p), \\ W^q(x) &= 0 \qquad (x \in \omega_q), \end{aligned}$$

where $\xi(x) = 1$ for $x \in S_\varepsilon(g)$, $\xi(x) = 0$ for $x \notin S_{2\varepsilon}(g)$, $\xi \in \dot{C}^\infty(\mathbb{R}^n)$. Therefore, since $\beta_n - m + c \geq m$ (see (12.15)), we have $\mathcal{D}^\kappa_n W^p(x)|_{x_n=0} = 0$ for $\kappa = 0, \ldots, m-1$.

Consider the system of equations (12.13) for each $j = 0, \ldots, m+k-1$. There is a unique solution of this system $Y^j(x) \in \dot{C}^{\infty, J_0}(\gamma)$. Evidently there exists a vector-valued function $\mathcal{Z} \in \dot{C}^{k+2m, J_0}(S_a(0))$ such that

$$\begin{aligned} \mathcal{D}^\kappa_n \mathcal{Z}(x)|_{x_n=0} &= 0 && (x \in \gamma,\ \kappa = 0, \ldots, m-1), \\ \mathcal{D}^{m+j}_n \mathcal{Z}(x)|_{x_n=0} &= Y^j(x) && (x \in \gamma,\ j = 0, \ldots, m+k-1). \end{aligned}$$

Let

$$\begin{aligned} u(x) &= U^1_t(x - h_{pt}) && \text{for } x \in Q_{pt} \cap S_a(y^t) \quad (t = 1, \ldots, N(p)), \\ u(x) &= 0 && \text{for } x \in Q \setminus \Big\{\bigcup_t (Q_{pt} \cap S_a(y^t))\Big\}, \end{aligned}$$

where $U^1 = (\mathcal{Z}_1, \ldots, \mathcal{Z}_{J_0}, W^p_1, \ldots, W^p_{N(p)-J_0})$. By virtue of (12.13), $u \in \mathcal{D}(\mathcal{A}_R)$ and $\mathcal{A}_R u \in W^k(Q)$.

Now we prove that $u \notin W^{k+2m}(S_a(y^l))$. By the definition of the functions $W^1(x), W^2(x)$, the system (12.13) for $j = c$ will have the form

$$B_p(x) Y^c(x) = \beta_1! \cdots \beta_{n-1}!(\beta_n - m + c)! B^r_{c\beta p}(x) + O(|x - g|) \qquad (x \in \gamma \cap S_\varepsilon(g)). \tag{12.17}$$

If $u \in W^{k+2m}(S_a(y^l))$, then

$$Y^c(x) = 0 \qquad (x \in \gamma \cap S_\varepsilon(g)). \tag{12.18}$$

By assumption, $B^r_{c\beta p}(x)$ is not a linear combination of the columns of $B_{pl}(x)$ for $|x - g| < \varepsilon$. Therefore, for sufficiently small $x - g$ $(x \in \gamma)$, the matrix and extended matrix of the system (12.17), (12.18) have ranks J_0 and $J_0 + 1$ respectively. Thus the function $u(x)$ does not satisfy the equation (12.18). □

Remark 12.1. If $J_0 = 1$, then in Theorem 12.1 necessary and sufficient conditions of smoothness of the solution in the ball $S_a(y^l)$ can be formulated in the following way: "The $1 \times (N(s)-1)$-matrices $B_{j\alpha s}(x)$ ($j = 0, \dots, m+k-1$, $|\alpha| \le 2m$, $2m - j \le \alpha_n$, $s = p, q$) are trivial for each $x \in \gamma$".

If $m = 1$, $k = 0$, then the conditions of Theorem 12.1 can be easily verified. In this case Theorem 12.1 will have the following form:

Theorem 12.2. *Let $Q \subset \mathbb{R}^n$ be a bounded domain with boundary $\partial Q \in C^\infty$ satisfying the condition 7.1 or a cylinder $(0,d) \times G$, where $G \subset \mathbb{R}^{n-1}$ is a bounded domain (with boundary $\partial G \in C^\infty$ if $n \ge 3$). Let the equation (9.1) be strongly elliptic in $\overline{Q}$.*

Then for a given l ($1 \le l \le J_0$), a generalized solution $u(x)$ of the boundary value problem (9.1), (9.2) belongs to $W^2(S_a(y^l))$ for all $f_0 \in L_2(Q)$ iff for every $x \in \gamma$ each column of the matrices $R_{\alpha s1}(x)$ ($\alpha = (0, \dots, 0, 2)$, $s = p, q$) is a linear combination of the columns of $B_{pl}(x)$.

Remark 12.2. If $J_0 = 1$, then in Theorem 12.2 necessary and sufficient conditions of smoothness of the solutions in the ball $S_a(y^l)$ can be formulated as the following: "The $1 \times (N(s)-1)$-matrices $R_{\alpha s1}(x)$ ($\alpha = (0, \dots, 0, 2)$, $s = p.q$) are trivial for each $x \in \gamma$".

Examples

We shall now give an example in which the smoothness of generalized solutions is disturbed on a boundary of neighboring subdomains.

Example 12.1. Consider the boundary value problem (10.5), (10.6). By Theorem 11.2, a generalized solution of the boundary value problem (10.5), (10.6) $u \in W^2(Q_{1l} \setminus \overline{\mathcal{K}^\varepsilon})$ for every $\varepsilon > 0$, where $Q_{1l} = (l-1, l) \times (0,1)$, $l = 1, 2$, $\mathcal{K} = \{(i,j) : i = 0, 1, 2,\ j = 1, 2\}$. Later we shall prove that $u \in W^2(Q_{1l})$ ($l = 1, 2$) and $R_Q u \in W^2(Q)$ (see Theorem 23.2).

Let us show that smoothness of a solution can be violated at the line $x_1 = 1$. We introduce a second class of subdomains $Q_{21} = Q_{12}$, $Q_{22} = Q_{11}$, reindexing subdomains of the first class (see Remark 11.1). Then $\{x : x_1 = 1,\ 0 < x_2 < 1\} \subset \partial Q_{11} \cap \partial Q_{21}$. It easy to see that

$$R_1 = R_2 = \begin{pmatrix} 2 & 1 \\ 1 & 2 \end{pmatrix}.$$

Denote by R_{11} (or R_{21}) the matrix obtained from R_1 (or R_2) by deleting the first column and the last row. Clearly, $R_{11} = R_{21} = (1) \ne (0)$. Hence, by virtue of Remark 12.2, there exists $f_0 \in L_2(Q)$ such that $u \notin W^2(Q)$. Since there is only one operator R_Q in equation (10.5), we omit the index α in the matrices $R_{\alpha s}$, $R_{\alpha s1}$.

Let us now construct a nonsmooth solution of the boundary value problem (10.5), (10.6) in an explicit form. For this it is sufficient to construct a function $u \in \mathring{W}^1(Q)$ such that $u \in W^2(Q_{1l})$ and

$$R_Q u_{x_1}|_{x_1=1-0} = R_Q u_{x_1}|_{x_1=1+0}, \tag{12.19}$$

$$u_{x_1}|_{x_1=1-0} \neq u_{x_1}|_{x_1=1+0} \tag{12.20}$$

(see (12.6), (12.16)). The condition (12.19) can be written in the form

$$2u_{x_1}|_{x_1=1-0} + u_{x_1}|_{x_1=2-0} = 2u_{x_1}|_{x_1=1+0} + u_{x_1}|_{x_1=0+0}. \tag{12.21}$$

We introduce a function $\xi \in \dot{C}^\infty(\mathbb{R}^2)$, $\xi(x) = 1$ for $|x| < 1/4$, $\xi(x) = 0$ for $|x| > 1/3$. Consider the function $u(x)$ defined by

$$u(x) = 2\xi(x_1, x_2 - 1/2)x_1 + \xi(x_1 - 1, x_2 - 1/2)(x_1 - 1) \quad \text{for } x \in Q_{11},$$
$$u(x) = 0 \quad \text{for } x \in Q_{12}.$$

Evidently, $u \in \mathring{W}^1(Q)$, $u \in W^2(Q_{1l})$ and the conditions (12.20), (12.21) are satisfied.

Now we shall demonstrate an example in which the solutions have a corresponding smoothness in the whole domain Q.

Example 12.2. Consider the equation

$$-(R_{1Q}u)_{x_1x_1} - (R_{2Q}u)_{x_2x_2} = f_0(x) \qquad (x \in Q) \tag{12.22}$$

with boundary conditions

$$u|_{\partial Q} = 0, \tag{12.23}$$

where $Q = (0,2)\times(0,2)$, $f_0 \in L_2(Q)$, $R_1u(x) = 2u(x)+u(x_1, x_2+1)+u(x_1, x_2-1)$, $R_2u(x) = 2u(x) + u(x_1+1, x_2) + u(x_1-1, x_2)$.

It is easy to see that the self-adjoint operators R_{iQ} are positive definite (see Example 8.3). Hence, the equation (12.22) is strongly elliptic (see Example 9.4). From this and from Theorem 10.1 it follows that there exists a unique generalized solution of the boundary value problem (12.22), (12.23). It is easy to demonstrate that the assumptions of Theorem 12.2 are satisfied for all neighboring subdomains Q_{1l}, Q_{1k}. Thus $u \in W^2(Q \setminus \overline{\mathcal{K}^\varepsilon})$ for every $\varepsilon > 0$, where $Q_{11} = (0,1) \times (0,1)$, $Q_{12} = (1,2) \times (0,1)$, $Q_{13} = (0,1) \times (1,2)$, $Q_{14} = (1,2) \times (1,2)$, $\mathcal{K} = \{(i,j) : i,j = 0,1,2\}$.

For example, let us show that $u_{x_2}|_{x_2=1-0} = u_{x_2}|_{x_2=1+0}$. We introduce a second class of subdomains reindexing subdomains of the first class (see Remark 11.1):

$$Q_{21} = Q_{13}, \quad Q_{22} = Q_{14}, \quad Q_{23} = Q_{11}, \quad Q_{24} = Q_{12}.$$

The matrices R_{21}, R_{22} will have the form

$$R_{21} = R_{22} = \begin{pmatrix} 2 & 1 & 0 & 0 \\ 1 & 2 & 0 & 0 \\ 0 & 0 & 2 & 1 \\ 0 & 0 & 1 & 2 \end{pmatrix}.$$

The matrix R_{210} is obtained from the matrix R_{21} by deleting the last two columns and lines. The matrices R_{211}, R_{221} are obtained from the matrix R_{21} by deleting the first two columns and the last two lines. Hence,

$$R_{210} = \begin{pmatrix} 2 & 1 \\ 1 & 2 \end{pmatrix}, \qquad R_{211} = R_{221} = \begin{pmatrix} 0 & 0 \\ 0 & 0 \end{pmatrix},$$

$$B_{11} = \begin{pmatrix} 1 \\ 2 \end{pmatrix}, \qquad B_{12} = \begin{pmatrix} 2 \\ 1 \end{pmatrix}.$$

Evidently, the columns of the matrices R_{211}, R_{221} are equal to $0 \cdot B_{11} = 0 \cdot B_{12}$. Thus, by virtue of Theorem 12.2,

$$u_{x_2}|_{x_2=1-0} = u_{x_2}|_{x_2=1+0}.$$

Now we shall prove that $u \in W^2(Q)$. We introduce the bounded operators $\mathcal{A}_0, \mathcal{A}_1 \colon W_0^2(Q) = \mathring{W}^1(Q) \cap W^2(Q) \to L_2(Q)$ by the formulas

$$\begin{aligned}
\mathcal{A}_0 u &= -2\Delta u, \\
\mathcal{A}_1 u(x) &= -u_{x_1x_1}(x_1, x_2+1) - u_{x_2x_2}(x_1+1, x_2) \qquad (x \in Q_{11}), \\
\mathcal{A}_1 u(x) &= -u_{x_1x_1}(x_1, x_2+1) - u_{x_2x_2}(x_1-1, x_2) \qquad (x \in Q_{12}), \\
\mathcal{A}_1 u(x) &= -u_{x_1x_1}(x_1, x_2-1) - u_{x_2x_2}(x_1+1, x_2) \qquad (x \in Q_{13}), \\
\mathcal{A}_1 u(x) &= -u_{x_1x_1}(x_1, x_2-1) - u_{x_2x_2}(x_1-1, x_2) \qquad (x \in Q_{14}).
\end{aligned}$$

It is well known that for every $f_0 \in L_2(Q)$ the equation $\mathcal{A}_0 u = f_0$ has a unique solution

$$u(x) = \sum_{i,j=1}^{\infty} \frac{2(f_0, u_{ij})_{L_2(Q)} u_{ij}(x)}{\pi^2(i^2 + j^2)}, \tag{12.24}$$

where

$$u_{ij}(x) = \sin(\pi i x_1/2) \cdot \sin(\pi j x_2/2),$$

while the series (12.24) converges in $W_0^2(Q)$ and

$$\|u\|^2_{W^2(Q)} \le c\|f_0\|^2_{L_2(Q)} \le c_1(\|u_{x_1x_1}\|^2_{L_2(Q)} + \|u_{x_2x_2}\|^2_{L_2(Q)}).$$

Thus, in the subspace $W_0^2(Q)$ of the space $W^2(Q)$ we can introduce the equivalent norm

$$\|u\|'_{W_0^2(Q)} = \{\|u_{x_1x_1}\|^2_{L_2(Q)} + \|u_{x_2x_2}\|^2_{L_2(Q)}\}^{1/2}.$$

By (12.24), $2\|u\|'_{W_0^2(Q)} \le \|f_0\|_{L_2(Q)}$. Hence, $\|\mathcal{A}_0^{-1}\| \le 1/2$. On the other hand, introducing new variables, we have

$$\begin{aligned}
\|\mathcal{A}_1 u\|^2_{L_2(Q)} &\le 2\Big\{ \int_{Q_{11}} (|u_{x_1x_1}(x_1, x_2+1)|^2 + |u_{x_2x_2}(x_1+1, x_2)|^2)dx + \cdots \\
&\quad + \int_{Q_{14}} (|u_{x_1x_1}(x_1, x_2-1)|^2 + |u_{x_2x_2}(x_1-1, x_2)|^2)dx \Big\} \\
&= 2(\|u\|'_{W_0^2(Q)})^2.
\end{aligned}$$

From this it follows that $\|\mathcal{A}_1\| \le \sqrt{2}$. Hence, the operator

$$\mathcal{A}_0 + \mathcal{A}_1 = \mathcal{A}_0(I + \mathcal{A}_0^{-1}\mathcal{A}_1)$$

has the bounded inverse operator $(\mathcal{A}_0 + \mathcal{A}_1)^{-1} \colon L_2(Q) \to \overset{\circ}{W}{}^2(Q)$.

Thus, for every $f_0 \in L_2(Q)$, there exists a unique solution of the boundary value problem (12.22), (12.23) as well as of the equation $(\mathcal{A}_0 + \mathcal{A}_1)u = f_0$. Now, it is sufficient to prove that every solution of the equation $(\mathcal{A}_0 + \mathcal{A}_1)u = f_0$ is a solution of the boundary value problem (12.22), (12.23). If the function $u(x)$ belongs to $\overset{\circ}{W}{}^2(Q)$ and satisfies the equation $(\mathcal{A}_0 + \mathcal{A}_1)u = f_0$, then

$$(\mathcal{A}_0 + \mathcal{A}_1)u = -R_{1Q}u_{x_1x_1} - R_{2Q}u_{x_2x_2} = f_0.$$

Since the equation (12.22) contain shifts and differentiation over the different variables, we have $R_{iQ}u_{x_ix_i} = (R_{iQ}u)_{x_ix_i}$ $(i = 1, 2)$. Therefore, $u(x)$ is a generalized solution of the boundary value problem (12.22), (12.23).

Remark 12.3. Generally speaking,

$$\mathcal{D}^\alpha R_{\alpha\beta Q}\mathcal{D}^\beta u \ne R_{\alpha\beta Q}\mathcal{D}^{\alpha+\beta}u \qquad (u \in \mathcal{D}(\mathcal{A}_R)).$$

Hence, the equation (17.1) is not equivalent to the equation

$$\sum_{|\alpha|,|\beta|\le m} R_{\alpha\beta Q}\mathcal{D}^{\alpha+\beta}u = f_0(x) \qquad (x \in Q). \tag{12.25}$$

Roughly speaking, the smoothness of the solution in the whole domain Q is the result of shifts of the arguments and differentiation over different variables.

However, sometimes there are disturbances in the smoothness when shifts of the arguments and differentiation are acting over different variables.

Example 12.3. Consider the equation

$$-u_{x_1x_1} - (R_{2Q}u)_{x_2x_2} = f_0(x) \qquad (x \in Q) \tag{12.26}$$

with boundary conditions

$$u|_{\partial Q} = 0, \tag{12.27}$$

where

$$Q = \{(0,1) \times (0,2)\} \cup \{(0,2) \times (0,1)\},$$
$$R_2u(x) = 2u(x) + au(x_1 + 1, x_2) + au(x_1 - 1, x_2), \qquad a \in \mathbb{R},\ |a| < 2.$$

Evidently the operator R_{2Q} is positive definite. Hence, by virtue of Example 9.4 and Theorem 10.1, there exists a unique generalized solution of the problem (12.26), (12.27). Denote $Q_{11} = (0,1) \times (0,1)$, $Q_{12} = (1,2) \times (0,1)$, $Q_{21} = (0,1) \times (1,2)$. We shall show that the smoothness of generalized solutions can be broken at the line $x_2 = 1$. Evidently $J_0 = 1$ and $\{x : 0 < x_1 < 1,\ x_2 = 1\} \subset \partial Q_{11} \cap \partial Q_{21}$. The matrices R_{21}, R_{22} have the following form

$$R_{21} = \begin{pmatrix} 1 & a \\ a & 1 \end{pmatrix}, \qquad R_{22} = (1).$$

Hence, $R_{211} = (a)$. By virtue of Remark 12.2, if $a \neq 0$, there exists $f_0 \in L_2(Q)$ such that $u \notin W^2_{\text{loc}}(Q)$.

Remark 12.4. All results of this chapter are valid if a domain Q satisfies the following conditions:

12.1. *Let $Q \subset \mathbb{R}^n$ be a bounded domain with boundary $\partial Q = \bigcup_i \overline{M_i}$ $(i = 1, \dots, N_0)$, where M_i are the $(n-1)$-dimensional manifolds of class C^∞, which are connected, open in the topology of ∂Q. In a neighborhood of each point $g \in \partial Q \setminus \cup_i M_i$ the domain Q is diffeomorphic to an n-dimensional dihedral angle $\Theta_b = \Theta_b(g) = \{x = (y, z) \in \mathbb{R}^n : |\varphi| < b,\ z \in \mathbb{R}^{n-2}\}$ if $n \geq 3$ and to a plane angle $\theta_b = \theta_b(g) = \{y \in \mathbb{R}^2 : |\varphi| < b\}$ if $n = 2$, where r, φ are the polar coordinates of y, $0 < b < \pi$.*

12.2. *$\partial Q \setminus \bigcup_i M_i \subset \mathcal{K}$ and $\mu_{n-1}(\mathcal{K} \cap \partial Q) = 0$.*

13 Elliptic Differential Equations with Nonlocal Conditions on Shifts of Boundary

Formulation of the Problem

Let us assume that a domain Q satisfies conditions 7.1, 8.1. Using the notation of Section 7, we consider the following differential equation

$$\sum_{|\alpha|,|\beta| \leq m} \mathcal{D}^\alpha a_{\alpha\beta}(x) \mathcal{D}^\beta w(x) = f_0(x) \qquad (x \in Q) \tag{13.1}$$

with nonlocal conditions

$$\left.\begin{aligned} \mathcal{D}_\nu^{\mu-1} w|_{\Gamma_{rl}} &= \sum_{j=1}^{J_0} \gamma_{lj}^r \mathcal{D}_\nu^{\mu-1} w|_{\Gamma_{rj}} && (r \in B,\ l = J_0 + 1, \dots, J), \\ \mathcal{D}_\nu^{\mu-1} w|_{\Gamma_{rl}} &= 0 && (r \notin B,\ l = 1, \dots, J). \end{aligned}\right\} \tag{13.2}$$

Here, $\mu = 1, \dots, m$, and the equation (13.1) is strongly elliptic in $\overline{Q}$, i.e.,

$$\sum_{|\alpha|,|\beta| = m} a_{\alpha\beta}(x) \xi^{\alpha+\beta} > 0$$

for all $x \in \overline{Q}$, $0 \neq \xi \in \mathbb{R}^n$, $a_{\alpha\beta} = a_{\beta\alpha} \in C^\infty(\mathbb{R}^n)$ are real-valued M-periodic in $\overline{Q}$ functions, $J_0 = J_0(r)$, $J = J(r)$; γ_{lj}^r are complex numbers, $B = \{r : J_0 > 0\}$ (see (8.17)), $f_0 \in L_2(Q)$.

We introduce the unbounded operator $\mathcal{A}_\gamma : L_2(Q) \to L_2(Q)$ with domain $\mathcal{D}(\mathcal{A}_\gamma) = \{u \in W_\gamma^m(Q) : \mathcal{A}_\gamma w \in L_2(Q)\}$ acting in the space of distributions $\mathcal{D}'(Q)$ by the formula

$$\mathcal{A}_\gamma w = \sum_{|\alpha|,|\beta| \le m} \mathcal{D}^\alpha a_{\alpha\beta}(x)\mathcal{D}^\beta w,$$

where $W_\gamma^m(Q)$ is the subspace of functions in $W^m(Q)$ satisfying the conditions (13.2).

Definition 13.1. A function w is called a *generalized solution* of the boundary value problem (13.1), (13.2) if $w \in \mathcal{D}(\mathcal{A}_\gamma)$ and

$$\mathcal{A}_\gamma w = f_0. \tag{13.3}$$

We can give an equivalent definition of a generalized solution.

Definition 13.2. Let $f_0 \in L_2(Q)$. A function w is called a *generalized solution* of the boundary value problem (13.1), (13.2) if $w \in W_\gamma^m(Q)$ and

$$\sum_{|\alpha|,|\beta| \le m} (a_{\alpha\beta}\mathcal{D}^\beta w, \mathcal{D}^\alpha v)_{L_2(Q)} = (f_0, v)_{L_2(Q)} \tag{13.4}$$

for all $v \in \mathring{W}^m(Q)$.

Theorem 8.1 enables us to establish a relation between the problem (13.1), (13.2) and the first boundary value problem for the strongly elliptic differential-difference equation investigated in Section 10. Now we shall formulate conditions for the existence of a difference operator corresponding to the boundary value problem (13.1), (13.2).

Let us introduce the set $\mathcal{M}_0 = \{h \in M : |h| \le \operatorname{diam} Q\}$. For each $s = 1, 2, \dots,$ we set up a relation between the set $\Lambda = \{a_h \in \mathbb{C} : h \in \mathcal{M}_0\}$ and the corresponding matrix $A_s(x)$ $(x \in \overline{Q_{s1}})$ of order $I \times I$ $(I = I(s,x))$ with elements

$$a_{ij}^s(x) = a_h \quad \text{if } x^j - x^i = h, \tag{13.5}$$

where $\{x^i\}$ $(i = 1, \dots, I(s,x))$ is the set of points $x + h \in \overline{Q}$ $(h \in M)$ renumbered so that $x^1 = x$, $x^i = x + h_{si}$ $(i = 1, \dots, N(s))$; h_{si} is determined by the condition $Q_{si} = Q_{s1} + h_{si}$. We introduce also the matrices R_s of order $N(s) \times N(s)$ obtained from the matrix A_s by deleting the last $I - N$ columns and rows.

By virtue of Lemma 7.7, for every $r = 1, 2, \dots$ there is a unique $s = s(r)$ such that $N(s) = J(r)$ and, after some renumbering, $\Gamma_{rl} \subset \partial Q_{sl}$ $(l = 1, \dots, N(s))$. Denote by $R_{s(r)}$ the matrix obtained from R_s $(s = s(r))$ by the appropriate renumbering of columns and rows. Let e_j^r $(j = 1, \dots, J(r))$ be the jth row of the $J \times J_0$-dimensional matrix obtained from the matrix $R_{s(r)}$ by deleting the last $J - J_0$ columns.

Definition 13.3. We say that for the boundary conditions (13.2) there are matrices A_s satisfying the condition 13.1 if the following condition is fulfilled:

13.1. *There is a set* Λ *such that, for all* $s = 1, 2, \dots,$ $x \in \overline{Q}_{s1}$*, the matrices* $A_s(x)+A_s^*(x)$ *are positive definite, and for every* $r \in B$ *and* $s = s(r)$ *the following relations are satisfied:*

$$e_l^r = \sum_{j=1}^{J_0} \gamma_{lj}^r e_j^r \qquad (l = J_0+1, \dots, J). \tag{13.6}$$

We note that condition 13.1 is purely algebraic. Its verification reduces to the solution of the system of homogeneous linear algebraic equations (13.6), for the unknowns a_h, and the subsequent verification that the matrices $A_s(x) + A_s^*(x)$ ($s = 1, 2, \dots,$ $x \in \overline{Q}_{s1}$), constructed from the solution $\{a_h\}$ corresponding to (13.5), are positive definite.

Discreteness of Spectrum

In this subsection we assume that there are matrices A_s satisfying condition 13.1.

Let us introduce the operator

$$R_Q = P_Q R I_Q, \tag{13.7}$$

where

$$Ru(x) = \sum_{h \in \mathcal{M}_0} a_h u(x+h), \tag{13.8}$$

and the coefficients a_h satisfy Definition 13.3.

Denote $R_Q^c = (R_Q + R_Q^*)/2$, $R_Q^k = (R_Q - R_Q^*)/2$.

Lemma 13.1. *There are numbers* $0 < \varepsilon < \pi$ *and* $q \geq 0$ *such that, for every* $\mu \in \Omega_{\varepsilon,q} = \{\mu \in \mathbb{C} : |\arg \mu| \geq \varepsilon,\ |\mu| \geq q\}$*, the space* $\mathring{W}^m(Q)$ *has the equivalent inner product*

$$(u,v)'_{\mathring{W}^m(Q)} = \sum_{|\alpha|,|\beta| \leq m} (a_{\alpha\beta} \mathcal{D}^\beta R_Q^c u, \mathcal{D}^\alpha v)_{L_2(Q)} - ((\operatorname{Re}\mu R_Q^c + i \operatorname{Im}\mu R_Q^k)u, u)_{L_2(Q)}. \tag{13.9}$$

Proof. By virtue of Lemmas 8.10, 8.14, the right-hand side of (13.9) defines a sesquilinear symmetric form $b[u,v]$. From the Schwarz inequality and Lemma 8.2 it follows that

$$b[u,u] \leq k_1 \|u\|^2_{W^m(Q)} \qquad (u \in \mathring{W}^m(Q)). \tag{13.10}$$

We prove that, for some $0 < \varepsilon < \pi$ and $q \geq 0$, the inequality

$$k_2 \|u\|^2_{W^m(Q)} \leq b[u,u] \qquad (u \in \mathring{W}^m(Q)) \tag{13.11}$$

holds for $\mu \in \Omega_{\varepsilon,q}$, where $k_2 = k_2(\mu) > 0$.

Since the equation (13.1) is strongly elliptic in $\overline{Q}$, the matrices $A_s + A_s^*$ are positive definite, and the coefficients $a_{\alpha\beta}(x)$ are M-periodic, we conclude that the conditions of Theorem 9.2 are satisfied. Hence,

$$\sum_{|\alpha|,|\beta|\le m} (a_{\alpha\beta}\mathcal{D}^\beta R_Q^c u, \mathcal{D}^\alpha u)_{L_2(Q)}$$
$$\ge c_1\|u\|^2_{W^m(Q)} - c_2\|u\|^2_{L_2(Q)} \qquad (u \in \mathring{W}^m(Q)), \quad (13.12)$$

where $c_1 > 0$ and $c_2 \ge 0$.

Since the matrices $A_s + A_s^*$ are positive definite, we conclude from Lemma 8.8 and Remark 9.1 that

$$(R_Q^c u, u)_{L_2(Q)} \ge k_3\|u\|^2_{L_2(Q)}. \qquad (13.13)$$

The inequality (13.11) follows from (13.12), (13.13). In addition, we choose ε and q so that $\pi/2 < \varepsilon < \pi$, $c_2/k_3 \cos(\pi-\varepsilon) \le q$ if $\|R_Q^k\| = 0$, and $0 < \pi - \varepsilon < \arctan(k_3/\|R_Q^k\|)$, $c_2/\{\|R_Q^k\|\cos(\pi-\varepsilon)\sin(\arctan(k_3/\|R_Q^k\|) - (\pi-\varepsilon))\} \le q$ if $\|R_Q^k\| \ne 0$. ☐

Lemma 13.2. *Let $0 < \varepsilon < \pi$ and $q \ge 0$ be as in Lemma 13.1.*

Then, for every $\mu \in \Omega_{\varepsilon,q}$, there is a linear bounded operator $K: \mathring{W}^m(Q) \to \mathring{W}^m(Q)$ such that $K^* = -K$ and

$$\sum_{|\alpha|,|\beta|\le m} (a_{\alpha\beta}\mathcal{D}^\beta R_Q^k u, \mathcal{D}^\alpha v)_{L_2(Q)} - ((\operatorname{Re}\mu R_Q^k + i \operatorname{Im}\mu R_Q^c)u, v)_{L_2(Q)}$$
$$= (Ku, v)'_{\mathring{W}^m(Q)} \quad (13.14)$$

for all $u, v \in \mathring{W}^m(Q)$.

Proof. By virtue of Lemmas 8.10, 8.14, the left hand side of (13.14) defines a sesquilinear skew-symmetric form $g[u,v]$ in $\mathring{W}^m(Q)$. Lemmas 8.2, 13.1 imply that for all $u, v \in \mathring{W}^m(Q)$

$$|g[u,v]| \le k_1\|u\|'_{\mathring{W}^m(Q)}\|v\|'_{\mathring{W}^m(Q)}.$$

Hence, by the Riesz theorem concerning a general form of a linear functional in a Hilbert space, for every $u \in \mathring{W}^m(Q)$ there exists a unique element $w(u) \in \mathring{W}^m(Q)$ such that

$$\|w(u)\|'_{\mathring{W}^m(Q)} \le k_1\|u\|'_{\mathring{W}^m(Q)}, \qquad (w(u), v)_{\mathring{W}^m(Q)} = g[u,v].$$

Thus, a correspondence $u \to w(u)$ defines a linear bounded operator $K: \mathring{W}^m(Q) \to \mathring{W}^m(Q)$. Since the sesquilinear form $g[u,v]$ is skew-symmetric, we have $K^* = -K$. ☐

Theorem 13.1. *Let the domain Q satisfy the conditions* 7.1, 8.1, *and let the equation* (13.1) *be strongly elliptic in $\overline{Q}$. Suppose that there exist matrices A_s satisfying condition* 13.1.

Then, the spectrum $\sigma(\mathcal{A}_\gamma)$ of the operator $\mathcal{A}_\gamma: L_2(Q) \to L_2(Q)$ is discrete and $\sigma(\mathcal{A}_\gamma) \subset \mathbb{C} \setminus \Omega_{\varepsilon,q}$, where $0 < \varepsilon < \pi$ and $0 \leq q$ are constants in Lemma 13.1. *If $\lambda \notin \sigma(\mathcal{A}_\gamma)$, then the resolvent $R(\lambda, \mathcal{A}_\gamma): L_2(Q) \to L_2(Q)$ is a compact operator.*

Proof. Let $\lambda \in \Omega_{\varepsilon,q}$. By virtue of Lemma 8.8, Theorem 8.1 and Remark 8.1, the operator R_Q, defined by (13.7), (13.8), maps $\mathring{W}^m(Q)$ onto $W^m_\gamma(Q)$ continuously and in a one-to-one manner. Hence, the equation

$$(\mathcal{A}_\gamma - \lambda I)w = f_0 \tag{13.15}$$

is equivalent to the equation

$$(\mathcal{A}_R - \lambda R_Q)u = f_0, \tag{13.16}$$

where $\mathcal{A}_R = \mathcal{A}_\gamma R_Q$. On the other hand, by virtue of Lemmas 13.1, 13.2, and the Riesz theorem concerning a general form of a linear functional in a Hilbert space, the equation (13.16) is equivalent to the integral identity

$$(u + Ku, v)'_{\mathring{W}^m(Q)} = (Af_0, v)'_{\mathring{W}^m(Q)} \qquad (v \in \mathring{W}^m(Q)), \tag{13.17}$$

where in (13.9), (13.14) we put $\mu = \lambda$, $A: L_2(Q) \to \mathring{W}^m(Q)$ is a linear bounded operator. From this we obtain that for $\lambda \in \Omega_{\varepsilon,q}$ the operator $\mathcal{A}_R - \lambda R_Q$ has a bounded inverse $(\mathcal{A}_R - \lambda R_Q)^{-1} = (I + K)^{-1}A: L_2(Q) \to \mathring{W}^m(Q)$. Hence, the operator $\mathcal{A}_\gamma - \lambda I$ has a bounded inverse $(\mathcal{A}_\gamma - \lambda I)^{-1} = R_Q(\mathcal{A}_R - \lambda R_Q)^{-1}: L_2(Q) \to W^m(Q)$. From this, compactness of the imbedding of $W^m(Q)$ into $L_2(Q)$ and Theorem A.8, Theorem 13.1 follows. □

By Theorem 22.2 concerning the elliptic equation (22.1) in a cylinder with nonlocal boundary conditions (22.2), for every $\varepsilon > 0$, there exists $q > 0$ such that $\sigma(\mathcal{A}_\gamma) \subset \mathbb{C} \setminus \Omega_{\varepsilon,q}$. In the case of the boundary value problem (13.1), (13.2), generally speaking, $0 < \varepsilon < \pi$ is not arbitrary. The following question is unanswered:

Problem 13.1. *Does Theorem* 13.1 *hold for $0 < \varepsilon \leq \pi/2$? Can we find, for every $\varepsilon > 0$, a number $q > 0$ such that $\sigma(\mathcal{A}_\gamma) \subset \mathbb{C} \setminus \Omega_{\varepsilon,q}$?*

Theorem 13.2. *Let the domain Q satisfy the conditions* 7.1, 8.1, *and let the equation* (13.1) *be strongly elliptic in $\overline{Q}$. Suppose that there exist matrices A_s satisfying condition* 13.1.

Then the operator $\mathcal{A}_\gamma: L_2(Q) \to L_2(Q)$ is Fredholm, and $\operatorname{ind} \mathcal{A}_\gamma = 0$.

The proof follows from the formula $\mathcal{A}_\gamma R_Q = \mathcal{A}_R$, strong ellipticity of the operator $\mathcal{A}_R$, Theorem 10.2 and Theorem A.1.

The Symmetric Case

In this subsection we assume that there are matrices A_s satisfying the conditions 13.1.

We note that after integration by parts of the expression $(\mathcal{A}_\gamma u, v)_{L_2(Q)}$, for functions $u, v \in C^\infty(\overline{Q})$ satisfying the nonlocal conditions (13.2), integrals of these functions and their derivatives arise on certain manifolds Γ_{rj} $(1 \leq j \leq J_0)$. Thus the operator $\mathcal{A}_\gamma$ is not self-adjoint. Below we establish sufficient conditions for the existence in $L_2(Q)$ of an equivalent inner product with respect to which $\mathcal{A}_\gamma$ is self-adjoint.

Lemma 13.3. *Let the operator $R_Q\colon L_2(Q) \to L_2(Q)$, defined by (13.7), (13.8), be self-adjoint and positive definite.*

Then there is a compact self-adjoint positive operator $T_R\colon \mathring{W}^m(Q) \to \mathring{W}^m(Q)$ such that

$$(R_Q u, v)_{L_2(Q)} = (T_R u, v)'_{\mathring{W}^m(Q)} \tag{13.18}$$

for all $u, v \in \mathring{W}^m(Q)$, where $(\cdot,\cdot)'_{\mathring{W}^m(Q)}$ is the equivalent inner product in $\mathring{W}^m(Q)$, defined by (13.9), in which $c_2 \geq 0$, $k_3 > 0$ are constants in (13.12), (13.13), $\mu \in \mathbb{R}$, $-\mu \geq c_2/k_3$.

The proof follows from Lemmas 8.2, 13.1, the Riesz theorem concerning the general form of a linear functional in a Hilbert space, and the compactness of the imbedding of $\mathring{W}^m(Q)$ into $L_2(Q)$.

Lemma 13.4. *Let the operator $R_Q\colon L_2(Q) \to L_2(Q)$, defined by (13.7), (13.8), be self-adjoint and positive definite.*

Then in the spaces $L_2(Q)$ and $W^m_\gamma(Q)$ we can introduce the equivalent inner products by the formulas

$$(u, v)'_{L_2(Q)} = (u, R_Q^{-1} v)_{L_2(Q)}, \tag{13.19}$$

$$(u, v)'_{W^m_\gamma(Q)} = \sum_{|\alpha|,|\beta| \leq m} (a_{\alpha\beta} \mathcal{D}^\beta u, \mathcal{D}^\alpha R_Q^{-1} v)_{L_2(Q)} - \mu (u, R_Q^{-1} v)_{L_2(Q)}, \tag{13.20}$$

where $c_2 \geq 0$, $k_3 > 0$ are constants in (13.12), (13.13), $\mu \in \mathbb{R}$, $-\mu \geq c_2/k_3$.

The proof follows from Lemmas 13.1, 8.10, 8.13 and Theorem 8.1.

Theorem 13.3. *Let the domain Q satisfy the conditions 7.1, 8.1, and let the equation (13.1) be strongly elliptic in $\overline{Q}$. Suppose that there are matrices A_s satisfying the condition 13.1 such that the matrices R_s are Hermitian for all $s = 1, 2, \ldots$.*

Then $\mathcal{A}_\gamma$ is self-adjoint in $L_2(Q)$ with inner product defined by (13.19). The spectrum $\sigma(\mathcal{A}_\gamma)$ consists of real isolated eigenvalues $\lambda_s > \mu$ of finite multiplicity. The eigenfunctions u_s of the operator $\mathcal{A}_\gamma$ form an orthonormal basis in $L_2(Q)$, and the functions $u_s/\sqrt{\lambda_s - \mu}$ form an orthonormal basis in $W^m_\gamma(Q)$ with inner product defined by the formula (13.20), where $\mu = -c_2/k_3$.

Proof. 1. We introduce the differential-difference operator $\mathcal{A}_R = \mathcal{A}_\gamma R_Q\colon L_2(Q) \to L_2(Q)$ with domain $\mathcal{D}(\mathcal{A}_R) = \{u \in \mathring{W}^m(Q) : \mathcal{A}_R u \in L_2(Q)\}$. By Theorem 9.2, the operator $\mathcal{A}_R$ is strongly elliptic. From Remark 9.1 and Lemmas 8.8, 8.12, it follows that the operator R_Q is self-adjoint positive definite. Hence, by virtue of Lemmas 8.10, 8.14, $(\mathcal{A}_R u, v)_{L_2(Q)} = (u, \mathcal{A}_R v)_{L_2(Q)}$ $(u, v \in \dot{C}^\infty(Q))$. Therefore, by Theorem 10.3, the operator $\mathcal{A}_R$ is self-adjoint. Thus, using the equivalent inner product in $L_2(Q)$ defined by (13.19), by virtue of Theorem 8.1, we have

$$
\begin{aligned}
(\mathcal{A}_\gamma u, v)'_{L_2(Q)} &= (\mathcal{A}_R \hat{u}, \hat{v})_{L_2(Q)} \\
&= (\hat{u}, \mathcal{A}_R \hat{v})_{L_2(Q)} = (u, \mathcal{A}_\gamma v)'_{L_2(Q)} \qquad \text{for all } u, v \in \mathcal{D}(\mathcal{A}_\gamma),
\end{aligned}
$$

where $\hat{u} = R_Q^{-1} u$, $\hat{v} = R_Q^{-1} v$. Hence, $\mathcal{A}_\gamma \subset \mathcal{A}_\gamma^*$.

We next prove that $\mathcal{A}_\gamma^* \subset \mathcal{A}_\gamma$. Let $v \in \mathcal{D}(\mathcal{A}_\gamma^*)$. Then

$$
(\mathcal{A}_\gamma u, v)'_{L_2(Q)} = (u, \mathcal{A}_\gamma^* v)'_{L_2(Q)}
$$

for all $u \in \mathcal{D}(\mathcal{A}_\gamma)$. On the other hand, by virtue of Theorem 8.1,

$$
(\mathcal{A}_\gamma u, v)'_{L_2(Q)} = (\mathcal{A}_R \hat{u}, \hat{v})_{L_2(Q)}, \qquad (u, \mathcal{A}_\gamma^* v)'_{L_2(Q)} = (\hat{u}, \mathcal{A}_\gamma^* R_Q \hat{v})_{L_2(Q)}.
$$

Thus $(\mathcal{A}_R \hat{u}, \hat{v})_{L_2(Q)} = (\hat{u}, \mathcal{A}_\gamma^* R_Q \hat{v})_{L_2(Q)}$ for all $\hat{u} \in \mathcal{D}(\mathcal{A}_R)$. Hence, since the operator $\mathcal{A}_R$ is self-adjoint, $\mathcal{A}_\gamma^* R_Q \subset \mathcal{A}_R$, i.e., $\mathcal{A}_\gamma^* \subset \mathcal{A}_\gamma$.

2. From the fact, that $\mathcal{A}_\gamma$ is self-adjoint, and from Theorem 13.1, it follows that the spectrum $\sigma(\mathcal{A}_\gamma)$ consists of real isolated eigenvalues of finite multiplicity.

3. By virtue of Theorem 8.1 and Lemmas 13.1, 13.3, the problem of eigenvalues $\mathcal{A}_\gamma u = \lambda u$ is equivalent to the integral identity

$$
(\hat{u}, v)'_{\mathring{W}^m(Q)} = (\lambda - \mu)(T_R \hat{u}, v)'_{\mathring{W}^m(Q)} \qquad (v \in \mathring{W}^m(Q)), \tag{13.21}
$$

where $(\cdot, \cdot)'_{\mathring{W}^m(Q)}$ is the inner product defined by (13.9), $\mu = -c_2/k_3$.

By the Hilbert–Schmidt theorem, there is an orthogonal basis in $\mathring{W}^m(Q)$ with inner product defined by (13.9), formed of the eigenfunctions $\hat{u}_s$ of the operator T_R corresponding to the eigenvalues $1/(\lambda_s - \mu)$ of this operator. Since the operator T_R is positive, $\lambda_s > \mu$. Suppose that $\|u_s\|'_{L_2(Q)} = 1$. From this, by virtue of Theorem 8.1 and formulas (13.9), (13.18)–(13.21), we have

$$
\begin{aligned}
(u_s, u_r)'_{L_2(Q)} &= (u_s, u_r)'_{W^m_\gamma(Q)}/(\lambda_s - \mu) \qquad (s \neq r), \\
(u_s/\sqrt{\lambda_s - \mu}, u_s/\sqrt{\lambda_s - \mu})'_{W^m_\gamma(Q)} &= (u_s, u_s)'_{L_2(Q)} = 1 \qquad (s = 1, 2, \ldots).
\end{aligned}
$$

Hence, a set of functions $u_s/\sqrt{\lambda_s - \mu}$ is the orthonormal basis in $W^m_\gamma(Q)$ with inner product defined by (13.20). In addition, a set of functions u_s is the orthonormal basis in $L_2(Q)$ with the inner product defined by (13.19). □

Equations of the Second Order

In some cases the conditions for the coefficients γ_{lj}^r can be weakened.

We consider the equation

$$A^0 w + a_0(x)w(x) = f_0(x) \qquad (x \in Q) \tag{13.22}$$

with nonlocal conditions

$$\left.\begin{aligned} w|_{\Gamma_{rl}} &= \sum_{j=1}^{J_0} \gamma_{lj}^r w|_{\Gamma_{rj}} && (r \in B,\ l = J_0+1, \dots, J), \\ w|_{\Gamma_{rl}} &= 0 && (r \notin B,\ l = 1, \dots, J). \end{aligned}\right\} \tag{13.23}$$

Here

$$A^0 = -\sum_{i,j=1}^{n} \frac{\partial}{\partial x_i} a_{ij} \frac{\partial}{\partial x_j}, \qquad a_{ij} = a_{ji},$$

$a_0 \in C^\infty(\mathbb{R}^n)$ are real-valued M-periodic in $\overline{Q}$ functions,

$$\sum_{i,j=1}^{n} a_{ij}(x)\xi_i\xi_j > 0 \quad \text{for all } x \in \overline{Q},\ 0 \neq \xi \in \mathbb{R}^n.$$

Remark 13.1. We assume that there are matrices R_s satisfying the following condition

13.2. *There is a set* Λ *such that for all* $s = 1, 2, \dots$ *the matrices* $R_s + R_s^*$ *are positive definite, and for each* $r \in B$ *and* $s = s(r)$ *the relations* (13.6) *are satisfied.*

Then Example 9.3 implies that for the boundary value problem (13.22), (13.23), Theorems 13.1–13.3 are valid.

Theorem 13.4. *Let the domain* Q *satisfy the conditions* 7.1, 8.1. *Let the equation* (13.22) *be strongly elliptic in* $\overline{Q}$, *and let* $a_0(x) \geq 0$ *for* $x \in \overline{Q}$. *Suppose that there are matrices* R_s *satisfying the condition* 13.2.

Then there is a unique generalized solution of the boundary value problem (13.22), (13.23).

Proof. By virtue of Example 9.3, we can put $c_2 = 0$ in (13.12). Hence, Lemma 13.1 remains valid for $\mu = 0$. Thus from the proof of Theorem 13.1 it follows that $0 \notin \sigma(\mathcal{A}_\gamma)$. □

Problem 13.2. *Do Theorems* 13.1–13.3 *remain in force if the existence of matrices* A_s *satisfying the condition* 13.1 *is replaced by the assumption that there are matrices* R_s *satisfying the condition* 13.2?

Example 13.1. Let $Q \subset \mathbb{R}^2$ be a bounded domain with boundary $\partial Q \in C^\infty$, which outside the disks $S_{1/8}((i4/3, j4/3))$ $(i, j = 0, 1)$ coincides with the boundary of the square $(0, 4/3) \times (0, 4/3)$. Suppose that a set M consists of vectors ph, where $h = (1, 1)$, $p = 0, \pm 1, \pm 2, \dots$. Then the set $\mathcal{K} \subset \partial Q$ and consists of four points $g^1 = (1/3, 0)$, $g^2 = (4/3, 1)$, $g^3 = (0, 1/3)$, $g^4 = (1, 4/3)$. We denote

$$\Gamma_{12} = \{x \in \partial Q : x_1 < 1/3,\ x_2 < 1/3\}, \qquad \Gamma_{11} = \Gamma_{12} + h,$$
$$\Gamma_{22} = \{x \in \partial Q : 1 < x_1,\ 1 < x_2\},$$
$$\Gamma_{21} = \Gamma_{22} - h, \qquad \Gamma_{31} = \partial Q \setminus (\overline{\Gamma}_{12} \cup \overline{\Gamma}_{22})$$

(see Example 11.2). Evidently $B = \{1, 2\}$.

We consider the boundary value problem

$$-\Delta w(x) = f_0(x) \qquad (x \in Q), \tag{13.24}$$

$$w|_{\Gamma_{12}} = \gamma_1 w|_{\Gamma_{11}}, \quad w|_{\Gamma_{22}} = \gamma_2 w|_{\Gamma_{21}}, \quad w|_{\Gamma_{31}} = 0, \tag{13.25}$$

where $\gamma_1, \gamma_2 \in \mathbb{R}$.

The decomposition $\mathcal{R}$ of the domain Q consists of two classes of subdomains: 1) Q_{11}, bounded by the curves $\overline{\Gamma}_{12}$, $\overline{\Gamma}_{21}$, and $Q_{12} = Q_{11} + h$; 2) $Q_{21} = Q \setminus (\overline{Q}_{11} \cup \overline{Q}_{12})$. We introduce the third class of subdomains, renumbering subdomains of the first class: $Q_{31} = Q_{12}$, $Q_{32} = Q_{11}$. Evidently $\Gamma_{1j} \subset \partial Q_{3j}$, $\Gamma_{2j} \subset \partial Q_{1j}$ $(j = 1, 2)$. Thus, $s(1) = 3$, $s(2) = 1$ (see Lemma 7.7). The set $\mathcal{M}_0$ consists of the vectors $(0, 0)$, h, $-h$. The matrices R_1, R_3 have the form

$$R_1 = \begin{pmatrix} a_0 & a_h \\ a_{-h} & a_0 \end{pmatrix}, \qquad R_3 = \begin{pmatrix} a_0 & a_{-h} \\ a_h & a_0 \end{pmatrix}.$$

The system of equations (13.6) will have the form

$$a_h = \gamma_1 a_0, \quad a_{-h} = \gamma_2 a_0. \tag{13.26}$$

The system (13.26) has one linear independent solution $a_0 = 1$, $a_h = \gamma_1$, $a_{-h} = \gamma_2$. If $|\gamma_1 + \gamma_2| < 2$, $a_0 = 1$, $a_h = \gamma_1$, $a_{-h} = \gamma_2$, then the matrix $R_1 + R_1^*$ is positive definite. The difference operator defined by (13.8) has the following form

$$Ru(x) = u(x) + \gamma_1 u(x_1 + 1, x_2 + 1) + \gamma_2 u(x_1 - 1, x_2 - 1).$$

Thus, by virtue of Remark 13.1 and Theorems 13.1, 13.4, if $|\gamma_1 + \gamma_2| < 2$, then the spectrum $\sigma(\mathcal{A}_\gamma)$ is discrete and $0 \notin \sigma(\mathcal{A}_\gamma)$. If $\gamma_1 = \gamma_2$, $|\gamma_1| < 1$, then, by Theorem 13.3 and Remark 13.1, the operator $\mathcal{A}_\gamma$ has a real discrete spectrum $\sigma(\mathcal{A}_\gamma)$ consisting of positive eigenvalues.

Example 13.2. We consider the boundary value problem

$$-\Delta w(x) = f_0(x) \qquad (x \in Q), \tag{13.27}$$

$$\begin{gathered} w|_{\Gamma_{12} \cup \Gamma_{22}} = \gamma_1 w|_{\Gamma_{11} \cup \Gamma_{21}}, \\ w|_{\Gamma_{42} \cup \Gamma_{52}} = \gamma_2 w|_{\Gamma_{41} \cup \Gamma_{51}}, \qquad w|_{\Gamma_{31} \cup \Gamma_{61}} = 0. \end{gathered} \tag{13.28}$$

Here $Q = \{x \in \mathbb{R}^2 : |x| < 1\}$, $\gamma_1, \gamma_2 \in \mathbb{R}^2$, Γ_{12}, Γ_{22}, Γ_{31}, Γ_{42}, Γ_{52}, Γ_{61} are the arcs of the circle $\partial Q : \{\pi(-2+j)/3 < \varphi < \pi(-1+j)/3\}$ $(j = 1, \dots, 6)$, $\Gamma_{11} = \Gamma_{12} - (1,0)$, $\Gamma_{21} = \Gamma_{22} - (1,0)$, $\Gamma_{41} = \Gamma_{42} + (1,0)$, $\Gamma_{51} = \Gamma_{52} + (1,0)$ (see Example 7.8).

The set M consists of the vectors ph, where $h = (1,0)$, $p = 0, \pm 1, \pm 2, \dots$, $B = \{1,2,4,5\}$. The set $\mathcal{M}_0$ consists of the vectors 0, $\pm h$, $\pm 2h$.

The decomposition $\mathcal{R}$ of the domain Q consists of three classes of subdomains: 1) Q_{11}, bounded by the curves $\overline{\Gamma}_{42}$, $\overline{\Gamma}_{52}$, $\overline{\Gamma}_{11}$, $\overline{\Gamma}_{21}$, and $Q_{12} = Q_{11} + (1,0)$; 2) Q_{21}, bounded by the curves $\overline{\Gamma}_{31}$, $\overline{\Gamma}_{21}$, $\overline{\Gamma}_{41}$; 3) Q_{31}, bounded by the curves $\overline{\Gamma}_{61}$, $\overline{\Gamma}_{51}$, $\overline{\Gamma}_{11}$. We introduce the fourth class of subdomains $Q_{41} = Q_{12}$, $Q_{42} = Q_{11}$. Evidently $\Gamma_{1j} \cup \Gamma_{2j} \subset \partial Q_{1j}$, $\Gamma_{4j} \cup \Gamma_{5j} \subset \partial Q_{4j}$ $(j = 1,2)$. Thus, $s(1) = s(2) = 1$, $s(4) = s(5) = 4$ (see Lemma 7.7). The matrices

$$R_1 = \begin{pmatrix} a_0 & a_h \\ a_{-h} & a_0 \end{pmatrix}, \qquad R_4 = \begin{pmatrix} a_0 & a_{-h} \\ a_h & a_0 \end{pmatrix}.$$

The system of equations (13.6) has the form

$$a_{-h} = \gamma_1 a_0, \qquad a_h = \gamma_2 a_0. \tag{13.29}$$

The system (13.29) has one linear independent solution $a_0 = 1$, $a_{-h} = \gamma_1$, $a_h = \gamma_2$. If $|\gamma_1 + \gamma_2| < 2$, $a_0 = 1$, $a_{-h} = \gamma_1$, $a_h = \gamma_2$, then the matrix $R_1 + R_1^* = R_4 + R_4^*$ is positive definite. The operator R defined by (13.8), has the form

$$Ru(x) = u(x) + \gamma_1 u(x-h) + \gamma_2 u(x+h).$$

Thus, if $|\gamma_1 + \gamma_2| < 2$, then the spectrum $\sigma(\mathcal{A}_\gamma)$ is discrete and $0 \notin \sigma(\mathcal{A}_\gamma)$. If $\gamma_1 = \gamma_2$, $|\gamma_1| < 1$, then the operator $\mathcal{A}_\gamma$ has a real discrete spectrum $\sigma(\mathcal{A}_\gamma)$ consisting of positive eigenvalues.

Nonlocal Problems in a Cylinder

We consider the equation

$$Aw = A^0 w + a_0(x) w(x) = f_0(x) \qquad (x \in Q) \tag{13.30}$$

with nonlocal conditions

$$w|_{x_1=0} = \sum_{i=1}^{k} \gamma_i^1 w|_{x_1=i}, \qquad w|_{x_1=d} = \sum_{i=1}^{k} \gamma_i^2 w|_{x_1=d-i}, \tag{13.31}$$
$$w|_{[0,d]\times\partial G} = 0.$$

Here $Q = (0,d) \times G$, $G \subset \mathbb{R}^{n-1}$ is a bounded domain (with boundary $\partial G \in C^\infty$ if $n \geq 3$), $d = k + \theta$, $0 < \theta \leq 1$, k is a natural number, the operator

$$A^0 = -\sum_{i,j} \frac{\partial}{\partial x_i} a_{ij} \frac{\partial}{\partial x_j} \tag{13.32}$$

is strongly elliptic in $\overline{Q}$, $a_{ij} = a_{ji}$, $a_0 \in C^\infty(\mathbb{R}^n)$ are real-valued 1-periodic in x_1 functions, γ_i^1, $\gamma_i^2 \in \mathbb{C}$ are constants, $f_0 \in L_2(Q)$.

Let us introduce the unbounded operator $\mathcal{A}_\gamma: L_2(Q) \to L_2(Q)$ with domain $\mathcal{D}(\mathcal{A}_\gamma) = \{w \in W_\gamma^1(Q) : \mathcal{A}_\gamma w \in L_2(Q)\}$ by the formula $\mathcal{A}_\gamma w = Aw$, where $W_\gamma^1(Q)$ is the subspace of the functions in $W^1(Q)$ satisfying the conditions (13.31).

Lemma 13.5. $\mathcal{D}(\mathcal{A}_\gamma) = W_\gamma^1(Q) \cap W^2(Q)$.

Proof. It is sufficient to prove that $\mathcal{D}(\mathcal{A}_\gamma) \subset W_\gamma^1(Q) \cap W^2(Q)$. Let $w \in \mathcal{D}(\mathcal{A}_\gamma)$. We introduce the functions $\eta(x_1)$ and $\xi(x)$ so that $\eta \in \dot{C}^\infty(\mathbb{R})$, $\eta(x_1) = 1$ for $x_1 \in (-\delta, \delta)$, $\eta(x_1) = 0$ for $x_1 \notin (-2\delta, 2\delta)$, $\xi(x) = \eta(x_1) \sum_{j=1}^k \gamma_j^1 w(x_1 + j, x_2, \dots, x_n)$ for $x \in (0, 2\delta) \times G$, $\xi(x) = \eta(x_1 - d) \sum_{j=1}^k \gamma_j^2 w(x_1 - j, x_2, \dots, x_n)$ for $x \in (d - 2\delta, d) \times G$, $\xi(x) = 0$ for $x \in [2\delta, d - 2\delta] \times G$, where $0 < 4\delta < \theta$.

Since $\partial G \in C^\infty$ and $w|_{[0,d] \times \partial G} = 0$, then, by Theorem C.5 and Lemma C.1, $w \in W^2((\varepsilon, d-\varepsilon) \times G)$ for every $\varepsilon > 0$. Hence, $\xi \in W^2(Q)$. Thus $A(w - \xi) \in L_2(Q)$ and by virtue of (13.31), $w - \xi \in \mathring{W}^1(Q)$. Using Theorem C.6 and Remark C.2, we obtain $w - \xi \in W^2(Q)$, i.e., $w \in W^2(Q)$. □

Remark 13.2. In Section 22, we shall show that, for any γ_i^1, γ_i^2, the spectrum $\sigma(\mathcal{A}_\gamma)$ is discrete, and for every $0 < \varepsilon < \pi$, all but perhaps a finite number of points of $\sigma(\mathcal{A}_\gamma)$ belong to the angle $|\arg \lambda| < \varepsilon$. The results of this section allow us to prove that under some assumptions the spectrum is real and $0 \notin \sigma(\mathcal{A}_\gamma)$.

We set up a relation between a vector $a = (a_{-k}, \dots, a_k) \in \mathbb{C}^{2k+1}$ and the corresponding $(k+1) \times (k+1)$-matrix R_1 with the elements $r_{ij}^1 = a_{j-i}$. Denote by $e_i(g_i)$ the ith row of the $(k+1) \times k$-matrix obtained from the matrix R_1 by deleting the first (last) column.

Theorem 13.5 follows from Theorem 13.4 and Lemma 13.5.

Theorem 13.5. *Let the equation* (13.30) *be strongly elliptic in* $\overline{Q}$, *and let* $a_0(x) \geq 0$ *for* $x \in \overline{Q}$. *Suppose that there is a matrix* R_1 *satisfying the condition*

13.3. $R_1 + R_1^*$ *is positive definite and*

$$e_1 = \sum_{i=1}^k \gamma_i^1 e_{i+1}, \qquad g_{k+1} = \sum_{i=1}^k \gamma_i^2 g_{k+1-i}. \tag{13.33}$$

Then there exists a unique generalized solution of the boundary value problem (13.30), (13.31).

Example 13.3 (cf. (0.3), (0.4)).Consider the boundary value problem

$$A^0 w = f_0(x) \qquad (x \in Q = (0,2) \times (0,1)), \tag{13.34}$$

$$\begin{gathered} w|_{x_1=0} = \gamma_1 w|_{x_1=1}, \qquad w|_{x_1=2} = \gamma_2 w|_{x_1=1}, \\ w|_{x_2=0} = w|_{x_2=1} = 0, \end{gathered} \tag{13.35}$$

where $\gamma_1, \gamma_2 \in \mathbb{R}$.

The set M consists of vectors ph, where $h=(0,1)$, $p=0,\pm1,\pm2,\ldots$. The decomposition $\mathcal{R}$ of the domain Q consists of two subdomains $Q_{11}=(0,1)\times(0,1)$, $Q_{12}=(1,2)\times(0,1)$ (see Example 8.4). The matrix $R_1=\begin{pmatrix} a_0 & a_1 \\ a_{-1} & a_0 \end{pmatrix}$. The condition (13.33) holds for $a_0=1$, $a_1=\gamma_1$, $a_{-1}=\gamma_2$. If $|\gamma_1+\gamma_2|<2$, then by virtue of Theorem 13.5, the boundary value problem (13.34), (13.35) has a unique generalized solution (cf. Examples 13.1, 22.2).

Theorem 13.6. *Let the equation (13.30) be strongly elliptic in $\overline{Q}$. Suppose that the following condition holds:*

13.4. $\gamma_i^1=\gamma_i^2\in\mathbb{R}$ *and there exists a vector* $a\in\mathbb{R}^{2k+1}$ *such that* $a_i=a_{-i}$ $(i=1,\ldots,k)$, *the matrix* R_1 *is positive definite and*

$$e_1=\sum_{i=1}^{k}\gamma_i^1 e_{i+1}. \tag{13.36}$$

Then the spectrum $\sigma(\mathcal{A}_\gamma)$ *is real.*

The proof follows from Theorem 13.3, Lemma 13.5 and the symmetry of the matrix R_1 with elements $r_{ij}^1=a_{|j-i|}$ about the principal and auxiliary diagonals.

Example 13.4. Let us reconsider Example 13.3 assuming in addition that $\gamma_1=\gamma_2\in\mathbb{R}$. If $|\gamma_1|<1$, then by Theorem 13.6, $\sigma(\mathcal{A}_\gamma)\subset\mathbb{R}$. If $A^0=-\Delta$, then, using separation of variables, one can show that for $|\gamma_1|\le 1$ the spectrum $\sigma(\mathcal{A}_\gamma)$ is real, and for $|\gamma_1|>1$ the spectrum $\sigma(\mathcal{A}_\gamma)$ contains a countable set of complex eigenvalues.

Example 13.5. Consider the boundary value problem

$$A^0w(x)=f_0(x) \qquad (x\in Q=(0,3)\times(0,1)), \tag{13.37}$$

$$\begin{gathered} w|_{x_1=0}=\gamma_1 w|_{x_1=1}+\gamma_2 w|_{x_1=2}, \\ w|_{x_1=3}=\gamma_1 w|_{x_1=2}+\gamma_2 w|_{x_1=1}, \\ w|_{x_2=0}=w|_{x_2=1}=0, \quad \gamma_1=39/28, \quad \gamma_2=-6/7. \end{gathered} \tag{13.38}$$

The matrix $R_1=\begin{pmatrix} a_0 & a_1 & a_2 \\ a_1 & a_0 & a_1 \\ a_2 & a_1 & a_0 \end{pmatrix}$. The equations (13.36) for the unknowns a_0, a_1, a_2 have the form

$$\left.\begin{aligned} \gamma_1 a_0+(\gamma_2-1)a_1 &=0, \\ \gamma_2 a_0+\gamma_1 a_1-a_2 &=0. \end{aligned}\right\} \tag{13.39}$$

The system of equations (13.39) has one linearly independent solution $a_0=1$, $a_1=3/4$, $a_2=3/16$. The matrix R_1 is positive definite in this case. Thus, by Theorem 13.6, the spectrum $\sigma(\mathcal{A}_\gamma)$ is real.

Remark 13.3. The system of equations (13.36) for the unknowns $a_0, \dots, a_k$ can be written in the form

$$\Gamma_1 C = 0, \tag{13.40}$$

where Γ_1 is the matrix of order $k \times (k+1)$ obtained from the matrix $\Gamma = \|\gamma_{ij}\|_{i,j=0}^k$ by deleting the first row, $\gamma_{ij} = \gamma_{i+j}^1 + \gamma_{i-j}^1$ ($\gamma_0^1 = -1$, $\gamma_i^1 = 0$ for $i > k$, $i < 0$), $C = (a_0/2, a_1, \dots, a_k)$. One can show that, if the condition 13.4 holds, then $\operatorname{rank} \Gamma_1 = k$. Thus, in order to verify the condition 13.4, one has to calculate $\operatorname{rank} \Gamma_1$. If $\operatorname{rank} \Gamma_1 < k$, then the condition 13.4 does not hold. If $\operatorname{rank} \Gamma_1 = k$, then the system of equations (13.36) has only one linearly independent solution. Now it remains to verify positive definiteness of the matrix R_1 defined to within a constant multiplier.

Notes

The results of this chapter were obtained by A. L. Skubachevskiĭ [2, 5, 8, 10, 12].

Solvability of the first boundary value problem for a strongly elliptic equation of the second order with arbitrary transformations of arguments was studied by G. A. Kamenskiĭ, A. D. Myshkis, A. L. Skubachevskiĭ [1].

The second and the third boundary value problems for strongly elliptic differential-difference equations were considered in the papers of A. L. Skubachevskiĭ and E. L. Tsvetkov [1] and E. L. Tsvetkov [1].

Chapter III

Applications to the Mechanics of a Deformable Body

In this chapter we consider a linear model of a mechanical system. We reduce this model to a boundary value problem for a strongly elliptic system of differential-difference equations.

Section 14 deals with an elastic system consisting of two parallel plates connected by a regular system of vertical and slanting ribs. We reduce the model of this type of system to a variational problem for the functional of total potential energy of a three-layer plate.

In Section 15, we consider this variational problem and the corresponding boundary value problem for a strongly elliptic system of differential-difference equations. We prove that the spectrum of the appropriate operator is discrete and consists of real, positive eigenvalues. The convergence of the Ritz method is stated.

Section 16 is devoted to the study of the smoothness of the generalized solutions. We prove the theorems on the smoothness of the solutions in subdomains and on the boundary of adjacent subdomains.

In order to demonstrate the methods of this chapter, in Section 17 we consider the elastic model in the one-dimensional case.

14 The Elastic Model

Continuous Model of Sandwich Plate

Figure III.1 shows an elastic system consisting of two parallel plates connected by a regular system of vertical and slanting ribs (at an angle α to the vertical), all oriented in the same direction.

It is natural to reduce this discrete-continuous system to a continuous model, "spreading out" both the vertical ribs (0-braces) and the slanting ribs (α-braces) in the space between the plates. To do this, we must introduce kinematically independent continuous fields of elastic displacements of the 0-braces and α-braces

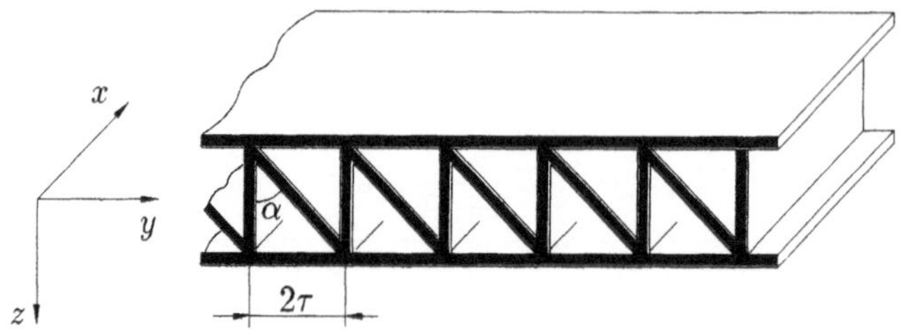

Fig. III.1

uniformly distributed in the space between the plates. As a result, we arrive at a three-layer plate with a "two-phase" model of a filler, which combines in itself a medium of 0-braces and one of α-braces.

We introduce a unified system of Cartesian coordinates x, y, z, making the middle surfaces of the plates coincide with the planes $z = \pm h$ in such a way that the ribs will be directed along the axis $0x$. We also introduce the local Cartesian coordinates x_β, 0_β, y_β ($\beta = 0, \alpha$) in the planes of the ribs, making the axis $0_\beta x_\beta$ coincide with the line of intersection of the corresponding rib and the plane $x0y$. We shall assume for the sake of simplicity that the plates themselves and the ribs, both vertical and slanting, are moment-free (zero rigidity out of the plane); the ribs offer no resistance to tension or compression in the longitudinal direction but are absolutely rigid in the transverse direction (in the plane of the ribs).

Remark 14.1. The hypothesis concerning absolute rigidity of the braces seems to be rather strong. In fact, in order to obtain a system of functional differential equations having the simplest form, we use this hypothesis. Consideration of the extensibility of the braces leads to a more complicated system of functional differential equations. In this book we shall not consider such a system. However, it should be noted that the methods stated here allow us to solve this problem.

Let us examine a single rib. The equation of the plane of the rib in the unified system of coordinates has the form

$$y - z\tan\beta = k \qquad (\beta = 0, \alpha). \tag{14.1}$$

Obviously, by virtue of the simplifications we have made, the cross-section of the rib, $x_\beta = \text{const}$, is displaced in the plane of the rib like a rigid body. Consequently, the elastic displacements u_β, v_β of an arbitrary point of the rib in the direction of the axes x_β, y_β can be represented in the form

$$\begin{aligned} u_\beta(x_\beta, y_\beta, k) &= \varphi_\beta(x_\beta, k) - \psi_\beta(x_\beta, k)y_\beta, \\ v_\beta(x_\beta, y_\beta, k) &= v_\beta(x_\beta, k), \end{aligned} \tag{14.2}$$

where φ_β, ψ_β are the translational displacement in the direction of the axis $0_\beta x_\beta$ and the rotation in the plane $x_\beta 0_\beta y_\beta$ of the rib cross-section $x_\beta = \text{const}$, respectively.

Having established the explicit relation (14.2) between the displacements of an individual rib and the coordinate y_β, we turn to the continuous model. The local coordinates are connected with the unit coordinates by the formulas,

$$x_\beta = x, \quad y_\beta = z/\cos\beta \qquad (\beta = 0, \alpha). \tag{14.3}$$

Spreading out the ribs in the space between the plates, we must introduce the continuous fields $U_\beta(x,y,z)$, $V_\beta(x,y,z)$ of the elastic displacements of the β-braces ($\beta = 0, \alpha$). To do this, after first passing from the local coordinates to the unified system of coordinates, we must "extend" the expressions (14.2), which relate to an individual rib, in accordance with the condition

$$\begin{aligned} U_\beta(x, k + z\tan\beta, z) &= u_\beta(x_\beta, y_\beta, k), \\ V_\beta(x, k + z\tan\beta, z) &= v_\beta(x_\beta, y_\beta, k), \end{aligned} \tag{14.4}$$

to the entire space between the plates.

From (14.1)–(14.4) we have

$$\begin{aligned} U_\beta(x,y,z) &= \varphi_\beta(x, y - z\tan\beta) - \psi_\beta(x, y - z\tan\beta)z/\cos\beta, \\ V_\beta(x,y,z) &= v_\beta(x, y - z\tan\beta). \end{aligned} \tag{14.5}$$

The expressions (14.5) for $\beta = 0, \alpha$ represent kinematically independent continuous fields of elastic displacements of the two-phase filler of a three-layer plate. These fields must be subjected to the conditions of the kinematic connection of the filler with the supporting layers

$$\begin{aligned} U_\beta(x,y,\pm h) &= u^\pm(x,y), \\ V_\beta(x,y,\pm h) &= v^\pm(x,y)\sin\beta + w^\pm(x,y)\cos\beta, \end{aligned} \tag{14.6}$$

where $u^\pm$, $v^\pm$, $w^\pm$ are the displacements of the surfaces of the lower plate ($z = h$) and the upper plate ($z = -h$), $\beta = 0, \alpha$.

From (14.5), (14.6), it follows that the field of elastic displacements of a three-layer plate with a two-phase filler is determined by 12 functions of two independent variables u^-, v^-, w^-, u^+, v^+, w^+, φ_α, ψ_α, v_α, φ_0, ψ_0, v_0, connected by the eight relations

$$\begin{aligned} &u^\pm(x,y) = \varphi_\beta(x, y \mp h\tan\beta) \mp \frac{\psi_\beta(x, y \mp h\tan\beta)h}{\cos\beta}, \\ &v^\pm(x,y)\sin\beta + w^\pm(x,y)\cos\beta = v_\beta(x, y \mp h\tan\beta) \end{aligned} \qquad (\beta = 0, \alpha). \tag{14.7}$$

Potential Energy of a Three-Layer Plate

The boundary value problem corresponding to the proposed adequate continuous interpretation of the discrete-continuous system under consideration can be naturally formulated on the basis of the Lagrange principle.

As our main unknown, we introduce the four-dimensional vector-valued function $u = (u^1, u^2, u^3, u^4)$ of two variables:

$$u^1 = \varphi_\alpha, \quad u^2 = \psi_\alpha, \quad u^3 = v_\alpha, \quad u^4 = v_0. \tag{14.8}$$

Then
$$\begin{aligned} &u^\pm = u^1_{\mp\tau} \mp h \sec\alpha u^2_{\mp\tau}, \quad v^\pm = \csc\alpha u^3_{\mp\tau} - \cot\alpha u^4, \\ &w^\pm = u^4, \quad \varphi_0 = (u^1_{+\tau} + u^1_{-\tau})/2 + h\sec\alpha(u^2_{+\tau} - u^2_{-\tau})/2, \\ &\psi_0 = (u^1_{+\tau} - u^1_{-\tau})/(2h) + \sec\alpha(u^2_{+\tau} + u^2_{-\tau})/2, \end{aligned} \tag{14.9}$$

where $\pm\tau$ denotes the shift of the second argument by $\tau = h\tan\alpha$, e.g. $u^i_{\pm\tau} = u^i(x, y \pm \tau)$.

We assume that the plates are rectangular $\{0 \le x \le a,\ 0 \le y \le b\}$. Let 0-media be included between the plates in a rectangular parallelepiped $V^0 = \{(x,y,z) : 0 \le x \le a,\ 0 \le y \le b,\ -h \le z \le h\}$, and let α-media be included in a parallelepiped $V^\alpha = \{(x,y,z) : 0 \le x \le a,\ -\tau \le y - z\tan\beta \le b + \tau,\ -h \le z \le h\}$.

We shall assume that in addition to the internal connections, the system is subjected to absolutely rigid external geometric connections (homogeneous geometric boundary conditions). We set:

a) for $x = 0, a$ there are no displacements of the β-media ($\beta = 0, \alpha$), i.e., for $x = 0, a$, $U_\beta(x,y,z) = V_\beta(x,y,z) = 0$, and hence, by virtue of (14.5), (14.8), we obtain

$$\begin{aligned} u^1(x,y) &= u^2(x,y) = u^3(x,y) = 0 \\ &\qquad ((x,y) \in \{(x,y) : x = 0, a; -\tau \le y \le b + \tau\}), \\ u^4(x,y) &= 0 \qquad ((x,y) \in \{(x,y) : x = 0, a; 0 \le y \le b\}), \end{aligned} \tag{14.10}$$

b) for $(x,y,z) \in \{(x,y,z) : 0 \le x \le a,\ -\tau \le y - z\tan\alpha \le \tau,\ -h \le z \le h\} \cup \{(x,y,z) : 0 \le x \le a,\ b - \tau \le y - z\tan\alpha \le b + \tau,\ -h \le z \le h\}$ there are no displacements of the α-braces, i.e., $U_\alpha(x,y,z) = V_\alpha(x,y,z) = 0$, and for $(x,y,z) \in \{(x,y,z) : 0 \le x \le a,\ y = 0, b; -h \le z \le h\}$ there are no displacements of the 0-braces, i.e., $U_0(x,y,z) = V_0(x,y,z) = 0$. From this, by (14.5), (14.8), we obtain

$$\begin{aligned} u^1(x,y) &= u^2(x,y) = u^3(x,y) = 0 \qquad ((x,y) \in G_1 \cup G_2), \\ u^4(x,y) &= 0 \qquad ((x,y) \in \{(x,y) : 0 \le x \le a,\ y = 0, b\}), \end{aligned} \tag{14.11}$$

where $G_1 = \{(x,y) : 0 < x < a,\ -\tau < y < \tau\}$, $G_2 = \{(x,y) : 0 < x < a,\ b - \tau < y < b + \tau\}$.

The potential energy of a thin elastic plate M has the form

$$E_M = \int_M \frac{1}{2}(\sigma_\xi \varepsilon_\xi + \sigma_\eta \varepsilon_\eta + \tau_{\xi\eta}\gamma_{\xi\eta})\, dM, \tag{14.12}$$

where ξ, η are the local Cartesian coordinates in the plane of the plate, ε_ξ, ε_η are the linear deformations, γ_{xy} is the shear, σ_ξ, σ_η are the normal stresses, and τ_{xy} is the tangent stress.

It is known that

$$\varepsilon_\xi = u_\xi, \quad \varepsilon_\eta = v_\eta, \quad \gamma_{\xi\eta} = u_\eta + v_\xi, \tag{14.13}$$

$$\sigma_\xi = \frac{2G}{1-\nu}(\varepsilon_\xi + \nu\varepsilon_\eta), \quad \sigma_\eta = \frac{2G}{1-\nu}(\varepsilon_\eta + \nu\varepsilon_\xi), \quad \tau_{\xi\eta} = G\gamma_{\xi\eta}, \tag{14.14}$$

where G, ν are the shear modulus and the Poisson coefficient, respectively, u, v are the elastic displacements in the direction of the axes ξ, η.

By virtue of (14.12)–(14.14), the potential energy of the plate $\Gamma^i = \{(x,y,z) : 0 \le x \le a,\ 0 \le y \le b,\ z = ih\}$ has the form

$$E_{\Gamma^i} = \int_0^a \int_0^b G\delta\left(\frac{(u_x^i)^2}{1-\nu} + \frac{(u_y^i)^2}{2} + \frac{(v_y^i)^2}{1-\nu} + \frac{(v_x^i)^2}{2} + \frac{2\nu}{1-\nu} u_x^i v_y^i + u_y^i v_x^i\right) dx dy$$
$$(i = 1, -1). \tag{14.15}$$

Here δ is the thickness of the plate.

The energy of the external forces is given by the formula

$$E_{F^i} = -\int_0^a \int_0^b (X^i u^i + Y^i v^i + Z^i w^i)\, dx\, dy \qquad (i = 1, -1), \tag{14.16}$$

where X^i, Y^i, Z^i are the components of the external load on the upper and lower plates in the unified system of coordinates.

By virtue of our assumptions, $\varepsilon_{y_\beta} = 0$, $\sigma_{x_\beta} = 0$. Thus, spreading out the ribs in V^β and using (14.12)–(14.14), we obtain the potential energy of the β-braces

$$E_{V^\beta} = \int_{V^\beta} \frac{G}{2} \mu_\beta (u_{\beta y_\beta} + v_{\beta x_\beta})^2\, dV^\beta, \tag{14.17}$$

where μ_β $(\beta = 0, \alpha)$ are the volumetric content of the β-braces in a unit volume V^β after spreading out.

From (14.1)–(14.3), (14.11) it follows that

$$E_{V^0} = \int_0^a \int_0^b Gh\mu_0 (v_{0x} - \psi_0)^2\, dx\, dy, \tag{14.18}$$

$$\begin{aligned} E_{V^\alpha} &= \frac{G\mu_\alpha}{2} \int_0^a dx \int_{-h}^h dz \int_{z\tan\alpha-\tau}^{z\tan\alpha+b+\tau} (v_{\alpha x}(x, y - z\tan\alpha) \\ &\qquad\qquad -\psi_\alpha(x, y - z\tan\alpha))^2 dy \\ &= \frac{G\mu_\alpha}{2} \int_0^a dx \int_{-h}^h dz \int_{-\tau}^{b+\tau} (v_{\alpha x}(x,t) - \psi_\alpha(x,t))^2 dt \\ &= Gh\mu_\alpha \int_0^a \int_0^b (v_{\alpha x}(x,t) - \psi_\alpha(x,t))^2\, dx\, dy. \end{aligned} \tag{14.19}$$

The functional of the total potential energy of the three-layer plate with a two-phase filler is given by the formula

$$E(u) = \sum_{i=1,-1} (E_{\Gamma^i} + E_{F^i}) + E_{V^0} + E_{V^\alpha}. \tag{14.20}$$

By virtue of (14.8), (14.9), (14.15), (14.16), (14.18), (14.19), we obtain

$$
\begin{aligned}
E(u) = \int_0^a \int_0^b \Big\{ \sum_{i=1,-1} G\delta \Big[& \frac{((u^1_{i\tau})_x + ih\sec\alpha(u^2_{i\tau})_x)^2}{1-\nu} \\
& + \frac{1}{2}((u^1_{i\tau})_y + ih\sec\alpha(u^2_{i\tau})_y)^2 + \frac{(\csc\alpha(u^3_{i\tau})_y - \cot\alpha u^4_y)^2}{1-\nu} \\
& + \frac{1}{2}(\csc\alpha(u^3_{i\tau})_x - \cot\alpha u^4_x)^2 \\
& + \frac{2\nu}{1-\nu}((u^1_{i\tau})_x + ih\sec\alpha(u^2_{i\tau})_x)(\csc\alpha(u^3_{i\tau})_y - \cot\alpha u^4_y) \\
& \qquad + ((u^1_{i\tau})_y + ih\sec\alpha(u^2_{i\tau})_y)(\csc\alpha(u^3_{i\tau})_x - \cot\alpha u^4_x) \Big] \\
& + Gh\mu_\alpha(u^3_x - u^2)^2 \\
& + Gh\mu_0 \Big[\frac{u^1_\tau - u^1_{-\tau}}{2h} + \frac{\sec\alpha(u^2_\tau + u^2_{-\tau})}{2} - u^4_x \Big]^2 \\
& - \sum_{i=1,-1} \big[X^{-i}(u^1_{i\tau} + ih\sec\alpha u^2_{i\tau}) \\
& \qquad + Y^{-i}(\csc\alpha u^3_{i\tau} - \cot\alpha u^4) + Z^i u^4 \big] \Big\} dx\, dy. \qquad (14.21)
\end{aligned}
$$

Thus the proposed continuous model can be described on the basis of the variational problem

$$E(u) \to \min \qquad (14.22)$$

with the boundary conditions (14.10), (14.11).

15 Variational and Boundary Value Problems

Necessary and Sufficient Conditions for Minimum

We introduce the real spaces of vector-valued functions

$$
\begin{aligned}
L^4_2 &= L_2(Q_1) \times L_2(Q_1) \times L_2(Q_1) \times L_2(Q_2), \\
W^{k,4} &= W^k(Q_1) \times W^k(Q_1) \times W^k(Q_1) \times W^k(Q_2) \qquad (k \geq 1), \\
\mathring{W}^{1,4} &= \mathring{W}^1(Q_1) \times \mathring{W}^1(Q_1) \times \mathring{W}^1(Q_1) \times \mathring{W}^1(Q_2),
\end{aligned}
$$

where $Q_1 = \{(x,y) : 0 < x < a,\ \tau < y < b - \tau\}$, $Q_2 = \{(x,y) : 0 < x < a,\ 0 < y < b\}$. Suppose that $X^i, Y^i, Z^i \in L_2(Q_2)$ $(i = 1, -1)$.

We shall find the extremum of the functional (14.21) with the boundary conditions (14.10), (14.11) in the space $\mathring{W}^{1,4}$ setting $u^i(x,y) = 0$ for $(x,y) \in G_1 \cup G_2$ $(i = 1, 2, 3)$.

Assume that the vector-valued function $u \in \mathring{W}^{1,4}$ yields a minimum of the functional (14.21) with the boundary conditions (14.10), (14.11). Then for each $v \in \mathring{W}^{1,4}$

$$\left.\frac{dE(u+tv)}{dt}\right|_{t=0} = 0. \tag{15.1}$$

We subdivide the resulting integral into the sum of integrals of the form

$$-\iint_{Q_2}\left[X^{-i}\left(v^1_{i\tau}+\frac{ih}{\cos\alpha}v^2_{i\tau}\right)+Y^{-i}\left(\frac{1}{\sin\alpha}v^3_{i\tau}-\cot\alpha v^4\right)+Z^iv^4\right]dx\,dy$$

and

$$\iint_{Q_2} k^{ij}_{rp}\mathcal{B}_r u^i \mathcal{B}_p v^j\,dx\,dy,$$

where $\mathcal{B}_1u^i = u^i$, $\mathcal{B}_2u^i = u^i_x$, $\mathcal{B}_3u^i = u^i_y$, $\mathcal{B}_4u^i = u^i_\tau$, $\mathcal{B}_5u^i = (u^i_\tau)_x$, $\mathcal{B}_6u^i = (u^i_\tau)_y$, $\mathcal{B}_7u^i = u^i_{-\tau}$, $\mathcal{B}_8u^i = (u^i_{-\tau})_x$, $\mathcal{B}_9u^i = (u^i_{-\tau})_y$, k^{ij}_{rp} are constants $(i,j = 1,\dots,4,\ r,p = 1,\dots,9)$. In the integrals containing the functions $v^i_{\pm\tau}$, $(v^i_{\pm\tau})_x$, $(v^i_{\pm\tau})_y$ $(i = 1,2,3)$ we make the change of variables $x' = x$, $y' = y \pm \tau$, thereby passing from integrals over the region Q_2 to integrals over the region $Q_1 \cup G_2$ if $y' = y+\tau$, and over the region $Q_1 \cup G_1$ if $y' = y-\tau$. Reducing similar terms and setting $v^j(x,y) = 0$ for $(x,y) \in G_1 \cup G_2$, we obtain

$$\begin{aligned}
&\iint_{Q_1}\Big\{G\delta\Big[\frac{4}{1-\nu}u^1_x+\frac{4\nu}{\sin\alpha(1-\nu)}u^3_y-\cot\alpha\,\frac{2\nu}{1-\nu}(R_1u^4)_y\Big]v^1_x\\
&\qquad+G\delta\Big[2u^1_y+\frac{2}{\sin\alpha}u^3_x-\cot\alpha(R_1u^4)_x\Big]v^1_y\\
&\qquad+G\Big[\mu_0(R_2u^4)_x+\frac{\mu_0}{2h}R_4u^1-\frac{\mu_0}{2\cos\alpha}R_3u^2\Big]v^1\Big\}dx\,dy\\
&+\iint_{Q_1}\Big\{G\delta\Big[\frac{4h^2}{\cos^2\alpha(1-\nu)}u^2_x+\frac{2h\nu}{\sin\alpha(1-\nu)}(R_2u^4)_y\Big]v^2_x\\
&\qquad+G\delta\Big[\frac{2h^2}{\cos^2\alpha}u^2_y+\frac{h}{\sin\alpha}(R_2u^4)_x\Big]v^2_y\\
&\qquad+G\Big[-2h\mu_\alpha u^3_x-\frac{h\mu_0}{\cos\alpha}(R_1u^4)_x+\frac{\mu_0}{2\cos\alpha}R_3u^1\\
&\qquad\qquad+\frac{h\mu_0}{2\cos^2\alpha}R_5u^2+2h\mu_\alpha u^2\Big]v^2\Big\}dx\,dy\\
&+\iint_{Q_1}\Big\{G\delta\Big[\frac{2}{\sin\alpha}u^1_y+\frac{2}{\sin^2\alpha}u^3_x-\frac{\cot\alpha}{\sin\alpha}(R_1u^4)_x\Big]v^3_x\\
&\qquad+G\delta\Big[\frac{4\nu}{\sin\alpha(1-\nu)}u^1_x+\frac{4}{\sin^2\alpha(1-\nu)}u^3_y-\frac{2\cot\alpha}{\sin\alpha(1-\nu)}(R_1u^4)_y\Big]v^3_y\\
&\qquad\qquad-2Gh\mu_\alpha(u^2-u^3_x)v^3_x\Big\}dx\,dy
\end{aligned}$$

$$+\iint_{Q_2}\Big\{G\delta\Big[-\cot\alpha(R_1u^1)_y-\frac{h}{\sin\alpha}(R_2u^2)_y$$
$$-\frac{\cot\alpha}{\sin\alpha}(R_1u^3)_x+2\cot^2\alpha u_x^4\Big]v_x^4$$
$$+G\delta\Big[-\cot\alpha\frac{2\nu}{1-\nu}(R_1u^1)_x-\frac{2h\nu}{\sin\alpha(1-\nu)}(R_2u^2)_x$$
$$-\frac{2\cot\alpha}{\sin\alpha(1-\nu)}(R_1u^3)_y+\cot^2\alpha\frac{4}{1-\nu}u_y^4\Big]v_y^4$$
$$+2Gh\mu_0\Big[u_x^4-\frac{1}{2h}R_2u^1-\frac{1}{2\cos\alpha}R_1u^2\Big]v_x^4\Big\}dx\,dy$$
$$=\iint_{Q_1}(X_\tau^1+X_{-\tau}^{-1})v^1dx\,dy$$
$$+\iint_{Q_1}\frac{h}{\cos\alpha}(X_{-\tau}^{-1}-X_\tau^1)v^2dx\,dy+\iint_{Q_1}\frac{1}{\sin\alpha}(Y_\tau^1+Y_{-\tau}^{-1})v^3dx\,dy$$
$$+\iint_{Q_2}[(Z^1+Z^{-1})-\cot\alpha(Y^1+Y^{-1})]v^4dx\,dy \quad (15.2)$$

for every $v\in\mathring{W}^{1,4}$, where $R_i\colon L_2(\mathbb{R}^2)\to L_2(\mathbb{R}^2)$ are the difference operators defined by

$$R_1w=w_\tau+w_{-\tau},\qquad R_2w=w_\tau-w_{-\tau},\qquad R_3w=w_{2\tau}-w_{-2\tau},$$
$$R_4w=2w-w_{2\tau}-w_{-2\tau},\qquad R_5w=2w+w_{2\tau}+w_{-2\tau}. \quad (15.3)$$

Denote by $I_i=I_{Q_i}\colon L_2(Q_i)\to L_2(\mathbb{R}^2)$ the operator of extension of functions from $L_2(Q_i)$ by zero in $\mathbb{R}^2\setminus Q_i$ $(i=1,2)$. Denote by $P_i=P_{Q_i}\colon L_2(\mathbb{R}^2)\to L_2(Q_i)$ the operator of restriction of functions from $L_2(\mathbb{R}^2)$ to Q_i $(i=1,2)$.

Lemma 15.1. *The operators $R_i\colon L_2(\mathbb{R}^2)\to L_2(\mathbb{R}^2)$ are bounded $(i=1,\dots,5)$. Moreover, $R_i^*=R_i$ $(i=1,4,5)$, $R_j^*=-R_j$ $(j=2,3)$.*

Lemma 15.2. *The operator $P_kR_iI_j\colon L_2(Q_j)\to L_2(Q_k)$ $(k,j=1,2,\ i=1,\dots,5)$ are bounded. Moreover, $(P_1R_iI_1)^*=P_1R_iI_1$ $(i=4,5)$, $(P_1R_3I_1)^*=-P_1R_3I_1$, $(P_1R_1I_2)^*=P_2R_1I_1$, $(P_1R_2I_2)^*=-P_2R_2I_1$.*

The proofs follow from Lemmas 8.1, 8.2.

Lemma 15.3. *The operators $P_kR_iI_j$ $(k,j=1,2,\ i=1,\dots,5)$ map $\mathring{W}^1(Q_j)$ into $W^1(Q_k)$ continuously.*

The proof is similar to that of Lemma 8.13.

The integral identity (15.2) with the boundary conditions $u^1(x,y) = u^2(x,y) = u^3(x,y) = 0$ for $(x,y) \in G_1 \cup G_2$ is equivalent to the identity

$$\begin{aligned}
&\iint_{Q_1} \Big\{ \delta\Big[\frac{4}{1-\nu}\, u_x^1 + \frac{4\nu}{\sin\alpha(1-\nu)}\, u_y^3 - \cot\alpha\,\frac{2\nu}{1-\nu}\,(P_1R_1I_2u^4)_y\Big] v_x^1 \\
&\qquad + \delta\Big[2u_y^1 + \frac{2}{\sin\alpha}\, u_x^3 - \cot\alpha (P_1R_1I_2u^4)_x\Big] v_y^1 \\
&\qquad + \Big[\mu_0 (P_1R_2I_2u^4)_x + \frac{\mu_0}{2h}\, P_1R_4I_1u^1 - \frac{\mu_0}{2\cos\alpha}\, P_1R_3I_1u^2\Big] v^1 \Big\} dx\,dy \\
&+ \iint_{Q_1} \Big\{ \delta\Big[\frac{4h^2}{\cos^2\alpha(1-\nu)}\, u_x^2 + \frac{2h\nu}{\sin\alpha(1-\nu)}\,(P_1R_2I_2u^4)_y\Big] v_x^2 \\
&\qquad + \delta\Big[\frac{2h^2}{\cos^2\alpha}\, u_y^2 + \frac{h}{\sin\alpha}\,(P_1R_2I_2u^4)_x\Big] v_y^2 \\
&\qquad + \Big[- 2h\mu_\alpha u_x^3 - \frac{h\mu_0}{\cos\alpha}\,(P_1R_1I_2u^4)_x \\
&\qquad + \frac{\mu_0}{2\cos\alpha}\, P_1R_3I_1u^1 + \frac{h\mu_0}{2\cos^2\alpha}\, P_1R_5I_1u^2 + 2h\mu_\alpha u^2\Big] v^2 \Big\} dx\,dy \\
&+ \iint_{Q_1} \Big\{ \delta\Big[\frac{2}{\sin\alpha}\, u_y^1 + \frac{2}{\sin^2\alpha}\, u_x^3 - \frac{\cot\alpha}{\sin\alpha}\,(P_1R_1I_2u^4)_x\Big] v_x^3 \\
&\qquad + \delta\Big[\frac{4\nu}{\sin\alpha(1-\nu)}\, u_x^1 + \frac{4}{\sin^2\alpha(1-\nu)}\, u_y^3 \\
&\qquad\qquad - \frac{2\cot\alpha}{\sin\alpha(1-\nu)}\,(P_1R_1I_2u^4)_y\Big] v_y^3 \\
&\qquad\qquad\qquad -2h\mu_\alpha (u^2 - u_x^3) v_x^3 \Big\} dx\,dy \\
&+ \iint_{Q_2} \Big\{ \delta\Big[- \cot\alpha (P_2R_1I_1u^1)_y - \frac{h}{\sin\alpha}\,(P_2R_2I_1u^2)_y \\
&\qquad\qquad - \frac{\cot\alpha}{\sin\alpha}\,(P_2R_1I_1u^3)_x + 2\cot^2\alpha u_x^4\Big] v_x^4 \\
&\qquad + \delta\Big[- \cot\alpha\,\frac{2\nu}{1-\nu}\,(P_2R_1I_1u^1)_x - \frac{2h\nu}{\sin\alpha(1-\nu)}\,(P_2R_2I_1u^2)_x \\
&\qquad\qquad - \frac{2\cot\alpha}{\sin\alpha(1-\nu)}\,(P_2R_1I_1u^3)_y + \cot^2\alpha\,\frac{4}{1-\nu}\, u_y^4\Big] v_y^4 \\
&\qquad + 2h\mu_0 \Big[u_x^4 - \frac{1}{2h}\, P_2R_2I_1u^1 - \frac{1}{2\cos\alpha}\, P_2R_1I_1u^2\Big] v_x^4 \Big\} dx\,dy \\
&\qquad\qquad = \sum_{i=1}^{3} \iint_{Q_1} f^i v^i\, dx\,dy + \iint_{Q_2} f^4 v^4\, dx\,dy \qquad (15.4)
\end{aligned}$$

for every $v \in \mathring{W}^{1,4}$, where $f^1 = (X^1_\tau + X^{-1}_{-\tau})/G$, $f^2 = h(X^{-1}_{-\tau} - X^1_\tau)/(G\cos\alpha)$, $f^3 = (Y^1_\tau + Y^{-1}_{-\tau})/(G\sin\alpha)$, $f^4 = [(Z^1 + Z^{-1}) - \cot\alpha(Y^1 + Y^{-1})]/G$. Obviously, $f^i \in L_2(Q_1)$ $(i = 1, 2, 3)$, $f^4 \in L_2(Q_2)$.

A vector-valued function $u \in \mathring{W}^{1,4}$ satisfies the identity (15.4) for each $v \in \mathring{W}^{1,4}$ if and only if u satisfies the following system of differential-difference equations in the sense of distributions:

$$
\begin{aligned}
&-2\delta\left(\Delta u^1 + \frac{1+\nu}{1-\nu}u^1_{xx}\right) - \frac{2\delta(1+\nu)}{\sin\alpha(1-\nu)}u^3_{xy}\\
&\quad+\delta\cot\alpha\frac{1+\nu}{1-\nu}(P_1R_1I_2u^4)_{xy} + \mu_0(P_1R_2I_2u^4)_x\\
&\quad\quad+\frac{\mu_0}{2h}P_1R_4I_1u^1 - \frac{\mu_0}{2\cos\alpha}P_1R_3I_1u^2 = f^1,\\
&-\frac{2\delta h^2}{\cos^2\alpha}\left(\Delta u^2 + \frac{1+\nu}{1-\nu}u^2_{xx}\right) - \frac{\delta h(1+\nu)}{\sin\alpha(1-\nu)}(P_1R_2I_2u^4)_{xy}\\
&\quad-2h\mu_\alpha u^3_x - \frac{h\mu_0}{\cos\alpha}(P_1R_1I_2u^4)_x + \frac{\mu_0}{2\cos\alpha}P_1R_3I_1u^1\\
&\quad\quad+\frac{h\mu_0}{2\cos^2\alpha}P_1R_5I_1u^2 + 2h\mu_\alpha u^2 = f^2,\\
&-\frac{2\delta(1+\nu)}{\sin\alpha(1-\nu)}u^1_{xy} - \frac{2\delta}{\sin^2\alpha}\left(\Delta u^3 + \frac{1+\nu}{1-\nu}u^3_{yy}\right)\\
&\quad+\frac{\delta\cot\alpha}{\sin\alpha}\left[\Delta(P_1R_1I_2u^4) + \frac{1+\nu}{1-\nu}(P_1R_1I_2u^4)_{yy}\right]\\
&\quad\quad-2h\mu_\alpha u^3_{xx} + 2h\mu_\alpha u^2_x = f^3,\\
&\delta\cot\alpha\frac{1+\nu}{1-\nu}(P_2R_1I_1u^1)_{xy} + \frac{\delta h(1+\nu)}{\sin\alpha(1-\nu)}(P_2R_2I_1u^2)_{xy}\\
&\quad+\frac{\delta\cot\alpha}{\sin\alpha}\left[\Delta(P_2R_1I_1u^3) + \frac{1+\nu}{1-\nu}(P_2R_1I_1u^3)_{yy}\right]\\
&\quad\quad-2\delta\cot^2\alpha\left(\Delta u^4 + \frac{1+\nu}{1-\nu}u^4_{yy}\right) - 2h\mu_0u^4_{xx}\\
&\quad\quad\quad+\mu_0(P_2R_2I_1u^1)_x + \frac{h\mu_0}{\cos\alpha}(P_2R_1I_1u^2)_x = f^4,
\end{aligned}
\tag{15.5}
$$

where the differential operators in the first three equations are acting in $\mathcal{D}'(Q_1)$, and the differential operators in the last equation are acting in $\mathcal{D}'(Q_2)$.

We rewrite the system of equations (15.5) in the form

$$\mathcal{A}_R u = f, \tag{15.6}$$

where $\mathcal{D}(\mathcal{A}_R) = \{u \in \mathring{W}^{1,4} : \mathcal{A}_R u \in L_2^4\}$ is the domain of the operator $\mathcal{A}_R$, $f = (f^1, f^2, f^3, f^4)$.

A vector-valued function u is called a generalized solution of the system (15.5) with the boundary conditions

$$u^1|_{\partial Q_1} = u^2|_{\partial Q_1} = u^3|_{\partial Q_1} = 0, \qquad u^4|_{\partial Q_2} = 0 \tag{15.7}$$

if $u \in \mathcal{D}(\mathcal{A}_R)$ and satisfies the equation (15.6).

Theorem 15.1. *The vector-valued function* $u \in \mathring{W}^{1,4}$ *yields a minimum of the functional* $E(u)$ *with the boundary conditions* (14.10), (14.11) *if and only if it is a generalized solution of the system* (15.5) *with the boundary conditions* (15.7).

Proof. The necessity was proved above.

Let $u \in \mathcal{D}(\mathcal{A}_R)$ be a solution of the system (15.6). We consider $E(u+v)$ setting $u^i(x,y) = v^i(x,y) = 0$ for $(x,y) \in G_1 \cup G_2$ $(i = 1,2,3)$, where $v \in \mathring{W}^{1,4}$ is an arbitrary vector-valued function. Clearly

$$E(u+v) = E(u) + \left.\frac{dE(u+tv)}{dt}\right|_{t=0} + J(v), \tag{15.8}$$

where

$$\begin{aligned} J(v) = \iint_{Q_2} \Bigg\{ \sum_{i=1,-1} G\delta \Bigg[&\frac{1}{1-\nu}\left((v^1_{i\tau})_x + \frac{ih}{\cos\alpha}\,(v^2_{i\tau})_x \right)^2 \\ &+ \frac{2\nu}{1-\nu}\left((v^1_{i\tau})_x + \frac{ih}{\cos\alpha}(v^2_{i\tau})_x \right)\left(\frac{1}{\sin\alpha}(v^3_{i\tau})_y - \cot\alpha v^4_y \right) \\ &+ \frac{1}{1-\nu}\left(\frac{1}{\sin\alpha}\,(v^3_{i\tau})_y - \cot\alpha v^4_y \right)^2 + \frac{1}{2}\left((v^1_{i\tau})_y + \frac{ih}{\cos\alpha}\,(v^2_{i\tau})_y \right)^2 \\ &+ \left((v^1_{i\tau})_y + \frac{ih}{\cos\alpha}(v^2_{i\tau})_y \right)\left(\frac{1}{\sin\alpha}\,(v^3_{i\tau})_x - \cot\alpha v^4_x \right) \\ &+ \frac{1}{2}\left(\frac{1}{\sin\alpha}\,(v^3_{i\tau})_x - \cot\alpha v^4_x \right)^2 \Bigg] + Gh\mu_\alpha(v^2 - v^3_x)^2 \\ &+ Gh\mu_0\left[\frac{1}{2h}\,(v^1_\tau - v^1_{-\tau}) + \frac{1}{2\cos\alpha}\,(v^2_\tau + v^2_{-\tau}) - v^4_x \right]^2 \Bigg\} dx\,dy. \end{aligned}$$

Since $0 < \nu < 1$, we have

$$\begin{aligned} J(v) \geq \iint_{Q_2} \Bigg\{ \sum_{i=1,-1} \frac{G\delta}{1-\nu} \Bigg[&(v^1_{i\tau})_x + \frac{ih}{\cos\alpha}\,(v^2_{i\tau})_x \\ &+\nu\left(\frac{1}{\sin\alpha}\,(v^3_{i\tau})_y - \cot\alpha v^4_y \right) \Bigg]^2 \Bigg\} dx\,dy \geq 0. \end{aligned} \tag{15.9}$$

The vector-valued function u is a solution of the system (15.6). Hence

$$\left.\frac{dE(u+tv)}{dt}\right|_{t=0} = 0 \tag{15.10}$$

for each $v \in \mathring{W}^{1,4}$. Thus, by virtue of (15.8)–(15.10), $E(u+v) \geq E(u)$ for each $v \in \mathring{W}^{1,4}$. □

Solvability of the Boundary Value Problem

Denote by $b[u,v]$ the left-hand side of (15.4), assuming that $u, v \in \mathring{W}^{1,4}$ are arbitrary.

Lemma 15.4. *The formula*

$$(u,v)'_{\mathring{W}^{1,4}} = b[u,v] \tag{15.11}$$

gives the equivalent inner product in the space $\mathring{W}^{1,4}$.

Proof. Since the set $\dot{C}^{\infty,4} = \dot{C}^{\infty}(Q_1) \times \dot{C}^{\infty}(Q_1) \times \dot{C}^{\infty}(Q_1) \times \dot{C}^{\infty}(Q_2)$ is dense in $\mathring{W}^{1,4}$, it is sufficient to prove that for all $u, v \in \dot{C}^{\infty,4}$

$$b[u,v] = b[v,u], \tag{15.12}$$

$$c_1(u,u)_{\mathring{W}^{1,4}} \le b[u,u] \le c_2(u,u)_{\mathring{W}^{1,4}}, \tag{15.13}$$

where $c_1, c_2 > 0$.

By virtue of Lemma 15.2, we have (15.12).

The right part of (15.13) follows from Lemma 15.3.

Now we prove the left part of (15.13). Denote $\varphi^j = I_1 u^j$ $(j = 1,2,3)$, $\varphi^4 = I_2 u^4$, where $u \in \dot{C}^{\infty,4}$. Since $P_i I_i$ is the identity operator in $L_2(Q_i)$ $(i = 1,2)$, by virtue of Lemma 8.1, we obtain

$$\begin{aligned}
b[u,u] &= (\mathcal{A}_R u, u)_{L_2^4} \\
&= \iint_{\mathbb{R}^2} \Bigg\{ \Bigg[-2\delta\Big(\Delta\varphi^1 + \frac{1+\nu}{1-\nu}\varphi^1_{xx}\Big) - \frac{2\delta(1+\nu)}{\sin\alpha(1-\nu)}\,\varphi^3_{xy} \\
&\quad + \delta\cot\alpha\,\frac{1+\nu}{1-\nu}\,(R_1\varphi^4)_{xy} + \mu_0(R_2\varphi^4)_x + \frac{\mu_0}{2h}\,R_4\varphi^1 - \frac{\mu_0}{2\cos\alpha}\,R_3\varphi^2\Bigg]\varphi^1 \\
&\quad + \Bigg[-\frac{2\delta h^2}{\cos^2\alpha}\Big(\Delta\varphi^2 + \frac{1+\nu}{1-\nu}\,\varphi^2_{xx}\Big) - \frac{\delta h(1+\nu)}{\sin\alpha(1-\nu)}\,(R_2\varphi^4)_{xy} - 2h\mu_\alpha\varphi^3_x \\
&\quad - \frac{h\mu_0}{\cos\alpha}\,(R_1\varphi^4)_x + \frac{\mu_0}{2\cos\alpha}\,R_3\varphi^1 + \frac{h\mu_0}{2\cos^2\alpha}\,R_5\varphi^2 + 2h\mu_\alpha\varphi^2\Bigg]\varphi^2 \\
&\quad + \Bigg[-\frac{2\delta(1+\nu)}{\sin\alpha(1-\nu)}\,\varphi^1_{xy} - \frac{2\delta}{\sin^2\alpha}\Big(\Delta\varphi^3 + \frac{1+\nu}{1-\nu}\,\varphi^3_{yy}\Big) \\
&\quad + \frac{\delta\cot\alpha}{\sin\alpha}\Big(\Delta(R_1\varphi^4) + \frac{1+\nu}{1-\nu}\,(R_1\varphi^4)_{yy}\Big) - 2h\mu_\alpha\varphi^3_{xx} + 2h\mu_\alpha\varphi^2_x\Bigg]\varphi^3 \\
&\quad + \Bigg[\delta\cot\alpha\,\frac{1+\nu}{1-\nu}\,(R_1\varphi^1)_{xy} + \frac{\delta h(1+\nu)}{\sin\alpha(1-\nu)}\,(R_2\varphi^2)_{xy} \\
&\quad + \frac{\delta\cot\alpha}{\sin\alpha}\Big(\Delta(R_1\varphi^3) + \frac{1+\nu}{1-\nu}\,(R_1\varphi^3)_{yy}\Big) \\
&\quad - 2\delta\cot^2\alpha\Big(\Delta\varphi^4 + \frac{1+\nu}{1-\nu}\varphi^4_{yy}\Big) \\
&\quad - 2h\mu_0\varphi^4_{xx} + \mu_0(R_2\varphi^1)_x + \frac{h\mu_0}{\cos\alpha}\,(R_1\varphi^2)_x\Bigg]\varphi^4\Bigg\}\,dx\,dy. \qquad (15.14)
\end{aligned}$$

Denote by $\psi^j(\xi) = \widehat{\varphi^j}(\xi)$ the Fourier transform of the function $\varphi^j(x)$, where

$$\widehat{\varphi^j}(\xi) = \frac{1}{2\pi}\iint_{R^2} \varphi^j(x,y)e^{-i(\xi_1 x+\xi_2 y)}\,dx\,dy \qquad (j=1,\dots,4).$$

Clearly for each $w \in \dot{C}^\infty(\mathbb{R}^2)$

$$(\widehat{w_x})(\xi) = (i\xi_1)\widehat{w}(\xi), \qquad (\widehat{w_y})(\xi) = (i\xi_2)\widehat{w}(\xi), \tag{15.15}$$

$$\left.\begin{aligned}
&(\widehat{R_1 w})(\xi) = 2\cos(\xi_2\tau)\widehat{w}(\xi),\\
&(\widehat{R_2 w})(\xi) = 2i\sin(\xi_2\tau)\widehat{w}(\xi),\\
&(\widehat{R_3 w})(\xi) = 2i\sin(2\xi_2\tau)\widehat{w}(\xi),\\
&(\widehat{R_4 w})(\xi) = 2(1-\cos(2\xi_2\tau))\widehat{w}(\xi),\\
&(\widehat{R_5 w})(\xi) = 2(1+\cos(2\xi_2\tau))\widehat{w}(\xi).
\end{aligned}\right\} \tag{15.16}$$

From (15.14)–(15.16) and from the Plancherel theorem it follows that

$$b(u,u) = \iint_{\mathbb{R}^2}\left\{\sum_{j=1}^{5}\Phi_j(\xi)\right\}d\xi, \tag{15.17}$$

where

$$\Phi_1(\xi) = 2\delta(\xi_1^2+\xi_2^2)\left(\psi^1\overline{\psi^1} + \frac{h^2}{\cos^2\alpha}\,\psi^2\overline{\psi^2}\right), \tag{15.18}$$

$$\begin{aligned}\Phi_2(\xi) = 2\delta(\xi_1^2+\xi_2^2)\Big\{&\frac{1}{\sin^2\alpha}\,\psi^3\overline{\psi^3} - \frac{\cot\alpha}{\sin\alpha}\cos(\xi_2\tau)(\psi^4\overline{\psi^3}+\psi^3\overline{\psi^4})\\
&+\cot^2\alpha\psi^4\overline{\psi^4}\Big\},\end{aligned} \tag{15.19}$$

$$\begin{aligned}\Phi_3(\xi) = 2\delta\,\frac{1+\nu}{1-\nu}\Big\{&\xi_1^2\psi^1\overline{\psi^1} + \frac{h^2}{\cos^2\alpha}\,\xi_1^2\psi^2\overline{\psi^2} + \frac{\xi_2^2}{\sin^2\alpha}\,\psi^3\overline{\psi^3} + \xi_2^2\cot^2\alpha\psi^4\overline{\psi^4}\\
&+\frac{\xi_1\xi_2}{\sin\alpha}(\psi^3\overline{\psi^1}+\psi^1\overline{\psi^3}) - \xi_1\xi_2\cot\alpha\cos(\xi_2\tau)(\psi^4\overline{\psi^1}+\psi^1\overline{\psi^4})\\
&+\frac{ih}{\sin\alpha}\xi_1\xi_2\sin(\xi_2\tau)(\psi^4\overline{\psi^2}-\psi^2\overline{\psi^4})\\
&-\frac{\cot\alpha}{\sin\alpha}\,\xi_2^2\cos(\xi_2\tau)(\psi^4\overline{\psi^3}+\psi^3\overline{\psi^4})\Big\},\end{aligned} \tag{15.20}$$

$$\Phi_4(\xi) = 2h\mu_\alpha(\psi^2\overline{\psi^2} - i\xi_1(\psi^3\overline{\psi^2}-\psi^2\overline{\psi^3}) + \xi_1^2\psi^3\overline{\psi^3}), \tag{15.21}$$

$$\begin{aligned}\Phi_5(\xi) = 2h\mu_0\Big\{&\frac{1-\cos(2\xi_2\tau)}{2h^2}\,\psi^1\overline{\psi^1} + \frac{1+\cos(2\xi_2\tau)}{2\cos^2\alpha}\,\psi^2\overline{\psi^2} + \xi_1^2\psi^4\overline{\psi^4}\\
&-\frac{i\sin(2\xi_2\tau)}{2h\cos\alpha}(\psi^2\overline{\psi^1}-\psi^1\overline{\psi^2}) - \frac{\xi_1}{h}\sin(\xi_2\tau)(\psi^4\overline{\psi^1}+\psi^1\overline{\psi^4})\\
&-\frac{i\xi_1}{\cos\alpha}\cos(\xi_2\tau)(\psi^4\overline{\psi^2}-\psi^2\overline{\psi^4})\Big\}.\end{aligned} \tag{15.22}$$

It is easy to see that

$$\Phi_2(\xi) = 2\delta(\xi_1^2+\xi_2^2)\left\{\left|\frac{\psi^3}{\sin\alpha} - \cot\alpha\cos(\xi_2\tau)\psi^4\right|^2 + \frac{1}{2}\cot^2\alpha(1-\cos(2\xi_2\tau))\psi^4\overline{\psi^4}\right\}, \tag{15.23}$$

$$\Phi_3(\xi) = 2\delta\frac{1+\nu}{1-\nu}\left\{\left|\xi_1\psi^1 + \frac{\xi_2}{\sin\alpha}\psi^3 - \xi_2\cot\alpha\cos(\xi_2\tau)\psi^4\right|^2 + \left|\frac{h\xi_1}{\cos\alpha}\psi^2 + i\xi_2\cot\alpha\sin(\xi_2\tau)\psi^4\right|^2\right\}, \tag{15.24}$$

$$\Phi_4(\xi) = 2h\mu_\alpha|i\psi^2+\xi_1\psi^3|^2, \tag{15.25}$$

$$\Phi_5(\xi) = 2h\mu_0\left|\frac{\sin(\xi_2\tau)}{h}\psi^1 - \frac{i\cos(\xi_2\tau)}{\cos\alpha}\psi^2 - \xi_1\psi^4\right|^2. \tag{15.26}$$

From (15.23) it follows that

$$\Phi_2(\xi) \geq \delta\cot^2\alpha(\xi_1^2+\xi_2^2)(1-\cos(2\xi_2\tau))\psi^4\overline{\psi^4}. \tag{15.27}$$

Similarly, we can prove that

$$\Phi_2(\xi) \geq \frac{\delta}{\sin^2\alpha}(\xi_1^2+\xi_2^2)(1-\cos(2\xi_2\tau))\psi^3\overline{\psi^3}. \tag{15.28}$$

Summing (15.27), (15.28), we obtain

$$\Phi_2(\xi) \geq \frac{\delta}{2\sin^2\alpha}(\xi_1^2+\xi_2^2)(1-\cos(2\xi_2\tau))(\psi^3\overline{\psi^3}+\cos^2\alpha\psi^4\overline{\psi^4}). \tag{15.29}$$

From (15.17), (15.18), (15.29), and (15.24)–(15.26), it follows that

$$b[u,u] \geq \iint_{\mathbb{R}^2}\left\{2\delta(\xi_1^2+\xi_2^2)\psi^1\overline{\psi^1} + \frac{2\delta h^2}{\cos^2\alpha}(\xi_1^2+\xi_2^2)\psi^2\overline{\psi^2} + \frac{\delta}{2\sin^2\alpha}(\xi_1^2+\xi_2^2)(1-\cos(2\xi_2\tau))\psi^3\overline{\psi^3} + \frac{\delta\cot^2\alpha}{2}(\xi_1^2+\xi_2^2)(1-\cos(2\xi_2\tau))\psi^4\overline{\psi^4}\right\}d\xi_1\,d\xi_2. \tag{15.30}$$

Hence, by virtue of the Plancherel theorem, the formulas (15.15), (15.16) and Lemma 8.1, we have

$$b[u,u] \geq \iint_{\mathbb{R}^2}\left\{2\delta[(\varphi_x^1)^2+(\varphi_y^1)^2] + \frac{2\delta h^2}{\cos^2\alpha}[(\varphi_x^2)^2+(\varphi_y^2)^2] + \frac{\delta}{4\sin^2\alpha}[(R_4\varphi_x^3)\varphi_x^3+(R_4\varphi_y^3)\varphi_y^3] + \frac{\delta\cot^2\alpha}{4}[(R_4\varphi_x^4)\varphi_x^4+(R_4\varphi_y^4)\varphi_y^4]\right\}dx\,dy$$

$$= \iint_{Q_1} 2\delta[(u_x^1)^2 + (u_y^1)^2]\,dx\,dy + \iint_{Q_1} \frac{2\delta h^2}{\cos^2\alpha}[(u_x^2)^2 + (u_y^2)^2]\,dx\,dy$$
$$+ \iint_{Q_1} \frac{\delta}{4\sin^2\alpha}[(P_1R_4I_1u_x^3)u_x^3 + (P_1R_4I_1u_y^3)u_y^3]\,dx\,dy$$
$$+ \iint_{Q_2} \frac{\delta\cot^2\alpha}{4}[(P_2R_4I_2u_x^4)u_x^4 + (P_2R_4I_2u_y^4)u_y^4]\,dx\,dy. \qquad (15.31)$$

Lemmas 15.1, 8.11, 8.3, 8.4, and 8.9 imply that the operators $P_iR_4I_i\colon L_2(Q_i) \to L_2(Q_i)$ are self-adjoint and positive definite. From this, using Theorem B.11 concerning the equivalent norms, we obtain

$$b[u,u] \geq c_0\Big\{\sum_{j=1}^{3}\iint_{Q_1}[(u_x^j)^2 + (u_y^j)^2]\,dx\,dy + \iint_{Q_2}[(u_x^4)^2 + (u_y^4)^2]\,dx\,dy\Big\}$$
$$\geq c_1\|u\|^2_{\mathring{W}^{1,4}}. \quad \square$$

Theorem 15.2. *The system of differential-difference equations* (15.5) *with the boundary conditions* (15.7) *has a unique generalized solution* $u \in \mathcal{D}(\mathcal{A}_R)$ *for each* $f \in L_2^4$ *and*

$$\|u\|_{\mathring{W}^{1,4}} \leq c\|f\|_{L_2^4}, \qquad (15.32)$$

where $c > 0$ *is a constant.*

Proof. A vector-valued function $u \in \mathring{W}^{1,4}$ is a solution of the system (15.6) if and only if it satisfies the integral identity

$$b[u,v] = (f,v)_{L_2^4} \qquad (15.33)$$

for all $v \in \mathring{W}^{1,4}$.

By the Riesz theorem and Lemma 15.1, there is a linear bounded operator $A\colon L_2^4 \to \mathring{W}^{1,4}$ such that

$$(f,v)_{L_2^4} = (Af,v)'_{\mathring{W}^{1,4}} \qquad (15.34)$$

for all $f \in L_2^4$ and $v \in \mathring{W}^{1,4}$. Hence, we can rewrite (15.33) in the following form:

$$(u,v)'_{\mathring{W}^{1,4}} = (Af,v)'_{\mathring{W}^{1,4}} \qquad (15.35)$$

for every $v \in \mathring{W}^{1,4}$. Thus, for each $f \in L_2^4$ there is a unique function $u = Af \in \mathring{W}^{1,4}$ satisfying (15.33) for every $v \in \mathring{W}^{1,4}$ and

$$\|u\|_{\mathring{W}^{1,4}} \leq \|A\| \cdot \|f\|_{L_2^4}. \qquad \square$$

Application of the Ritz Method

In order to apply variational methods to the solution of the boundary value problem (15.5), (15.7), we consider the functional

$$I(v) = b[v,v] - 2(f,v)_{L_2^4} \qquad (v \in \mathring{W}^{1,4}), \tag{15.36}$$

where $f \in L_2^4$.

By virtue of Lemma 15.4, we can rewrite (15.36) in the following form:

$$I(v) = (v,v)'_{\mathring{W}^{1,4}} - 2(f,v)_{L_2^4} \qquad (v \in \mathring{W}^{1,4}). \tag{15.37}$$

From (15.37) and from Theorem A.17 it follows that there is a unique vector-valued function $u \in \mathring{W}^{1,4}$ which yields a minimum of the functional I on $\mathring{W}^{1,4}$. It is easy to see that this function u satisfies the integral identity

$$b[u,v] - (f,v)_{L_2^4} = 0 \tag{15.38}$$

for all $v \in \mathring{W}^{1,4}$. Hence u is a generalized solution of the boundary value problem (15.5), (15.7).

Let $\{\varphi^k\}$ be an arbitrary linearly independent system of vector-valued functions, whose linear span is dense everywhere in $\mathring{W}^{1,4}$. Then, by virtue of Theorem A.18, we have the following assertion:

Theorem 15.3. *The Ritz sequence of the functional I constructed with respect to a system $\{\varphi^k\}$ converges in $\mathring{W}^{1,4}$ to the generalized solution of the boundary value problem* (15.5), (15.7).

Spectrum of a Differential-Difference Operator

Now let the spaces L_2^4, $\mathring{W}^{1,4}$ be complex. Denote by $b[u,v]$ the left side of (15.4) in which the functions $v^1,\dots.v^4$ are replaced by $\overline{v^1},\dots,\overline{v^4}$. Then Lemmas 15.1–15.4 and Theorem 15.2 hold.

From Lemma 15.4 and Theorem A.16 we obtain:

Theorem 15.4. *The unbounded operator $\mathcal{A}_R\colon L_2^4 \to L_2^4$ is self-adjoint, the spectrum $\sigma(\mathcal{A}_R)$ consists of real isolated eigenvalues $\lambda_s > 0$ of finite multiplicity. The eigenfunctions v_s of the operator $\mathcal{A}_R$ form an orthonormal basis in L_2^4. Moreover, the functions $v_s/\sqrt{\lambda_s}$ form an orthonormal basis in $\mathring{W}^{1,4}$ with inner product given by the formula* (15.11).

Remark 15.1. Let $A_R\colon L_2^4 \to L_2^4$ be the unbounded operator given by

$$\left.\begin{aligned} &\mathcal{D}(A_R) = \dot{C}^{\infty,4}, \\ &A_R u = \mathcal{A}_R u \quad (u \in \mathcal{D}(A_R)) \end{aligned}\right\}. \tag{15.39}$$

Then $\mathcal{A}_R$ is the Friedrichs extension of A_R.

16 Smoothness of Solutions

Differentiability of Solutions in Subdomains

Let N_α be the number of slanting ribs, and let $N = 2N_\alpha$. Then $b = N\tau$. Denote $Q_{1l} = (0,a) \times ((l-1)\tau, l\tau)$ $(l = 1,\dots,N)$. We introduce the set

$$\mathcal{K} = \{(x,y) : x = 0, a,\ y = l\tau\ (l = 0,\dots,N)\} \tag{16.1}$$

(cf. (7.1)).

As in the proof of Theorem 11.3, we obtain the following:

Theorem 16.1. *Let u be a generalized solution of the boundary value problem (15.5), (15.7), and let $f \in W^{k,4}$.*

Then $u^i \in W^{k+2}(Q_{1l} \setminus \overline{\mathcal{K}^\varepsilon})$ $(i = 1,2,3,\ l = 2,\dots,N-1)$, $u^4 \in W^{k+2}(Q_{1l} \setminus \overline{\mathcal{K}^\varepsilon})$ $(l = 1,\dots,N)$ for each $\varepsilon > 0$.

Differentiability of Solutions on a Boundary of Neighboring Subdomains

We define the matrices A_{11}, A_{12}, A_{21}, A_{22} of order $(N-3)\times(N-3)$, $(N-3)\times(N-1)$, $(N-1)\times(N-3)$, $(N-1)\times(N-1)$, respectively, by the formulas

$$A_{11} = \begin{pmatrix} -2 & 0 & 0 & \dots & 0 \\ 0 & -2 & 0 & \dots & 0 \\ 0 & 0 & -2 & \dots & 0 \\ \vdots & \vdots & \vdots & \ddots & \vdots \\ 0 & 0 & 0 & \dots & -2 \end{pmatrix}, \quad A_{12} = \begin{pmatrix} 1 & 0 & 1 & 0 & 0 & \dots & 0 & 0 & 0 \\ 0 & 1 & 0 & 1 & 0 & \dots & 0 & 0 & 0 \\ 0 & 0 & 1 & 0 & 1 & \dots & 0 & 0 & 0 \\ \vdots & \vdots & \vdots & \vdots & \vdots & \ddots & \vdots & \vdots & \vdots \\ 0 & 0 & 0 & 0 & 0 & \dots & 1 & 0 & 1 \end{pmatrix},$$

$$A_{21} = \begin{pmatrix} 1 & 0 & 0 & \dots & 0 \\ 0 & 1 & 0 & \dots & 0 \\ 1 & 0 & 1 & \dots & 0 \\ 0 & 1 & 0 & \dots & 0 \\ 0 & 0 & 1 & \dots & 0 \\ \vdots & \vdots & \vdots & \ddots & \vdots \\ 0 & 0 & 0 & \dots & 1 \\ 0 & 0 & 0 & \dots & 0 \\ 0 & 0 & 0 & \dots & 1 \end{pmatrix}, \quad A_{22} = \begin{pmatrix} -2 & 0 & 0 & 0 & \dots & 0 & 0 & 0 \\ 0 & -2 & 0 & 0 & \dots & 0 & 0 & 0 \\ 0 & 0 & -2 & 0 & \dots & 0 & 0 & 0 \\ 0 & 0 & 0 & -2 & \dots & 0 & 0 & 0 \\ 0 & 0 & 0 & 0 & \dots & 0 & 0 & 0 \\ \vdots & \vdots & \vdots & \vdots & \ddots & \vdots & \vdots & \vdots \\ 0 & 0 & 0 & 0 & \dots & -2 & 0 & 0 \\ 0 & 0 & 0 & 0 & \dots & 0 & -2 & 0 \\ 0 & 0 & 0 & 0 & \dots & 0 & 0 & -2 \end{pmatrix}.$$

Clearly $A_{21} = A_{12}^*$.

Let A be a matrix of order $(2N-4)\times(2N-4)$ given by

$$A = \begin{pmatrix} A_{11} & A_{12} \\ A_{21} & A_{22} \end{pmatrix}. \tag{16.2}$$

It is easy to prove the following:

Lemma 16.1. $\det A \neq 0$.

Denote by A_l the $(2N-4)\times(2N-5)$-matrix obtained from A by deleting the lth column. We define the columns B_j $(j=\overline{N-1,2N-5})$ of dimension $2N-4$ with elements $b_s^j=\delta_{js}$, where

$$\delta_{js}=\begin{cases}1 & \text{if } s=j,\\ 0 & \text{if } s\neq j.\end{cases}$$

Let $\Omega_l=(0,a)\times((l-1)\tau,(l+1)\tau)$ $(l=1,\dots,N-1)$.

Theorem 16.2. *Let u be a generalized solution of the boundary value problem (15.5), (15.7), and let $f\in L_2^4$.*

Then for each $\varepsilon>0$ we have:

1) $u^i\in W^2(Q_1\setminus\overline{\mathcal{K}^\varepsilon})$ $(i=1,2)$ for all $f\in L_2^4$.

2) For a given $2\le l\le N-2$, u^3 belongs to $W^2(\Omega_l\setminus\overline{\mathcal{K}^\varepsilon})$ for all $f\in L_2^4$ if and only if the columns B_j $(j=\overline{N-1,2N-5})$ are linear combinations of the columns of A_{l-1}.

3) For a given $1\le l\le N-1$, u^4 belongs to $W^2(\Omega_l\setminus\overline{\mathcal{K}^\varepsilon})$ for all $f\in L_2^4$ if and only if the columns B_j $(j=\overline{N-1,2N-5})$ are linear combinations of the columns of A_{N-3+l}.

Proof. 1. Since $u\in\mathring{W}^{1,4}$, by Lemma 15.3 all the terms in the left-hand side of (15.5) containing the functions $u^1,\dots,u^4$ and their derivatives of the first order belong to L_2. On the other hand, by virtue of Theorem 16.1, all the terms containing the derivatives $u^1_{xy},\dots,u^4_{xy}$ and $u^1_{xx},\dots,u^4_{xx}$ belong to L_2. Therefore, since $f\in L_2^4$, the sums of terms containing the derivatives $u^1_{yy},\dots,u^4_{yy}$ belong to L_2. From this and from Theorem 16.1 it follows that

$$u^i_y|_{y=l\tau+0}=u^i_y|_{y=l\tau-0}\qquad(i=1,2,\ l=2,\dots,N-2),\tag{16.3}$$

$$\begin{aligned}&-2u^3_y|_{y=l\tau+0}+\cos\alpha(P_1R_1I_2u^4)_y|_{y=l\tau+0}\\&\quad=-2u^3_y|_{y=l\tau-0}+\cos\alpha(P_1R_1I_2u^4)_y|_{y=l\tau-0}\qquad(l=2,\dots,N-2),\end{aligned}\tag{16.4}$$

$$\begin{aligned}&(P_2R_1I_1u^3)_y|_{y=l\tau+0}-2\cos\alpha u^4_y|_{y=l\tau+0}\\&\quad=(P_2R_1I_1u^3)_y|_{y=l\tau-0}-2\cos\alpha u^4_y|_{y=l\tau-0}\qquad(l=1,\dots,N-1).\end{aligned}\tag{16.5}$$

Theorem 16.1 and the equalities (16.3) imply that $u^i\in W^2(Q_1\setminus\overline{\mathcal{K}^\varepsilon})$ $(i=1,2)$. Thus we proved the first part of the theorem.

2. We set $\varphi_l^+=u^3_y|_{y=l\tau+0}$ $(l=1,\dots,N-2)$, $\varphi_l^-=u^3_y|_{y=l\tau-0}$ $(l=2,\dots.N-1)$, $\psi_l^+=u^4_y|_{y=l\tau+0}$ $(l=0,\dots,N-1)$, $\psi_l^-=u^4_y|_{y=l\tau-0}$ $(l=1,\dots,N)$. Then the relations (16.4), (16.5) will have the form

$$\left.\begin{aligned}&-2\varphi_2^+ +\cos\alpha\psi_1^+ +\cos\alpha\psi_3^+ =-2\varphi_2^- +\cos\alpha\psi_1^- +\cos\alpha\psi_3^-,\\&-2\varphi_3^+ +\cos\alpha\psi_2^+ +\cos\alpha\psi_4^+ =-2\varphi_3^- +\cos\alpha\psi_2^- +\cos\alpha\psi_4^-,\\&\dots\dots\dots\dots\dots\dots\dots\dots\dots\dots\dots\dots\\&-2\varphi_{N-2}^+ +\cos\alpha\psi_{N-3}^+ +\cos\alpha\psi_{N-1}^+\\&\qquad=-2\varphi_{N-2}^- +\cos\alpha\psi_{N-3}^- +\cos\alpha\psi_{N-1}^-,\end{aligned}\right\}\tag{16.6}$$

$$\left.\begin{aligned}
&\varphi_2^+ - 2\cos\alpha\psi_1^+ = \varphi_2^- - 2\cos\alpha\psi_1^-,\\
&\varphi_1^+ + \varphi_3^+ - 2\cos\alpha\psi_2^+ = \varphi_3^- - 2\cos\alpha\psi_2^-,\\
&\varphi_2^+ + \varphi_4^+ - 2\cos\alpha\psi_3^+ = \varphi_2^- + \varphi_4^- - 2\cos\alpha\psi_3^-,\\
&\cdots\cdots\cdots\cdots\cdots\cdots\cdots\cdots\cdots\cdots\cdots\cdots\\
&\varphi_{N-4}^+ + \varphi_{N-2}^+ - 2\cos\alpha\psi_{N-3}^+ = \varphi_{N-4}^- + \varphi_{N-2}^- - 2\cos\alpha\psi_{N-3}^-,\\
&\varphi_{N-3}^+ - 2\cos\alpha\psi_{N-2}^+ = \varphi_{N-3}^- + \varphi_{N-1}^- - 2\cos\alpha\psi_{N-2}^-,\\
&\varphi_{N-2}^+ - 2\cos\alpha\psi_{N-1}^+ = \varphi_{N-2}^- - 2\cos\alpha\psi_{N-1}^-.
\end{aligned}\right\} \tag{16.7}$$

We introduce the vector-valued function of dimension $2N-4$ by the formula

$$\Phi = (\varphi_2^+ - \varphi_2^-, \ldots, \varphi_{N-2}^+ - \varphi_{N-2}^-, \cos\alpha(\psi_1^+ - \psi_1^-), \ldots, \cos\alpha(\psi_{N-1}^+ - \psi_{N-1}^-)).$$

Then the system (16.6), (16.7) can be written as

$$A\Phi = -B_{N-1}\varphi_1^+ + B_{2N-5}\varphi_{N-1}^-. \tag{16.8}$$

Now we prove the second part of the theorem.

3. Sufficiency. By Theorem 16.1, $u^3 \in W^2(\Omega_l \setminus \overline{\mathcal{K}^\varepsilon})$ $(2 \le l \le N-2)$ if and only if

$$\Phi_{l-1} = 0. \tag{16.9}$$

By what has been proved above, the solution of the boundary value problem (15.5), (15.7) satisfies the equation (16.8). Lemma 16.1 implies that $\det A \neq 0$. Hence there is a unique solution of the system (16.8). Suppose that B_{N-1}, B_{2N-5} are linear combinations of the columns of A_{l-1}. Then the matrix of the system (16.8), (16.9) and the extended matrix have the same rank $2N-4$. Hence a solution Φ of the system (16.8) also satisfies the equation (16.9). Thus $u^3 \in W^2(\Omega_l \setminus \overline{\mathcal{K}^\varepsilon})$.

4. Necessity. Assume to the contrary that B_{N-1} is not a linear combination of the columns of A_{l-1}. Then we shall prove that there is $u \in \mathcal{D}(\mathcal{L}_R)$ such that $u^3 \notin W^2_{\text{loc}}(\Omega_l)$.

Denote $w(x,y) = y\xi(x-a/2, y)$, where $8\delta < \min\{a, \tau\}$, $\xi \in \dot{C}^\infty(\mathbb{R}^2)$, $\xi(x,y) = 1$ for $x \in S_\delta(0)$, $\operatorname{supp}\xi \in S_{2\delta}(0)$. Consider the system

$$A\Phi = -B_{N-1}w_y|_{y=0}. \tag{16.10}$$

There is a unique solution of the system (16.10) $\Phi \in \dot{C}^{\infty,2N-4}(a/2-2\delta, a/2+2\delta)$. Clearly there is a vector-valued function $\mathcal{Z} \in \dot{C}^{2,2N-4}(\mathbb{R}^2)$ such that $\operatorname{supp}\mathcal{Z} \subset S_{4\delta}(g)$ and

$$\mathcal{Z}|_{y=0} = 0, \qquad \mathcal{Z}_y|_{y=0} = \Phi, \tag{16.11}$$

where $g = (a/2, 0)$.

Let

$$\begin{aligned}
u^i(x,y) &\equiv 0 && (i = 1,2,\ (x,y) \in Q_1),\\
u^3(x,y) &= w(x, y-\tau) && ((x,y) \in Q_{12}),\\
u^3(x,y) &= \mathcal{Z}_l(x, y-(l+1)\tau) && ((x,y) \in Q_{1,l+2},\ l = 1,\ldots,N-3),\\
u^4(x,y) &= 0 && ((x,y) \in Q_{11}),\\
u^4(x,y) &= \mathcal{Z}_{N-3+l}(x, y-l\tau) && ((x,y) \in Q_{1,l+1},\ l = 1,\ldots,N-1).
\end{aligned}$$

By virtue of (16.11), $u \in \mathring{W}^{1,4}$. The system (16.10) implies that $u(x,y)$ satisfies (16.8). Therefore $\mathcal{A}_R u \in L_2^4$. On the other hand, B_{N-1} is not a linear combination of the columns of A_{l-1}. Hence, the matrix and the extended matrix of the system (16.10), (16.9) have the ranks $2N-4$ and $2N-3$, respectively. Thus the function $u(x,y)$ does not satisfy the equation (16.9).

Similarly we can prove the third part of the theorem. □

Example 16.1. Let $N = 4$. Then $Q_1 = (0,a) \times (\tau, 3\tau)$, $Q_2 = (0,a) \times (0, 4\tau)$, $\mathcal{K} = \{(x,y) : x = 0, a;\ y = l\tau\ (l = 0, \dots, 4)\}$. The matrix A has the form

$$A = \begin{pmatrix} -2 & 1 & 0 & 1 \\ 1 & -2 & 0 & 0 \\ 0 & 0 & -2 & 0 \\ 1 & 0 & 0 & -2 \end{pmatrix}.$$

Clearly $N - 1 = 2N - 5 = 3$. Therefore

$$B_3 = \begin{pmatrix} 0 \\ 0 \\ 1 \\ 0 \end{pmatrix}.$$

The column B_3 is a linear combination of the columns of A_i $(i = 1,2,4)$. On the other hand, B_3 is not a linear combination of the columns of A_3. Then, by Theorem 16.2, $u^i \in W^2(Q_1 \setminus \overline{\mathcal{K}^\varepsilon})$ $(i = 1,2)$, $u^3 \in W^2(\Omega_2 \setminus \overline{\mathcal{K}^\varepsilon})$, $u^4 \in W^2(\Omega_1 \setminus \overline{\mathcal{K}^\varepsilon})$, $u^4 \in W^2(\Omega_3 \setminus \overline{\mathcal{K}^\varepsilon})$, but generally speaking, $u^4 \notin W^2(\Omega_2 \setminus \overline{\mathcal{K}^\varepsilon})$.

Problem 16.1. *Does Theorem* 16.2 *hold for* $\varepsilon = 0$?

17 The One-Dimensional Case

The One-Dimensional Elastic Model

In order to demonstrate the methods of this chapter, we also consider the one-dimensional case. The one-dimensional analog of a three-layer plate with a two-phase filler which is under consideration here is the continuous interpretation of a two-belt rod system — a truss with a regular set of absolutely rigid vertical and diagonal braces (Fig. III.2). A representation of the field of elastic displacements of this model can be obtained from the corresponding expressions for a three-layer plate, assuming that the displacements take place only in the plane $y0z$ and are independent of the coordinate x. Using (14.5), (14.7), we have

$$V_\beta(y,z) = v_\beta(y - z\tan\beta), \tag{17.1}$$

$$v^\pm(y)\sin\beta + w^\pm(y)\cos\beta = v_\beta(y \mp h\tan\beta) \qquad (\beta = 0, \alpha), \tag{17.2}$$

from which it follows that the field of displacements of a three-layer beam with a two-phase filler is determined by six functions of one variable $(v^-, w^-, v^+, w^+, v_\alpha, v_0)$ connected by four relations (17.2).

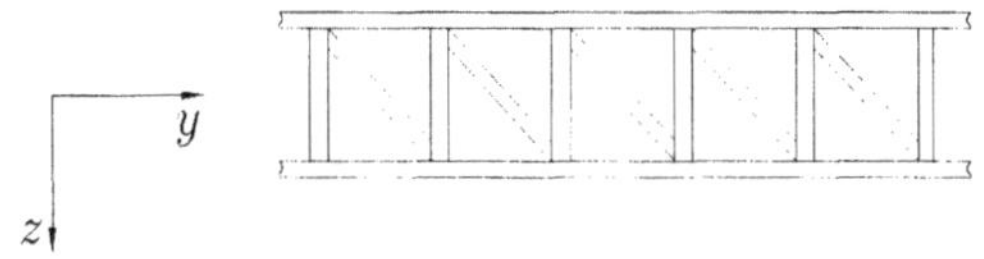

Fig. III.2

We introduce the two-dimensional vector-valued function $u = (u^1, u^2)$ of one variable:

$$u^1 = v_\alpha, \qquad u^2 = v_0. \tag{17.3}$$

Then we obtain

$$v^\pm = \csc\alpha u^1_{\mp\tau} - \cot\alpha u^2, \qquad w^\pm = u^2, \tag{17.4}$$

where $u^i_{\pm\tau} = u^i(y \pm \tau)$.

The Variational and Boundary Value Problem in the One-Dimensional Case

Using (17.1)–(17.4), in a manner analogous to the previous case, we arrive at a problem involving the minimum of the functional

$$J(u) = \int_0^b \sum_{i=1,-1} \big[EF(1-\nu)^{-1}(\csc\alpha(u^1_{i\tau})' - \cot\alpha(u^2)')^2$$
$$-Y^{-i}(\csc\alpha u^1_{i\tau} - \cot\alpha u^2) - Z^i u^2\big]dy \tag{17.5}$$

with the boundary conditions

$$\left.\begin{aligned} u^1(y) &= 0 \quad (y \in [-\tau,\tau] \cup [b-\tau, b+\tau]),\\ u^2(y) &= 0 \quad (y = 0, b), \end{aligned}\right\} \tag{17.6}$$

where E is the Young modulus, F is the cross-sectional area of the belts, Y^i, Z^i are the components of the external load on the belts.

Denote by $I_i = I_{Q_i}\colon L_2(Q_i) \to L_2(\mathbb{R})$ the operator of extension of functions from $L_2(Q_i)$ by zero in $\mathbb{R} \setminus Q_i$ $(i = 1,2)$, where $Q_1 = (\tau, b-\tau)$, $Q_2 = (0,b)$. Denote by $P_i = P_{Q_i}\colon L_2(\mathbb{R}) \to L_2(Q_i)$ the operator of restriction of functions from $L_2(\mathbb{R})$ to Q_i $(i = 1,2)$.

Assume that $Y^i, Z^i \in L_2(0,b)$ $(i = 1,2)$. Then

$$f^1 = (1-\nu)\sin\alpha(2EF)^{-1}(Y^{-1}_{-\tau} + Y^1_{+\tau}) \in L_2(\tau, b-\tau),$$
$$f^2 = (1-\nu)\sin\alpha(2EF)^{-1}[(Z^{-1} + Z^1)\sin\alpha - (Y^{-1} + Y^1)\cos\alpha] \in L_2(0,b).$$

We consider the system of differential-difference equations

$$\left.\begin{aligned} -2(u^1)'' + \cos\alpha(P_1R_1I_2u^2)'' &= f^1(y) \qquad (y \in (\tau, b-\tau)),\\ \cos\alpha(P_2R_1I_1u^1)'' - 2\cos^2\alpha(u^2)'' &= f^2(y) \qquad (y \in (0,b)) \end{aligned}\right\} \tag{17.7}$$

with the boundary conditions

$$u^1(\tau) = u^1(b-\tau) = 0, \qquad u^2(0) = u^2(b) = 0, \tag{17.8}$$

where $R_1 w = w_\tau + w_{-\tau}$.

A vector-valued function $u \in \mathring{W}^{1,2} = \mathring{W}^1(\tau, b-\tau) \times \mathring{W}^1(0,b)$ is called a generalized solution of the problem (17.7), (17.8) if it satisfies the system (17.7) in a sense of distributions.

As in Section 15, we can show that the vector-valued function $u \in \mathring{W}^{1,2}$ yields a minimum of the functional (17.5) with boundary conditions (17.6) if and only if it is a generalized solution of the problem (17.7), (17.8).

We rewrite the system of equations (17.7) in the form

$$\mathcal{A}_R u = f, \tag{17.9}$$

where $\mathcal{D}(\mathcal{A}_R) = \{u \in \mathring{W}^{1,2} : \mathcal{A}_R u \in L_2^2\}$ is the domain of the operator $\mathcal{A}_R$, which acts in the space of distributions. Here $L_2^2 = L_2(\tau, b-\tau) \times L_2(0,b)$, $f = (f^1, f^2)$.

It can be shown that as in the two-dimensional case:

1) the solution of the boundary value problem (17.7), (17.8) exists and is unique;

2) the spectrum $\sigma(\mathcal{A}_R)$ consists of real isolated eigenvalues $\lambda_s > 0$ of finite multiplicity;

3) the Ritz method converges.

Differentiability of Solutions

As in Section 16, we write $b = N\tau$, $N = 2N_\alpha$. We introduce the $(2N-2)$-dimensional vector-valued function $W = (w_1, \dots, w_{2N-2})$ defined in the interval $(0,\tau)$:

$$w_j(y) = \begin{cases} u^1(y + j\tau) & (j = 1, \dots, N-2), \\ u^2(y + (j-N+1)\tau) & (j = N-1, \dots, 2N-2). \end{cases} \tag{17.10}$$

Then, passing to the variables (17.10), we reduce the system of differential-difference equations (17.7) to a system of $2N-2$ ordinary differential equations in the functions $w_1, \dots, w_{2N-2}$:

$$CW''(y) = F(y) \qquad (y \in (0,\tau)), \tag{17.11}$$

where $F(y)$ is the $(2N-2)$-dimensional vector-valued function with coordinates

$$F_j(y) = \begin{cases} f^1(y + j\tau) & (j = 1, \dots, N-2), \\ f^2(y + (j-N+1)\tau) & (j = N-1, \dots, 2N-2), \end{cases} \tag{17.12}$$

C is the matrix of order $(2N-2) \times (2N-2)$ given by

$$C = \begin{pmatrix} C_{11} & C_{12} \\ C_{21} & C_{22} \end{pmatrix}, \tag{17.13}$$

C_{11}, C_{12}, C_{21}, C_{22} are the matrices of order $(N-2)\times(N-2)$, $(N-2)\times N$, $N\times(N-2)$, $N\times N$, respectively, defined by the formulas:

$$C_{11}=\begin{pmatrix} -2 & 0 & 0 & \dots & 0\\ 0 & -2 & 0 & \dots & 0\\ 0 & 0 & -2 & \dots & 0\\ \vdots & \vdots & \vdots & \ddots & \vdots\\ 0 & 0 & 0 & \dots & -2 \end{pmatrix}, \quad C_{12}=\cos\alpha\begin{pmatrix} 1 & 0 & 1 & 0 & 0 & \dots & 0 & 0 & 0\\ 0 & 1 & 0 & 1 & 0 & \dots & 0 & 0 & 0\\ 0 & 0 & 1 & 0 & 1 & \dots & 0 & 0 & 0\\ \vdots & \vdots & \vdots & \vdots & \vdots & \ddots & \vdots & \vdots & \vdots\\ 0 & 0 & 0 & 0 & 0 & \dots & 1 & 0 & 1 \end{pmatrix},$$

$$C_{21}=\cos\alpha\begin{pmatrix} 1 & 0 & 0 & \dots & 0\\ 0 & 1 & 0 & \dots & 0\\ 1 & 0 & 1 & \dots & 0\\ 0 & 1 & 0 & \dots & 0\\ 0 & 0 & 1 & \dots & 0\\ \vdots & \vdots & \vdots & \ddots & \vdots\\ 0 & 0 & 0 & \dots & 1\\ 0 & 0 & 0 & \dots & 0\\ 0 & 0 & 0 & \dots & 1 \end{pmatrix}, \quad C_{22}=\cos^2\alpha\begin{pmatrix} -2 & 0 & 0 & \dots & 0 & 0 & 0\\ 0 & -2 & 0 & \dots & 0 & 0 & 0\\ 0 & 0 & -2 & \dots & 0 & 0 & 0\\ 0 & 0 & 0 & \dots & 0 & 0 & 0\\ 0 & 0 & 0 & \dots & 0 & 0 & 0\\ \vdots & \vdots & \vdots & \ddots & \vdots & \vdots & \vdots\\ 0 & 0 & 0 & \dots & -2 & 0 & 0\\ 0 & 0 & 0 & \dots & 0 & -2 & 0\\ 0 & 0 & 0 & \dots & 0 & 0 & -2 \end{pmatrix}.$$

Clearly $C_{21}=C_{12}^*$.

Since $\det C\neq 0$, we obtain the following assertion:

Theorem 17.1. *Let u be a generalized solution of the boundary value problem (17.7), (17.8), and let $f\in W^{k,2}=W^k(\tau,b-\tau)\times W^k(0,b)$.*

Then $u^1\in W^{k+2}(j\tau,(j+1)\tau)$ $(j=1,\dots,N-2)$, $u^2\in W^{k+2}(j\tau,(j+1)\tau)$ $(j=0,\dots,N-1)$.

The proof of Theorem 17.1 indicates a method of finding the solution of the boundary value problem (17.7), (17.8) in explicit form. We first find the general solution of a system of $2N-2$ equations dependent on $2(2N-2)$ arbitrary constants. The conditions for the continuity of the functions u^1, u^2 and the boundary conditions

$$\begin{aligned} u^1(j\tau-0) &= u^1(j\tau+0) \qquad (j=2,\dots,N-2),\\ u^1(\tau) &= u^1(b-\tau)=0, \end{aligned} \tag{17.14}$$

$$\begin{aligned} u^2(j\tau-0) &= u^2(j\tau+0) \qquad (j=1,\dots,N-1),\\ u^2(0) &= u^2(b)=0 \end{aligned} \tag{17.15}$$

enable us to eliminate $2N$ constants. By Theorem 17.1, the functions $r_1(y)=-2(u^1)'+\cos\alpha(P_1R_1I_2u^2)'$ and $r_2(y)=\cos\alpha(P_2R_1I_1u^1)'-2\cos^2\alpha(u^2)'$ are piecewise continuous. By the definition of a generalized solution, however, $r_1'(y)$, $r_2'(y)$ must not contain any terms of the δ-function type. The condition for this is the continuity of the functions $r_1(y)$ on $[\tau,b-\tau]$ and $r_2(y)$ on $[0,b]$

$$r_1(j\tau-0) = r_1(j\tau+0) \qquad (j=2,\dots,N-2), \tag{17.16}$$

$$r_2(j\tau-0) = r_2(j\tau+0) \qquad (j=1,\dots,N-1), \tag{17.17}$$

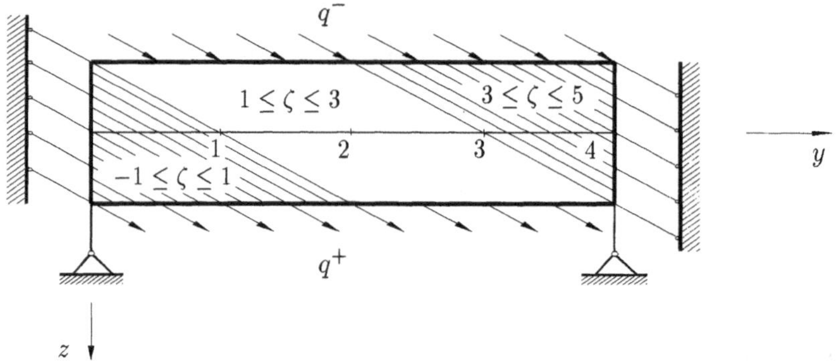

Fig. III.3

which enables us to eliminate the remaining $2N-4$ constants. By the existence and uniqueness theorem, the functions u^1, u^2 we have obtained are the solution of the boundary value problem (17.7), (17.8).

It must be noted that the last $2N-4$ constants are eliminated by the condition of the absence of δ-functions on the right-hand sides of the system of equations (17.7), and not by the condition of continuity of the first derivatives of the functions u^1 and u^2 at the corresponding points, which at first glance seems more natural.

Using the notation of Theorem 16.2, we obtain the following:

Theorem 17.2. *Let* u *be a generalized solution of the boundary value problem* (17.7), (17.8), *and let* $f \in L_2^2$.

Then:

1) *For a given* $2 \le l \le N-2$, u^1 *belongs to* $W^2((l-1)\tau, (l+1)\tau)$ *for all* $f \in L_2^2$ *if and only if the columns* B_j $(j = N-1, 2N-5)$ *are linear combinations of the columns of* A_{l-1}.

2) *For a given* $1 \le l \le N-1$, u^2 *belongs to* $W^2((l-1)\tau, (l+1)\tau)$ *for all* $f \in L_2^2$ *if and only if the columns* B_j $(j = N-1, 2N-5)$ *are linear combinations of the columns of* A_{N-3+l}.

Example 17.1. We consider the three-layer beam ($b = 4$, $\tau = 1$, $\alpha = \pi/3$) shown in Fig. III.3. The belts of this beam are loaded with the uniformly distributed load $q^{\pm}(y)$, applied in the direction of the oblique braces $f^1(y) \equiv 1$, $f^2(y) \equiv 0$.

The boundary value problem (17.7), (17.8) takes the form

$$\left.\begin{aligned} -4(u^1)'' + (P_1 R_1 I_2 u^2)'' &= 2 \quad (y \in (1,3)), \\ (P_2 R_1 I_1 u^1)'' - (u^2)'' &= 0 \quad (y \in (0,4)), \end{aligned}\right\} \tag{17.18}$$

$$u^1(1) = u^1(3) = 0, \qquad u^2(0) = u^2(4) = 0. \tag{17.19}$$

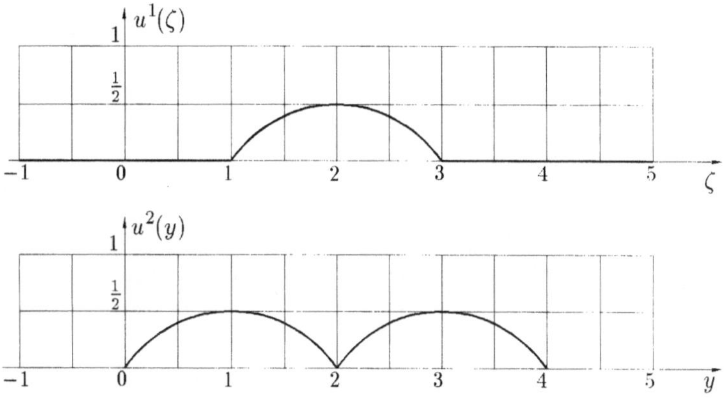

Fig. III.4

It easy to see that the solution of the problem (17.18), (17.19) is given by the functions

$$u^1(y) = (-(y-2)^2+1)/2 \qquad (y \in [1,3]), \tag{17.20}$$

$$u^2(y) = \begin{cases} (-(y-1)^2+1)/2 & (y \in [0,2]), \\ (-(y-3)^2+1)/2 & (y \in [2,4]) \end{cases} \tag{17.21}$$

(see Fig. III.4).

In fact, $u^1(1) = u^1(3) = 0$, $u^2(0) = u^2(2) = u^2(4) = 0$. Thus $u^1 \in \mathring{W}^1(1,3)$, $u^2 \in \mathring{W}^1(0,4)$. We can be convinced without difficulty that the functions $u^1(y)$, $u^2(y)$ satisfy the system (17.18).

We note that the function $(P_2R_1I_1u^1)'' = (u^2)''$ is singular, since $(u^2)'' = -1 + 2\delta(y-2)$.

Notes

The main results in Sections 14, 15, 17 are taken from G. G. Onanov and A. L. Skubachevskiĭ [1]. Theorems concerning the smoothness of solutions (see Sections 16, 17) are published here for the first time.

Chapter IV

Semi-Bounded Differential-Difference Operators with Degeneration

In this chapter we study elliptic differential-difference equations with degeneration.

In Section 18, we obtain a priori estimates for a semi-bounded differential-difference operator with degeneration A_R. Next we consider a self-adjoint Friedrichs extension $\mathcal{A}_R$ of A_R.

Section 19 deals with the spectrum of $\mathcal{A}_R$. We prove that the range $\mathcal{R}(\mathcal{A}_R)$ is closed in $L_2(Q)$, and $\operatorname{codim} \mathcal{R}(\mathcal{A}_R) = \dim \mathcal{N}(\mathcal{A}_R) = \infty$. The spectrum $\sigma(\mathcal{A}_R)$ consists of isolated eigenvalues, $\lambda_0 = 0$ of infinite multiplicity and λ_s of finite multiplicity.

In Section 20, we consider the smoothness of generalized solutions of a boundary value problem for a differential-difference equation with degeneration. It is shown that such a problem can have solutions which do not even belong to $W^1_{\text{loc}}(Q)$. However, the orthogonal projection of the solution onto the range of the difference operator R_Q already has the appropriate smoothness. The disturbance of the smoothness implies that the traces of the solution cannot be determined on some manifolds $\Gamma_{rl} \subset \partial Q$.

18 Self-Adjoint Extension of a Semi-Bounded Differential-Difference Operator

Notation

In Chapter IV, we consider the *differential-difference operator* A_R with domain $\mathcal{D}(A_R) = \dot{C}^\infty(Q)$ given by

$$A_R = AR_Q. \tag{18.1}$$

Here $R_Q = P_Q R I_Q$; $R: L_2(\mathbb{R}^n) \to L_2(\mathbb{R}^n)$ is the difference operator defined by the relation

$$Ru(x) = \sum_{h \in \mathcal{M}} a_h u(x+h),$$

$\mathcal{M}$ is a finite set of vectors with integer coordinates, and, if $h \in \mathcal{M}$, then $-h \in \mathcal{M}$, $a_h = a_{-h}$, a_h are real numbers,

$$A = \sum_{|\alpha|,|\beta|\le m} \mathcal{D}^\alpha a_{\alpha\beta}(x)\mathcal{D}^\beta,$$

$a_{\alpha\beta} = a_{\beta\alpha} \in C^\infty(\mathbb{R}^n)$ are real-valued M-periodic in $\overline{Q}$ functions, $Q \subset \mathbb{R}^n$ is a bounded domain with boundary $\partial Q \in C^\infty$ or a cylinder $(0,d) \times G$, where $G \subset \mathbb{R}^{n-1}$ is a bounded domain (with boundary $\partial G \in C^\infty$ if $n \ge 3$).

We introduce the matrices R_s of order $N \times N$ ($N = N(s)$, $s = 1,2,\dots$) with elements

$$r_{ij}^s = \begin{cases} a_h & \text{if } h = h_{sj} - h_{si} \in \mathcal{M}, \\ 0 & \text{if } h_{sj} - h_{si} \notin \mathcal{M}, \end{cases} \tag{18.2}$$

where h_{si} are such that $Q_{si} = Q_{s1} + h_{si}$ (see (8.6)).

Let $x \in \overline{Q}_{s1}$ be an arbitrary point. Consider all points $x^i \in \overline{Q}$ such that $x^i - x \in M$. Since the domain Q is bounded, the set $\{x^i\}$ consists of a finite number of points $I = I(s,x)$ ($I \ge N(s)$). We shall number all points x^i so that $x^i = x + h_{si}$ for $i = 1,\dots,N = N(s)$, $x^1 = x$. We introduce the $I \times I$-matrices $A_s = A_s(x)$ ($I = I(s,x)$, $s = 1,2,\dots$, $x \in \overline{Q}_{s1}$) with elements

$$a_{ij}^s(x) = \begin{cases} a_h & \text{if } h = x^j - x^i \in \mathcal{M}, \\ 0 & \text{if } x^j - x^i \notin \mathcal{M} \end{cases} \tag{18.3}$$

(cf. (9.12), (13.5)).

Clearly the matrices R_s and A_s are symmetric.

In this chapter we assume that the following conditions are fulfilled:

18.1. $\sum\limits_{|\alpha|,|\beta|=m} a_{\alpha\beta}(x)\xi^{\alpha+\beta} > 0$ *for all* $x \in \overline{Q}$, $0 \ne \xi \in \mathbb{R}^n$.

18.2. $a_0 > 0$, *and there exists* s_1 *such that* $\det R_{s_1} = 0$.

18.3. *If* $m = 1$, *the matrices* R_s *are non-negative*; *if* $m > 1$, *the matrices* A_s *are non-negative* ($s = 1,2,\dots$).

A Priori Estimates

By virtue of Lemmas 8.2, 8.10, 8.11, and 8.14, the operators A_R and R_Q are symmetric in the space $L_2(Q)$. In order to study a self-adjoint Friedrichs extension of A_R, we consider some inequalities.

Lemma 18.1. *There exist constants* $c_0 \ge 0$, $c_1 > 0$ *such that*

$$((A_R + c_0R_Q)u,u)_{L_2(Q)} \ge c_1\|R_Qu\|^2_{W^m(Q)} \tag{18.4}$$

for all $u \in \dot{C}^\infty(Q)$.

Proof. 1. From decomposition (8.3), Lemma 8.17 and formulas (8.5), (8.10), (8.36) it follows that

$$\begin{aligned}
\|R_Q u\|^2_{W^m(Q)} &= \sum_{s,l} \|P_s R_Q P^R u\|^2_{W^m(Q_{sl})} = \sum_s \|R_s U_s P_s P^R u\|^2_{W^{m,N}(Q_{s1})} \\
&\le k_1 \sum_s \|U_s P_s P^R u\|^2_{W^{m,N}(Q_{s1})} \\
&= k_1 \sum_s \sum_{|\alpha|\le m} \|P^R_s U_s P_s \mathcal{D}^\alpha u\|^2_{L_2^N(Q_{s1})} \qquad (18.5)
\end{aligned}$$

for all $u \in \dot{C}^\infty(Q)$. By virtue of Remark 9.1 and condition 18.3, the matrices R_s are non-negative for all $s = 1, 2, \ldots$. Thus, since the number of different matrices R_s is finite, we conclude from Lemmas 8.16, 8.17, and formulas (8.36), (8.5), (8.10), (8.3) that

$$\begin{aligned}
&k_1 \sum_s \sum_{|\alpha|\le m} \|P^R_s U_s P_s \mathcal{D}^\alpha u\|^2_{L_2^N(Q_{s1})} \\
&\quad \le k_2 \sum_s \sum_{|\alpha|\le m} (R_s P^R_s U_s P_s \mathcal{D}^\alpha u, P^R_s U_s P_s \mathcal{D}^\alpha u)_{L_2^N(Q_{s1})} \\
&\quad = k_2 \sum_{|\alpha|\le m} (R_Q \mathcal{D}^\alpha u, \mathcal{D}^\alpha u)_{L_2(Q)}. \qquad (18.6)
\end{aligned}$$

2. First let $m = 1$. The formulas (8.3), (8.5), (8.10), (8.14), condition 18.1 and Lemma 8.17 imply that

$$\begin{aligned}
&\sum_{|\alpha|=1} (R_Q \mathcal{D}^\alpha u, \mathcal{D}^\alpha u)_{L_2(Q)} \\
&\quad = \sum_{|\alpha|=1} \sum_s (R_s U_s P_s \mathcal{D}^\alpha u, U_s P_s \mathcal{D}^\alpha u)_{L_2^N(Q_{s1})} \\
&\quad = \sum_{|\alpha|=1} \sum_s (\sqrt{R_s} U_s P_s \mathcal{D}^\alpha u, \sqrt{R_s} U_s P_s \mathcal{D}^\alpha u)_{L_2^N(Q_{s1})} \\
&\quad \le k_3 \sum_{|\alpha|,|\beta|=1} \sum_s (a_{\alpha\beta} \sqrt{R_s} U_s P_s \mathcal{D}^\beta u, \sqrt{R_s} U_s P_s \mathcal{D}^\alpha u)_{L_2^N(Q_{s1})} \\
&\quad = k_3 \sum_{|\alpha|,|\beta|=1} (a_{\alpha\beta} \mathcal{D}^\beta R_Q u, P^R \mathcal{D}^\alpha u)_{L_2(Q)} \\
&\quad \le k_4 (A_R u, u)_{L_2(Q)} + k_5 \|R_Q u\|_{W^1(Q)} \|R_Q u\|_{L_2(Q)}.
\end{aligned}$$

From this and (18.5), (18.6), (9.6) it follows that

$$\|R_Q u\|^2_{W^1(Q)} \le k_6 \{(A_R u, u)_{L_2(Q)} + q^{-1} \|R_Q u\|^2_{W^1(Q)} + q \|R_Q u\|^2_{L_2(Q)}\}. \qquad (18.7)$$

By virtue of Lemmas 8.17, 8.7, we have

$$\|R_Q u\|^2_{L_2(Q)} \le k_7\|P^R u\|^2_{L_2(Q)} \le k_8(R_Q P^R u, P^R u)_{L_2(Q)}$$
$$= k_8(R_Q u, u)_{L_2(Q)}. \tag{18.8}$$

Putting $k_6 q^{-1} < 1/2$, from the inequalities (18.7), (18.8) we obtain (18.4) for $m = 1$.

3. Now let $m > 1$. Using the partition of unity in Lemma 9.1, Lemmas 8.10, 8.14, the Leibniz formula and Lemma 8.17, we obtain

$$\begin{aligned}
&\sum_{|\alpha|\le m} (R_Q\mathcal{D}^\alpha u, \mathcal{D}^\alpha u)_{L_2(Q)} \\
&\quad = \sum_{|\alpha|\le m}\sum_j (R_Q(\varphi_j\mathcal{D}^\alpha u), \varphi_j\mathcal{D}^\alpha u)_{L_2(Q)} \\
&\quad = \sum_{|\alpha|= m}\sum_j (R_Q\mathcal{D}^\alpha(\varphi_j u), \mathcal{D}^\alpha(\varphi_j u))_{L_2(Q)} \\
&\qquad + \sum_{|\alpha|\le m}\sum_j\sum_{\mu,\beta} b_{\alpha\beta\gamma}(R_Q(\mathcal{D}^{\alpha-\mu}\varphi_j\mathcal{D}^{\alpha-\beta}\varphi_j\mathcal{D}^\beta u), P^R\mathcal{D}^\mu u)_{L_2(Q)} \\
&\quad \le \sum_{|\alpha|= m}\sum_j (R_Q\mathcal{D}^\alpha(\varphi_j u), \mathcal{D}^\alpha(\varphi_j u))_{L_2(Q)} \\
&\qquad + k_9(a)\Big\{\|R_Q u\|_{W^m(Q)}\sum_{|\gamma|\le m-1}\|P^R\mathcal{D}^\gamma u\|_{L_2(Q)} \\
&\qquad\qquad + \|R_Q u\|_{W^{m-1}(Q)}\sum_{|\gamma|\le m}\|P^R\mathcal{D}^\gamma u\|_{L_2(Q)}\Big\} \\
&\quad \le \sum_{|\alpha|= m}\sum_j (R_Q\mathcal{D}^\alpha(\varphi_j u), \mathcal{D}^\alpha(\varphi_j u))_{L_2(Q)} \\
&\qquad + k_{10}(a)\|R_Q u\|_{W^{m-1}(Q)}\|R_Q u\|_{W^m(Q)},
\end{aligned} \tag{18.9}$$

where $b_{\alpha\beta\mu}$ are positive constants, and we sum over μ, β such that $\mu_i, \beta_i \le \alpha_i$, $|\mu| + |\beta| < 2m$.

We consider an arbitrary point $y^j \in \overline{Q}$ from Lemma 9.1. There exists a subdomain Q_{sl} such that $y^j \in \overline{Q}_{sl}$. Denote $z^j = y^j - h_{sl}$. Then $z^j \in \overline{Q}_{s1}$. Let us introduce the vector-valued function $W^j \in \dot{C}^{\infty,I}(S_\delta(z^j))$ with coordinates

$$W_i^j(x) = (\varphi_j u)(x + z^{ji} - z^j) \qquad (x \in S_\delta(z^j)), \tag{18.10}$$

where $i = 1, \dots, I = I(s, z^j)$, the points z^{ji}, corresponding to z^j, are defined in the manner described at the beginning of this section. The matrices $A_s(x)$ can have different order at different points $x \in S_\delta(z^j)$ (see Example 9.1). Therefore, we consider the auxiliary matrices A_s^j of order $I(s, z^j)\times I(s, z^j)$ defined by the formula $A_s^j = A_s(z^j)$. Since the matrices A_s^j are non-negative, there exist non-negative matrices $\sqrt{A_s^j}$. Then, using the ellipticity of the operator A, the M-periodicity of the coefficients $a_{\alpha\beta}(x)$ in $\overline{Q}$, Lemma 8.17 and the Plancherel theorem, we obtain

$$\begin{aligned}
&\sum_{|\alpha|=m}\sum_j (R_Q\mathcal{D}^\alpha(\varphi_j u), \mathcal{D}^\alpha(\varphi_j u))_{L_2(Q)} \\
&\quad= \sum_{|\alpha|=m}\sum_j (A_s^j\mathcal{D}^\alpha W^j, \mathcal{D}^\alpha W^j)_{L_2^I(\mathbb{R}^n)} \\
&\quad= \sum_{|\alpha|=m}\sum_j (\mathcal{D}^\alpha\sqrt{A_s^j}W^j, \mathcal{D}^\alpha\sqrt{A_s^j}W^j)_{L_2^I(\mathbb{R}^n)} \\
&\quad= \sum_j\sum_{i=1}^I\int_{\mathbb{R}^n}\sum_{|\alpha|=m}\xi^{2\alpha}|(\sqrt{A_s^j}\widehat{W}^j)_i|^2 d\xi \\
&\quad\le k_{11}\sum_j\sum_i\int_{\mathbb{R}^n}\sum_{|\alpha|,|\beta|=m} a_{\alpha\beta}(z^j)\xi^{\alpha+\beta}|(\sqrt{A_s^j}\widehat{W^j})_i|^2 d\xi \\
&\quad= k_{11}\sum_j\sum_{|\alpha|,|\beta|=m}(a_{\alpha\beta}(z^j)\mathcal{D}^\beta R_Q(\varphi_j u), \mathcal{D}^\alpha(\varphi_j u))_{L_2(Q)} \\
&\quad= (q_1+q_2)k_{11},
\end{aligned} \tag{18.11}$$

where

$$\begin{aligned}
q_1 &= \sum_j\sum_{|\alpha|,|\beta|=m}(a_{\alpha\beta}(x)\mathcal{D}^\beta R_Q(\varphi_j u), \mathcal{D}^\alpha(\varphi_j u))_{L_2(Q)}, \\
q_2 &= \sum_j\sum_{|\alpha|,|\beta|=m}((a_{\alpha\beta}(z^j)-a_{\alpha\beta}(x))\mathcal{D}^\beta R_Q(\varphi_j u), P^R\mathcal{D}^\alpha(\varphi_j u))_{L_2(Q)}.
\end{aligned}$$

Since the functions $a_{\alpha\beta}(x)$ are M-periodic in $\overline{Q}$ and uniformly continuous in $\overline{Q^1}$, we have

$$|q_2| \le \varepsilon(a)\sum_j\|R_Q(\varphi_j u)\|_{W^m(Q)}\sum_{|\alpha|=m}\|P^R\mathcal{D}^\alpha(\varphi_j u)\|_{L_2(Q)}, \tag{18.12}$$

where $\varepsilon(a)\to 0$ as $a\to 0$.

By virtue of the inequalities (18.11), (18.12), the Leibniz formula and Lemma 8.17, the following relations hold:

$$\begin{aligned}
&\sum_{|\alpha|=m}\sum_j (R_Q\mathcal{D}^\alpha(\varphi_j u), \mathcal{D}^\alpha(\varphi_j u))_{L_2(Q)} \\
&\quad\le k_{12}\sum_{|\alpha|,|\beta|\le m}(a_{\alpha\beta}(x)\mathcal{D}^\beta R_Q u, \mathcal{D}^\alpha u)_{L_2(Q)} + k_{13}\varepsilon(a)\|R_Q u\|^2_{W^m(Q)} \\
&\qquad\qquad + k_{14}(a)\|R_Q u\|_{W^m(Q)}\|R_Q u\|_{W^{m-1}(Q)}.
\end{aligned} \tag{18.13}$$

From the inequalities (18.5), (18.6), (18.9), (18.13), (9.6), (B.20) it follows that

$$\|R_Q u\|^2_{W^m(Q)} \le k_{15}(A_R u, u)_{L_2(Q)} + (k_{16}\varepsilon(a) + k_{17}(a)q^{-1})\|R_Q u\|^2_{W^m(Q)} + k_{18}(a)q^{2m-1}\|R_Q u\|^2_{L_2(Q)}. \quad (18.14)$$

Let $a > 0$ be such that $4k_{16}\varepsilon(a) < 1$. Then, choosing $q > 0$ so that $4k_{17}(a)q^{-1} < 1$, from inequalities (18.14), (18.8) we obtain (18.4) for $m > 1$. □

Lemma 18.2. *There exists a constant $c_2 > 0$ such that for all $u, v \in \dot{C}^\infty(Q)$*

$$((A_R + c_0 R_Q)u, v)_{L_2(Q)} \le c_2\|R_Q u\|_{W^m(Q)}\|R_Q v\|_{W^m(Q)}. \quad (18.15)$$

Proof. Integrating by parts and using Lemmas 8.10, 8.14, 8.17, we obtain

$$\begin{aligned} &((A_R + c_0 R_Q)u, v)_{L_2(Q)} \\ &= \sum_{|\alpha|,|\beta|\le m} (R_Q a_{\alpha\beta}\mathcal{D}^\beta u, P^R \mathcal{D}^\alpha v)_{L_2(Q)} + c_0(R_Q u, P^R v)_{L_2(Q)} \\ &\le k_1 \sum_{|\alpha|,|\beta|\le m} \|R_Q\mathcal{D}^\beta u\|_{L_2(Q)}\|R_Q\mathcal{D}^\alpha v\|_{L_2(Q)} \\ &\le k_2\|R_Q u\|_{W^m(Q)}\|R_Q v\|_{W^m(Q)}. \quad \square \end{aligned}$$

A Friedrichs Extension

By virtue of Lemmas 8.2, 8.11, 8.14, 8.10, and 18.1, we can introduce the inner product

$$(u, v)_{A_R} = ((A_R + c_0 R_Q + I)u, v)_{L_2(Q)}$$

in $\dot{C}^\infty(Q)$. Denote by W_{A_R} the set of elements $u \in L_2(Q)$, for which there is a sequence $\{u_p\} \subset \dot{C}^\infty(Q)$, such that

$$\lim_{p\to\infty} \|u_p - u\|_{L_2(Q)} = 0, \qquad \lim_{p,q\to\infty} \|u_p - u_q\|_{A_R} = 0.$$

We introduce the norm

$$\|u\|_{A_R} = \lim_{p\to\infty} \|u_p\|_{A_R},$$

where $u_p \in \dot{C}^\infty(Q), u_p \to u$ in $L_2(Q)$ and $\lim_{p,q\to\infty}\|u_p - u_q\|_{A_R} = 0$. This norm does not depend on the choice of the sequence $\{u_p\}$. The space W_{A_R} with such an inner product is complete.

We introduce the unbounded operators $\mathcal{A}_R, \mathcal{A}_{Rc_0}: L_2(Q) \to L_2(Q)$ acting in the space of distributions $\mathcal{D}'(Q)$ by formulas $\mathcal{A}_R = AR_Q$, $\mathcal{A}_{Rc_0} = AR_Q + c_0 R_Q$, with domain

$$\mathcal{D}(\mathcal{A}_R) = \mathcal{D}(\mathcal{A}_{Rc_0}) = \{u \in W_{A_R} : \mathcal{A}_R u \in L_2(Q)\}.$$

Theorem 18.1. *Let the conditions* 18.1–18.3 *be fulfilled.*

Then the operators $\mathcal{A}_R, \mathcal{A}_{Rc_0}: L_2(Q) \to L_2(Q)$ *are the self-adjoint Friedrichs extensions of* A_R, $A_R + c_0 R_Q$, *respectively. Furthermore the operator* $\mathcal{A}_{Rc_0}$ *is non-negative.*

The proof follows from Lemma 18.1 and Theorem A.13.

The following question remains open:

Problem 18.1. *Do Lemma* 18.1 *and Theorem* 18.1 *hold, when condition* 18.3 *in them is replaced by condition* 18.4*?*

18.4. *The matrices* R_s *are non-negative* $(s = 1, 2, \ldots)$.

19 The Spectrum of Semi-Bounded Differential-Difference Operators

Some Properties of the Space W_{A_R}

Lemma 19.1. $\dim \mathcal{N}(R_Q) = \infty$ *and* $\mathcal{N}(R_Q) \subset W_{A_R}$.

Proof. Let $u \in \mathcal{N}(R_Q)$. We prove that there exists a sequence of functions $\{u_p\} \subset \dot{C}^\infty(Q)$ such that $u_p \in \mathcal{N}(R_Q)$, $\lim_{p\to\infty} \|u_p - u\|_{L_2(Q)} = 0$.

By virtue of (8.3), for each $p > 0$ there exists N_0 such that

$$\sum_{s>N_0} \sum_{l=1}^{N} \|u\|^2_{L_2(Q_{sl})} < \frac{1}{2p}.$$

From (8.10) it follows that $P_s u \in \mathcal{N}(R_Q)$, i.e., $U_s P_s u \in \mathcal{N}(R_{Qs})$. If $\det R_s = 0$, then a subspace $\mathcal{N}(R_{Qs})$ consists of vector-valued functions $V \in L_2^N(Q_{s1})$ such that their $i_1, \ldots, i_j$th coordinates are linear combinations of the remaining $N - j$ coordinates. From this it follows that $\dim \mathcal{N}(R_Q) = \infty$. Since $\dot{C}^\infty(Q_{s1})$ is dense in $L_2(Q_{s1})$, we conclude that for each $s = 1, \ldots, N_0$ there is $V_{sp} \in \mathcal{N}(R_{Qs})$, for which

$$V_{sp} \in \dot{C}^{\infty,N}(Q_{s1}) \quad \text{and} \quad \sum_{s=1}^{N_0} \|V_{sp} - U_s P_s u\|^2_{L_2^N(Q_{s1})} < \frac{1}{2p}.$$

By construction, $U_s^{-1} V_{sp} \in \mathcal{N}(R_Q)$. Let $u_p = \sum_{s=1}^{N_0} U_s^{-1} V_{sp}$. Then $u_p \in \mathcal{N}(R_Q)$, $u_p \equiv 0$ for $x \in \bigcup_{s>N_0} \bigcup_{l=1}^{N} Q_{sl}$, $u_p \in \dot{C}^\infty(Q)$, and

$$\begin{aligned}\|u_p - u\|^2_{L_2(Q)} &= \sum_{s=1}^{N_0} \sum_{l=1}^{N} \|u_p - u\|^2_{L_2(Q_{sl})} + \sum_{s>N_0} \sum_{l=1}^{N} \|u\|^2_{L_2(Q_{sl})} \\ &< \sum_{s=1}^{N_0} \|V_{sp} - U_s P_s u\|^2_{L_2^N(Q_{s1})} + \frac{1}{2p} < \frac{1}{p}. \quad \square\end{aligned}$$

Denote by W_R the closure of the linear manifold $R_Q(\dot{C}^\infty(Q))$ in $W^m(Q)$.

Lemma 19.2. $W_R = R_Q(W_{A_R})$.

Proof. Lemma 18.1 implies that $R_Q(W_{A_R}) \subset W_R$. It remains to prove the reverse inclusion.

Let $w \in W_R$, i.e., $w \in W^m(Q)$ and there is a sequence $\{v_p\} \subset \dot{C}^\infty(Q)$ such that $\lim_{p\to\infty} \|R_Q v_p - w\|_{W^m(Q)} = 0$. Thus $\lim_{p,q\to\infty} \|R_Q v_p - R_Q v_q\|_{W^m(Q)} = 0$. Hence from Lemma 18.2 and inequality (8.34), we have

$$\lim_{p,q\to\infty} ((A_R + c_0 R_Q)(v_p - v_q), v_p - v_q)_{L_2(Q)} = 0, \tag{19.1}$$

$$\lim_{p,q\to\infty} \|P^R v_p - P^R v_q\|_{L_2(Q)} = 0. \tag{19.2}$$

By virtue of (8.3), for each $p > 0$ there is N_0 such that

$$\sum_{s>N_0} \sum_{l=1}^{N} \int_{Q_{sl}} |(I - P^R) v_p(x)|^2 dx < \frac{1}{2p}. \tag{19.3}$$

Let $\theta_{sp} \in C^\infty(Q_{\alpha 1})$ be the functions defined as follows: $0 \le \theta_{sp}(x) \le 1$,

$$\theta_{sp}(x) = \begin{cases} 1, & x \in Q_{s1} \setminus (Q_{s1})_\varepsilon, \\ 0, & x \in (Q_{s1})_{2\varepsilon} \end{cases}$$

$(s = 1, \dots, N_0)$, $\theta_{sp}(x) \equiv 1$ for $x \in Q_{s1}$ $(s > N_0)$. Here $(Q_{sl})_\delta = \{x \in Q_{sl} : \rho(x, \partial Q_{sl}) > \delta\}$, $\delta = \varepsilon, 2\varepsilon$; $\varepsilon = \varepsilon(p)$ is such that

$$\sum_{s=1}^{N_0} \sum_{l=1}^{N} \int_{Q_{sl} \setminus (Q_{sl})_{2\varepsilon}} |(I - P^R) v_p(x)|^2 dx < \frac{1}{2p}. \tag{19.4}$$

Let $v_p^0 = \sum_s U_s^{-1} \theta_{sp} U_s P_s (I - P^R) v_p \in L_2(Q)$. Clearly, if $V \in \mathcal{N}(R_{Qs})$, then $\theta_{sp} V \in \mathcal{N}(R_{Qs})$. Hence it follows from (8.10) and the boundedness of R_Q in $L_2(Q)$ that $v_p^0 \in \mathcal{N}(R_Q)$.

Denote $u_p(x) = P^R v_p(x) + v_p^0(x)$. By construction, $v_p^0(x) = (I - P^R) v_p(x)$ for $x \in \Omega_p$, where

$$\Omega_p = \left[\bigcup_{s=1}^{N_0} \bigcup_{l=1}^{N} (Q_{sl} \setminus (Q_{sl})_\varepsilon) \right] \cup \left[\bigcup_{s>N_0} \bigcup_{l=1}^{N} Q_{sl} \right],$$

i.e., $u_p(x) = v_p(x)$ for $x \in \Omega_p$. But, since $v_p \in \dot{C}^\infty(Q)$, (8.35) implies that $P^R v_p$, $v_p^0 \in C^\infty(Q_{sl})$ $(s = 1, 2, \dots,\ l = 1, \dots, N)$. Thus $u_p \in \dot{C}^\infty(Q)$.

By virtue of (19.3), (19.4), $\lim_{p\to\infty} \|v_p^0\|_{L_2(Q)} = 0$. Hence from (19.2) we obtain $\lim_{p,q\to\infty} \|u_p - u_q\|_{L_2(Q)} = 0$, i.e., $\lim_{p\to\infty} \|u_p - u\|_{L_2(Q)} = 0$, where $u \in L_2(Q)$.

But $v_p^0 \in \mathcal{N}(R_Q)$, and so Lemma 8.17 and formula (19.1) imply that

$$\lim_{p,q\to\infty}((A_R + c_0 R_Q)(u_p - u_q), (u_p - u_q))_{L_2(Q)} = \lim_{p,q\to\infty}((A_R + c_0 R_Q)(v_p - v_q), (v_p - v_q))_{L_2(Q)} = 0.$$

Thus, $u \in W_{A_R}$ and $\lim_{p\to\infty}\|u_p - u\|_{A_R} = 0$. Hence, by virtue of Lemma 18.1, $R_Q u \in W^m(Q)$ and $\lim_{p\to\infty}\|R_Q u_p - R_Q u\|_{W^m(Q)} = 0$. On the other hand, Lemma 8.17 implies that

$$\lim_{p\to\infty}\|R_Q u_p - w\|_{W^m(Q)} = \lim_{p\to\infty}\|R_Q v_p - w\|_{W^m(Q)} = 0.$$

The uniqueness of the limit implies that $w = R_Q u$, i.e., $w \in R_Q(W_{A_R})$. Therefore $W_R \subset R_Q(W_{A_R})$. □

Properties of Spectrum

Let $\mathcal{A}_R^0$ and $\mathcal{A}_{Rc_0}^0$ denote the restrictions of $\mathcal{A}_R$ and $\mathcal{A}_{Rc_0}$, respectively, on $\mathcal{D}(\mathcal{A}_R) \cap \mathcal{R}(R_Q)$.

Lemma 19.3. *The operator $\mathcal{A}_{Rc_0}^0 \colon \mathcal{R}(R_Q) \to \mathcal{R}(R_Q)$ is self-adjoint, and has a compact inverse operator defined on the whole of $\mathcal{R}(R_Q)$.*

Proof. By virtue of Lemma 19.1, $\mathcal{N}(R_Q) \subset \mathcal{N}(\mathcal{A}_{Rc_0}) \subset \mathcal{D}(\mathcal{A}_{Rc_0})$. Hence $\mathcal{R}(R_Q)$ is an invariant subspace of the operator $\mathcal{A}_{Rc_0}$, i.e., $\mathcal{A}_{Rc_0}^0$ is self-adjoint.

Lemma 8.17 implies that

$$(\mathcal{A}_{Rc_0}u, v)_{L_2(Q)} = ((A + c_0 I)R_Q u, P^R v)_{L_2(Q)} = ((A + c_0 I)R_Q u, (R_Q^R)^{-1} R_Q v)_{L_2(Q)}$$

for $u, v \in \dot{C}^\infty(Q)$. On $R_Q(\dot{C}^\infty(Q)) \times R_Q(\dot{C}^\infty(Q))$ we introduce the symmetric sesquilinear form $b_1[R_Q u, R_Q v] = (\mathcal{A}_{Rc_0}u, v)_{L_2(Q)}$. By virtue of Lemma 18.2, it can be continued to a bounded symmetric sesquilinear form defined on $W_R \times W_R$. Hence, by the Riesz theorem, for all $w_1, w_2 \in W_R$

$$b_1[w_1, w_2] = (B_1 w_1, w_2)_{W^m(Q)},$$

where $B_1 \colon W_R \to W_R$ is a bounded self-adjoint operator.

Take arbitrary $u \in W_{A_R}$ and $v \in \dot{C}^\infty(Q)$. Let $u_p \in \dot{C}^\infty(Q)$, and assume that $\lim_{p\to\infty}\|u_p - u\|_{A_R} = 0$. Then, by virtue of Lemma 18.1, $R_Q u \in W^m(Q)$ and $\lim_{p\to\infty}\|R_Q u_p - R_Q u\|_{W^m(Q)} = 0$. Hence $\|R_Q u_p - R_Q u\|_{L_2(Q)} \to 0$, i.e., $\mathcal{A}_{Rc_0}u_p \to \mathcal{A}_{Rc_0}u$ in the space $\mathcal{D}'(Q)$. Thus,

$$\langle \mathcal{A}_{Rc_0}u, \overline{v}\rangle = \lim_{p\to\infty}(\mathcal{A}_{Rc_0}u_p, v)_{L_2(Q)} = \lim_{p\to\infty}(B_1(R_Q u_p), R_Q v)_{W^m(Q)} = (B_1(R_Q u), R_Q v)_{W^m(Q)}.$$

Thus, for all $u \in W_{A_R}$ and $v \in \dot{C}^\infty(Q)$,

$$\langle \mathcal{A}_{Rc_0} u, \overline{v} \rangle = (B_1(R_Q u), R_Q v)_{W^m(Q)}. \tag{19.5}$$

On the other hand, by virtue of Lemma 8.17 for every $f_0 \in \mathcal{R}(R_Q)$ and $v \in \dot{C}^\infty(Q)$

$$(f_0, v)_{L_2(Q)} = (f_0, P^R v)_{L_2(Q)} = (f_0, (R_Q^R)^{-1} R_Q v)_{L_2(Q)}.$$

Hence, $\psi_{f_0}(R_Q v) = (f_0, v)_{L_2(Q)}$ is a semi-linear functional, defined on W_R and continuous on $W^m(Q)$. It follows from the Riesz theorem that there is a unique $F = F(f_0) \in W_R$ such that

$$\psi_{f_0}(R_Q v) = (F, R_Q v)_{W^m(Q)}, \qquad \|F\|_{W^m(Q)} \le \|(R_Q^R)^{-1}\| \cdot \|f_0\|_{L_2(Q)}.$$

The relation $F = F(f)$ therefore defines a bounded operator $B_2 \colon L_2(Q) \to W_R$ for which

$$(f_0, v)_{L_2(Q)} = (B_2 f_0, R_Q v)_{W^m(Q)}. \tag{19.6}$$

Formulas (19.5), (19.6) imply that $u \in \mathcal{D}(\mathcal{A}_{Rc_0})$ is a solution of the equation

$$\mathcal{A}_{Rc_0} u = f_0 \tag{19.7}$$

for $f_0 \in \mathcal{R}(R_Q)$, if and only if $u \in W_{A_R}$, and for every $v \in \dot{C}^\infty(Q)$

$$(B_1(R_Q u), R_Q v)_{W^m(Q)} = (B_2 f_0, R_Q v)_{W^m(Q)}. \tag{19.8}$$

By virtue of (19.5) and Lemma 18.1,

$$(B_1(R_Q u), R_Q u)_{W^m(Q)} \ge c_1 \|R_Q u\|^2_{W^m(Q)} \tag{19.9}$$

for $u \in W_{A_R}$. The operator B_1, therefore, has a bounded inverse B_1^{-1} on $R_Q(W_{A_R})$, and $\|B_1^{-1}\| \le 1/c_1$. But Lemma 19.2 implies that $R_Q(\dot{C}^\infty(Q))$ is dense in $R_Q(W_{A_R})$. We thus conclude that the integral identity (19.8) is equivalent to the equation

$$R_Q u = B_1^{-1} B_2 f_0, \tag{19.10}$$

moreover

$$\|R_Q u\|_{W^m(Q)} \le \frac{\|(R_Q^R)^{-1}\|}{c_1} \|f_0\|_{L_2(Q)}. \tag{19.11}$$

It follows from (19.10) and the existence of a bounded operator $(R_Q^R)^{-1} \colon \mathcal{R}(R_Q) \to \mathcal{R}(R_Q)$ that for every $f_0 \in \mathcal{R}(R_Q)$, the equation

$$\mathcal{A}^0_{Rc_0} u = f_0 \tag{19.12}$$

has a unique solution $u \in \mathcal{D}(\mathcal{A}_{Rc_0}) \cap \mathcal{R}(R_Q)$ and

$$u = (R_Q^R)^{-1} B_1^{-1} B_2 f_0. \tag{19.13}$$

Thus, $\mathcal{A}^0_{Rc_0}$ has a bounded inverse operator $(\mathcal{A}^0_{Rc_0})^{-1} = (R_Q^R)^{-1} B_1^{-1} B_2$ defined on the whole of $\mathcal{R}(R_Q)$. Furthermore, it follows from (19.11), the compactness of the imbedding operator from $W^m(Q)$ into $L_2(Q)$, and the boundedness of $(R_Q^R)^{-1} \colon \mathcal{R}(R_Q) \to \mathcal{R}(R_Q)$ that the operator $(\mathcal{A}^0_{Rc_0})^{-1} \colon \mathcal{R}(R_Q) \to \mathcal{R}(R_Q)$ is compact. □

Lemma 19.4. *The operator* $\mathcal{A}_R^0 \colon \mathcal{R}(R_Q) \to \mathcal{R}(R_Q)$ *is Fredholm, and* $\operatorname{ind} \mathcal{A}_R^0 = 0$. *The spectrum* $\sigma(\mathcal{A}_R^0)$ *is real and discrete.*

Proof. Since $\mathcal{R}(R_Q)$ is an invariant subspace of the operator $\mathcal{A}_R$, the operator $\mathcal{A}_R^0$ is self-adjoint. Hence the spectrum $\sigma(\mathcal{A}_R^0)$ is real. Let $\lambda \in \mathbb{C} \setminus \sigma(\mathcal{A}_R^0)$. Consider the equation

$$\mathcal{A}_R^0 u - \lambda u = f_0. \tag{19.14}$$

This equation has a unique solution $u \in \mathcal{R}(R_Q)$ and

$$\|u\|_{L_2(Q)} \leq k_1 \|f_0\|_{L_2(Q)}. \tag{19.15}$$

We rewrite the equation (19.14) in the form

$$\mathcal{A}_{Rc_0}^0 u = c_0 R_Q u + \lambda u + f_0. \tag{19.16}$$

From (19.15), (19.11) and the boundedness of $R_Q \colon L_2(Q) \to L_2(Q)$ it follows that

$$\|R_Q u\|_{W^m(Q)} \leq \frac{\|(R_Q^R)^{-1}\|}{c_1} \|c_0 R_Q u + \lambda u + f_0\|_{L_2(Q)} \leq k_2 \|f_0\|_{L_2(Q)}.$$

Hence, by virtue of Lemma 8.17, the operator $(\mathcal{A}_R^0 - \lambda I)^{-1} \colon \mathcal{R}(R_Q) \to \mathcal{R}(R_Q)$ is compact. Thus, by Theorem A.8, the spectrum $\sigma(\mathcal{A}_R^0)$ is discrete.

On the other hand, $\mathcal{A}_R^0 R(\lambda, \mathcal{A}_R^0) = I + \lambda R(\lambda, \mathcal{A}_R^0)$, i.e., by Theorem A.1, the operator $\mathcal{A}_R^0 \colon \mathcal{R}(R_Q) \to \mathcal{R}(R_Q)$ is Fredholm, and $\operatorname{ind} \mathcal{A}_R^0 = 0$. □

Theorem 19.1. *Let the conditions* 18.1–18.3 *be fulfilled.*

Then the operator $\mathcal{A}_R \colon L_2(Q) \to L_2(Q)$ *has a closed range* $\mathcal{R}(\mathcal{A}_R)$, $\mathcal{N}(R_Q) \subset \mathcal{N}(\mathcal{A}_R)$, *and* $\dim \mathcal{N}(\mathcal{A}_R) = \operatorname{codim} \mathcal{R}(\mathcal{A}_R) = \infty$. *The spectrum* $\sigma(\mathcal{A}_R)$ *consists of isolated eigenvalues,* $\lambda_0 = 0$ *of infinite multiplicity and* λ_s *of finite multiplicities;* $\lambda_s \to +\infty$ *as* $s \to +\infty$.

Proof. Lemmas 19.1, 19.4 imply that $\mathcal{N}(R_Q) \subset \mathcal{N}(\mathcal{A}_R)$, $\dim \mathcal{N}(\mathcal{A}_R) = \operatorname{codim} \mathcal{R}(\mathcal{A}_R) = \infty$, and $\mathcal{R}(\mathcal{A}_R)$ is closed in $L_2(Q)$. Clearly $\{0\} \cup \sigma(\mathcal{A}_R^0) \subset \sigma(\mathcal{A}_R)$. On the other hand, if $\lambda \notin \{0\} \cup \sigma(\mathcal{A}_R^0)$, it is easy to show that $(\mathcal{A}_R^0 - \lambda I)^{-1} P^R - \lambda^{-1}(I - P^R)$ is the inverse operator for $\mathcal{A}_R - \lambda I$. Thus, $\sigma(\mathcal{A}_R) = \{0\} \cup \sigma(\mathcal{A}_R^0)$. By virtue of Theorem 18.1 and the boundedness of the operator $R_Q \colon L_2(Q) \to L_2(Q)$, we have $(\mathcal{A}_R u, u)_{L_2(Q)} \geq -k_1 (u, u)_{L_2(Q)}$ for all $u \in \mathcal{D}(\mathcal{A}_R)$. Hence $\lambda_s \geq -k_1$. □

Theorem 19.2. *Let the conditions* 18.1–18.3 *be fulfilled. Assume*

$$A_1 \colon L_2(Q) \to L_2(Q)$$

is a bounded self-adjoint operator.

Then the operator $\mathcal{A}_R + A_1 \colon L_2(Q) \to L_2(Q)$ *is self-adjoint; the spectrum* $\sigma(\mathcal{A}_R + A_1) \subset [-\|A_1\|, +\infty)$, *and the set* $(\|A_1\|, +\infty) \cap \sigma(\mathcal{A}_R + A_1)$ *consists of isolated eigenvalues of finite multiplicity.*

The proof follows from Theorems 18.1, 19.1, and Theorem A.7.

20 Smoothness of Solutions of Equations with Degeneration

Smoothness inside Subdomains

We consider the equation

$$\mathcal{A}_R u = f_0, \tag{20.1}$$

where $f_0 \in L_2(Q)$. By virtue of Theorem 19.1, $\mathcal{N}(R_Q) \subset \mathcal{N}(\mathcal{A}_R)$. Therefore, the equation (20.1) can have solutions $u \in \mathcal{D}(\mathcal{A}_R)$ from $L_2(Q)$, not even in $W^1(Q)$. However, we shall establish below that an orthogonal projection of the solution onto $\mathcal{R}(R_Q)$ already has the appropriate smoothness.

Theorem 20.1. *Let the conditions* 18.1–18.3 *be fulfilled. Let* $u \in \mathcal{D}(\mathcal{A}_R)$ *be a solution of the equation* (20.1), *and let* $f_0 \in L_2(Q) \cap W^k_{\text{loc}}(Q)$ $(s = 1, 2, \dots, l = 1, \dots, N(s))$.

If s *is such that* $(\bigcup_l \partial Q_{sl}) \cap \partial Q \neq \emptyset$, *then* $P^R u \in W^{k+2m}_{\text{loc}}(Q_{sl})$ $(l = 1, \dots, N(s))$.

If s *is such that* $(\bigcup_l \partial Q_{sl}) \cap \partial Q = \emptyset$, *then* $P^R u \in W^{k+2m}(Q_{sl})$ $(l = 1, \dots, N(s))$.

Proof. By virtue of Theorem C.5 concerning the interior smoothness of generalized solutions, $R_Q u \in W^{k+2m}_{\text{loc}}(Q)$. Hence the theorem is a consequence of Lemma 8.19. □

Remark 20.1. The case in which $(\bigcup_l \partial Q_{sl}) \cap \partial Q = \emptyset$ seems unnatural. However it occurs if $Q = (0, 4/3) \times (0, 4/3)$, and the set $\mathcal{M}$ consists of vectors $(1, 0)$, $(0, 1)$. A class consisting of one subdomain $Q_{11} = (1/3, 1) \times (1/3, 1)$ is required.

Smoothness near Boundaries of Subdomains

By virtue of Lemma 7.7, for every $r = 1, 2, \dots$ there exists a unique $s = s(r)$ such that $N(s) = J(r)$, and after some reindexing of subdomains of the sth class, $\Gamma_{rl} \subset \partial Q_{sl}$ $(l = 1, \dots, N(s))$. For each $r \in B = \{r : J_0 > 0\}$ we denote by C_s the matrix of order $N \times J_0$ obtained from R_s by deleting the last $N - J_0$ columns, where $N = N(s)$, $J_0 = J_0(r)$, $s = s(r)$.

Lemma 20.1. *Each of the last* $i = J_0 + 1, \dots, N$ *rows of the matrix* C_s *is a linear combination of the first* $1, \dots, J_0$ *rows of* C_s.

Proof. Assume the contrary. For example, suppose that for some r and corresponding $s = s(r)$ the $(J_0 + 1)$th row of C_s is not a linear combination of the first $1, \dots, J_0$ rows of C_s. Let C'_s and C''_s be the matrices of order $J_0 \times J_0$ and $(J_0 + 1) \times (J_0 + 1)$, respectively, obtained from R_s by deleting the last $N - J_0$ and $N - J_0 - 1$ rows and columns. If $\det C'_s \neq 0$, then the rows of C'_s form a basis in $\mathbb{R}^{J_0}$, in contradiction to our assumption. This proves the lemma. Let $\det C'_s = 0$. Denote by $x = (x_1, \dots, x_{J_0})$ a vector in $\mathbb{R}^{J_0}$, and let $y = (x, y_1)$ be a vector in $\mathbb{R}^{J_0+1}$. Clearly, for $x \in \mathcal{N}(C'_s)$, we have $(C'_s x, x) = (C''_s y, y) = 0$, where

$y = (x, 0)$. But, since condition 18.3 and Remark 9.1 implies that $R_s \geq 0$, then $C''_s \geq 0$. Hence, by virtue of the extremal property of eigenvalues, $C''_s y = 0$ for each $y = (x, 0)$, where $x \in \mathcal{N}(C'_s)$. Thus $\dim \mathcal{N}(C'_s) \leq \dim \mathcal{N}(C''_s)$.

By assumption, the (J_0+1)th row of C_s is not a linear combination of rows of C'_s, and C''_s is symmetric. Hence $\operatorname{rang} C''_s = \operatorname{rang} C'_s + 2$. From this it follows that $\dim \mathcal{N}'(C_s) = \dim \mathcal{N}(C''_s) + 1$. The resulting contradiction proves the lemma. □

Lemma 20.2. *Let the conditions* 18.1–18.3 *be fulfilled. Suppose that a domain* Q *satisfies the condition* 7.1.

Then there is a set of real numbers $\gamma = \{\gamma^r_{lj}\}$ *such that* $R_Q(W_{\mathcal{A}_R}) \subset W^m_\gamma(Q)$, *where* $W^m_\gamma(Q)$ *is the subspace of functions in* $W^m(Q)$ *satisfying the conditions*

$$\left.\begin{aligned} &\mathcal{D}^{\mu-1}_\nu w|_{\Gamma_{rl}} = \textstyle\sum_{j=1}^{J_0} \gamma^r_{lj} \mathcal{D}^{\mu-1}_\nu w|_{\Gamma_{rj}} && (r \in B,\ l = J_0+1, \dots, J), \\ &\mathcal{D}^{\mu-1}_\nu w|_{\Gamma_{rl}} = 0 && (r \notin B,\ l = 1, \dots, J), \end{aligned}\right\} \tag{20.2}$$

where $\mu = 1, \dots, m$.

Proof. By virtue of Lemma 19.2, it is sufficient to prove that the conditions (20.2) hold for every function $w = R_Q u$, where $u \in \dot{C}^\infty(Q)$.

Evidently the conditions (20.2) are fulfilled in the case $r \notin B$.

Let $r \in B$, and let $s = s(r)$. By Lemma 20.1, real numbers γ^r_{lj} exist such that

$$e^r_l = \sum_{j=1}^{J_0} \gamma^r_{lj} e^r_j \qquad (l = J_0+1, \dots, N), \tag{20.3}$$

where e^r_i is the ith row of C_s. Hence, as in (8.19), the equality $\mathcal{D}^{\mu-1}_\nu u|_{\Gamma_{rl}} = 0$ $(l = J_0+1, \dots, N)$ implies that

$$\mathcal{D}^{\mu-1}_\nu (R_Q u)|_{\Gamma_{rl}} = \sum_{j=1}^{J_0} \gamma^r_{lj} \mathcal{D}^{\mu-1}_\nu (R_Q u)|_{\Gamma_{rj}} \qquad (l = J_0+1, \dots, J). \quad \square \tag{20.4}$$

Theorem 20.2. *Let* $Q \subset \mathbb{R}^n$ *be a bounded domain with boundary* $\partial Q \in C^\infty$ *satisfying the condition* 7.1, *or a cylinder* $(0, d) \times G$, *where* $G \subset \mathbb{R}^{n-1}$ *is a bounded domain (with boundary* $\partial G \in C^\infty$ *if* $n \geq 3$). *Let the conditions* 18.1–18.3 *be fulfilled. Let* $u \in \mathcal{D}(\mathcal{A}_R)$ *be a solution of the equation* (20.1), *and let* $f_0 \in W^k(Q)$.

Then $R_Q u \in W^{k+2m}(Q \setminus \overline{(\partial Q \cap \mathcal{K})^\varepsilon})$, *and* $P^R u \in W^{k+2m}(Q_{sl} \setminus \overline{\mathcal{K}^\varepsilon})$ *for each* $\varepsilon > 0$ $(s = 1, 2, \dots,\ l = 1, \dots, N(s))$.

Proof. Theorem C.5 concerning the interior smoothness of generalized solutions implies that $R_Q u \in W^{k+2m}_{\text{loc}}(Q)$. Hence, since $\Gamma_{rj} \subset Q$ $(j = 1, \dots, J_0,\ r \in B)$, we obtain

$$\mathcal{D}^{\mu-1}_\nu (R_Q u)\big|_{\Gamma_{rj} \setminus \overline{\mathcal{K}^{\varepsilon/2}}} \in W^{k+2m-\mu+1/2}(\Gamma_{rl} \setminus \overline{\mathcal{K}^{\varepsilon/2}}) \tag{20.5}$$

$(j = 1, \dots, J_0)$. Therefore, by virtue of (20.2), the relations (20.5) are valid for all $r = 1, 2, \dots,\ l = 1, \dots, J$. On the other hand, the function $w = R_Q u \in W^m(Q)$

is a solution of the equation $Aw = f_0$ in a sense of distributions. Thus Theorem B.6 and Lemma C.1 imply that

$$R_Q u \in W^{k+2m}(Q \setminus \overline{(\partial Q \cap \mathcal{K})^\varepsilon}).$$

Hence, by virtue of Lemma 8.19, $P^R u \in W^{k+2m}(Q_{sl} \setminus \overline{\mathcal{K}^\varepsilon})$. □

The following example shows that Theorem 20.2 cannot be generalized to the case when $\varepsilon = 0$.

Example 20.1. Let $A = -\Delta$, $Ru(x) = u(x) + u(x_1+1, x_2+1) + u(x_1-1, x_2-1)$. Assume that Q is a domain with boundary $\partial Q \in C^\infty$, which outside the disks $S_{1/8}((i4/3, j4/3))$ $(i,j=0,1)$ coincides with the boundary of the square $(0,4/3) \times (0,4/3)$ (see Example 11.2).

We denote $\Gamma_{12} = \{x \in \partial Q : x_1 < 1/3,\ x_2 < 1/3\}$, $\Gamma_{11} = \Gamma_{12} + (1,1)$, $\Gamma_{22} = \{x \in \partial Q : 1 < x_1,\ 1 < x_2\}$, $\Gamma_{21} = \Gamma_{22} - (1,1)$. The decomposition $\mathcal{R}$ consists of two classes: 1) Q_{11}, bounded by the curves $\overline{\Gamma_{12}}$ and $\overline{\Gamma_{21}}$; Q_{21}, bounded by the curves $\overline{\Gamma_{11}}$ and $\overline{\Gamma_{22}}$, and 2) $Q_{21} = Q \setminus (\overline{Q}_{11} \cup \overline{Q}_{12})$. The set $\mathcal{K} \subset \partial Q$ and consists of four points: $g^1 = (1/3, 0)$, $g^2 = (4/3, 1)$, $g^3 = (1/3, 0)$, $g^4 = (4/3, 1)$.

We introduce the function $u(x)$ by the formula

$$u(x) = \begin{cases} \frac{1}{2}\{u_1(x_1 - \frac{1}{3}, x_2) + u_2(x_1 - \frac{1}{3}, x_2)\} & (x \in Q_{11}), \\ \frac{1}{2}\{u_1(x_1 - \frac{4}{3}, x_2 - 1) - u_2(x_1 - \frac{4}{3}, x_2 - 1)\} & (x \in Q_{12}), \\ u_1(x_1 - \frac{1}{3}, x_2) + u_1(x_1 - \frac{4}{3}, x_2 - 1) & (x \in Q_{21}), \end{cases}$$

where $u_1(r,\varphi) = \xi(r) r^{2/3} \sin \frac{2}{3}\varphi$, $u_2(r,\varphi) = \xi(r) r^{2/3} \sin \frac{4}{3}\varphi$, $\xi(r) \in \dot{C}^\infty(\mathbb{R})$, $0 \le \xi(r) \le 1$, $\xi(r) = 1$ for $r \le 1/8$, $\xi(r) = 0$ for $r \ge 1/6$; r, φ are polar coordinates.

Hence $R_Q u(x) = u_1(x_1 - \frac{1}{3}, x_2) + u_1(x_1 - \frac{4}{3}, x_2 - 1)$,

$$P^R u = R_Q u(x) \quad (x \in Q_{21}), \qquad P^R u = \tfrac{1}{2} R_Q u(x) \quad (x \in Q_{11} \cup Q_{12}).$$

Clearly $u \in \mathring{W}^1(Q)$. Thus from Lemmas 8.13, 18.2 it follows that $u \in W_{A_R}$. Since $-\Delta R_Q u \in L_2(Q)$, we have $u \in \mathcal{D}(A_R)$. On the other hand, $R_Q u \notin W^2(Q \cap S_\delta(g^1))$, and $P^R u \notin W^2(Q_{s1} \cap S_\delta(g^1))$ for $\delta > 0$.

Smoothness of Solutions in a Cylinder

If Q is a cylinder, then, under certain assumptions concerning the operators A and R_Q, we prove Theorem 20.2 for $\varepsilon = 0$.

Suppose that

$$A = -\sum_{i,j=1}^{n} \frac{\partial}{\partial x_i} a_{ij}(x) \frac{\partial}{\partial x_j}, \tag{20.6}$$

$$Ru(x) = \sum_{i=-k}^{k} b_i u(x_1 + i, x_2, \dots, x_n), \tag{20.7}$$

and that $Q = (0, d) \times G$, where $d = k + \theta$, $0 < \theta \leq 1$, k is a natural number, $G \subset \mathbb{R}^{n-1}$ is a bounded domain (with boundary $\partial G \in C^\infty$ if $n \geq 3$). Here $a_{ij} = a_{ji} \in C^\infty(\mathbb{R}^n)$ are real-valued 1-periodic in x_1 functions, $b_i = b_{-i}$ are real numbers.

Then the decomposition $\mathcal{R}$ consists of one class of subdomains $\{Q_{1l}\}$ if $\theta = 1$, and of two classes of subdomains $\{Q_{1l}\}$, $\{Q_{2l}\}$ if $\theta < 1$. The matrices R_1 coincide in both cases and have the form (8.8).

Theorem 20.3. *Let the conditions* 18.1–18.3 *be fulfilled. Suppose that the operators* A, R *and the domain* Q *satisfy the conditions of this subsection. Let* $u \in \mathcal{D}(\mathcal{A}_R)$ *be a solution of the equation* (20.1), *and let* $f_0 \in L_2(Q)$.

Then $R_Q u \in W^2(Q)$, *and* $P^R u \in W^2(Q_{sl})$ (*if* $\theta = 1$, *then* $s = 1$, $l = 1, \dots, k+1$; *if* $\theta < 1$, *then* $s = 1, 2$, $l = 1, \dots, k+1$ *for* $s = 1$ *and* $l = 1, \dots, k$ *for* $s = 2$).

Proof. By virtue of Lemmas 13.5, 8.19, it is sufficient to prove that a function $w = R_Q u$ satisfies the conditions

$$w|_{x_1=0} = \sum_{j=1}^{k} \gamma_j^1 w|_{x_1=j}, \qquad w|_{x_1=d} = \sum_{j=1}^{k} \gamma_j^1 w|_{x_1=d-j}, \qquad w|_{[0,d]\times\partial G} = 0, \tag{20.8}$$

where $\gamma_j^1 \in \mathbb{R}$ are constants.

We define the isomorphism $U_1 \colon L_2(\bigcup_{l=1}^{k+1} Q_{1l}) \to L_2^{k+1}(Q_{11})$ by the formula

$$(U_1 u)_l(x) = u(x_1 + l - 1, x_2, \dots, x_n) \qquad (x \in Q_{11}) \tag{20.9}$$

($l = 1, \dots, k+1$), where $Q_{11} = (0, \theta) \times G$. Lemma 6.2 and the equality (8.10) implies that

$$\begin{aligned}(U_1 P_1 R_Q u)_1(x) &= (R_1 U_1 P_1 u)_1(x) = \sum_{j=1}^{k} \gamma_j^1 (R_1 U_1 P_1 u)_{j+1}(x) \\ &= \sum_{j=1}^{k} \gamma_j^1 (U_1 P_1 R_Q u)_{j+1}(x) \qquad (x \in Q_{11}),\end{aligned} \tag{20.10}$$

where γ_j^1 are real constants. Since the matrix R_1 is symmetric about the principal and auxiliary diagonals, we have

$$(U_1 P_1 R_Q u)_{k+1}(x) = \sum_{j=1}^{k} \gamma_j^1 (U_1 P_1 R_Q u)_{k+1-j}(x) \qquad (x \in Q_{11}). \tag{20.11}$$

From (20.9)–(20.11) it follows that

$$\begin{aligned}(R_Q u)(x) &= \sum_{j=1}^{k} \gamma_j^1 (R_Q u)(x_1 + j, x_2, \dots, x_n) \qquad (x \in Q_{11}),\\ (R_Q u)(x) &= \sum_{j=1}^{k} \gamma_j^1 (R_Q u)(x_1 - j, x_2, \dots, x_n) \qquad (x \in Q_{1,k+1}).\end{aligned} \tag{20.12}$$

By virtue of Lemma 18.1, $R_Q u \in W^1(Q)$. Hence the relations (20.8) follow from (20.12). □

Remark 20.2. Theorem 20.3 could be proved by applying Lemma 20.1. The proof we gave was based on the relations (20.12) obtained from Lemma 6.2. These relations also establish a connection between semi-bounded differential-difference operators with degeneration and elliptic problems with nonlocal conditions of A. V. Bitsadze and A. A. Samarskiĭ type (see problem (20.20)–(20.22)).

The Domain of $\mathcal{A}_R$

Let the operators A, R and the domain Q satisfy the conditions of previous subsection, as before. In this case Theorem 20.3 enables us to find a simple description of the domain $\mathcal{D}(\mathcal{A}_R)$.

Theorem 20.4. *Let the operators A and R have the form* (20.6) *and* (20.7), *respectively, and let $Q = (0,d) \times G$, where $G \subset \mathbb{R}^{n-1}$ is a bounded domain (with boundary $\partial G \in C^\infty$ if $n \geq 3$). Assume that the conditions* 18.1–18.3 *hold.*

Then

$$\mathcal{D}(\mathcal{A}_R) = \{u \in L_2(Q) : R_Q u \in W^2(Q),\ R_Q u|_{[0,d]\times\partial G} = 0\}.$$

Proof. 1. By virtue of Theorem 20.3, it is sufficient to show that $W_{\mathcal{A}_R} = H$, where $H = \{u \in L_2(Q) : R_Q u \in W^1(Q),\ R_Q u|_{[0,d]\times\partial G} = 0\}$. Clearly $R_Q u|_{[0,d]\times\partial G} = 0$ for every $u \in \dot{C}^\infty(Q)$. Hence, by virtue of Lemma 18.1, $W_{\mathcal{A}_R} \subset H$. In order to prove the reverse inclusion, it is sufficient to show that, for each $u \in H$, there is $u_0 \in \mathring{W}^1(Q)$ such that $R_Q u = R_Q u_0$.

In fact, assume that this assertion is true. Lemmas 8.13, 18.2 imply that $((A + c_0 I)R_Q v, v)_{L_2(Q)} \leq c(v,v)_{\mathring{W}^1(Q)}$ for $v \in \dot{C}^\infty(Q)$, where $c > 0$. Since $\dot{C}^\infty(Q)$ is dense in $\mathring{W}^1(Q)$, we obtain $u_0 \in W_{\mathcal{A}_R}$. But, by virtue of Lemma 19.1, $(u - u_0) \in \mathcal{N}(R_Q) \subset W_{\mathcal{A}_R}$, i.e., $u \in W_{\mathcal{A}_R}$.

2. Let us consider the case $\theta = 1$. Then the decomposition $\mathcal{R}$ consists of one class of subdomains $Q_{1l} = (l-1, l) \times G$ $(l = 1, \dots, k+1)$. For $x \in \{0\} \times G$ the

matrix $A_1(x)$, defined by (18.3), has the form

$$A_1(x) = \begin{pmatrix} b_0 & b_1 & \dots & 0 \\ b_1 & b_0 & \dots & b_k \\ b_2 & b_1 & \dots & b_{k-1} \\ \vdots & \vdots & \ddots & \vdots \\ 0 & b_k & \dots & b_0 \end{pmatrix}.$$

The matrix R_1 is obtained from A_1 by deleting the last column and the last row.

Let $S_0^1(G) \subset L_2(G)$ be a linear manifold of functions of $n-1$ arguments such that, for $v \in S_0^1(G)$, there is $u \in \mathring{W}^1((-\varepsilon,\varepsilon)\times G)$ for which $u|_{x_1=0} = v$, where $0 < \varepsilon < 1/4$. Denote by $S_0^r(G) \subset L_2^r(G)$ a linear manifold of vector-valued functions $V = (V_1, \dots, V_r)$ such that $V_i \in S_0^1(G)$ $(i = 1, \dots, r)$.

3. Let $u \in H$. By virtue of (8.38) we obtain $(U_1 P^R u)_l \in W^1(Q_{11})$ and

$$(U_1 P^R u)_l|_{(0,1)\times\partial G} = 0 \qquad (l = 1, \dots, k+1), \quad \text{i.e.,}$$
$$V' = U_1 P^R u|_{x_1=0} \in S_0^{k+1}(G), \quad V'' = U_1 P^R u|_{x_1=1} \in S_0^{k+1}(G).$$

Since $R_Q u \in W^1(Q)$, we have

$$K_1 V' = K_2 V'', \tag{20.13}$$

where K_1 and K_2 are the matrices of order $k \times (k+1)$, obtained from R_1 by deleting the first and the last row, respectively.

We prove that there is a function $u_1 \in \mathring{W}^1(Q)$ such that

$$R_Q u_1|_{x_1=l} = R_Q u|_{x_1=l} \qquad (l = 0, \dots, k+1). \tag{20.14}$$

For this, by virtue of the definition of $S_0^{k+2}(G)$, it is sufficient to show that there exists a solution $X \in S_0^{k+2}(G)$ of the system

$$A_1 X = F \tag{20.15}$$

such that $X_1 = X_{k+2} = 0$, where $F = ((R_1 V')_1, \dots, (R_1 V')_{k+1}, (R_1 V'')_{k+1}) \in S_0^{k+2}(G)$. This problem is equivalent to the following

$$K_3 Y = F, \tag{20.16}$$

where K_3 is the matrix of order $(k+2)\times k$, obtained from A_1 by deleting the first and the last column, $Y = (X_2, \dots, X_{k+1}) \in S_0^k(G)$. By virtue of Lemma 6.2, the first row of the matrix K_3 and of the extended matrix K_4 of the system (20.16) is a linear combination of the lth rows of these matrices with the same coefficients γ_{l-1}^1 $(l = 2, \dots, k+1)$. The relations (20.13) imply that $F = ((R_1 V')_1, (R_1 V'')_1, \dots, (R_1 V'')_{k+1})$.

From this, using Lemma 6.2 and the symmetry of the matrix R_1 about the principal and auxiliary diagonals, we obtain that the $(k+2)$th row of matrices

K_3 and K_4 is equal to a linear combination of the lth rows of these matrices with coefficients γ^1_{k+2-l} $(l = 2, \dots, k+1)$. Thus, $\operatorname{rank} K_3 = \operatorname{rank} K_3'$, $\operatorname{rank} K_4 = \operatorname{rank} K_4'$, where K_3' and K_4' are the matrices of order $k \times k$ and $k \times (k+1)$, respectively, obtained from K_3 and K_4 by deleting the first and last rows. By construction, K_3' is obtained from R_1 by deleting the last row and last column. On the other hand, the matrix K_4' can be constructed by adding the column $K_2 V''$ to K_3'. By Lemma 6.2, the last column of the matrix K_2 is a linear combination of the columns of K_3'. Hence $\operatorname{rank} K_3' = \operatorname{rank} K_4'$. Therefore the system of equations (20.16) is compatible.

4. We have thus proved that there is a function $u_1 \in \mathring{W}^1(Q)$ satisfying the conditions (20.14). From (20.14) and Lemma 8.19 we obtain

$$\begin{aligned} &R_Q(u - u_1)|_{x_1 = l} = 0 && (l = 0, \dots, k+1), \\ &P^R(u - u_1) \in W^1(Q_{1l}) && (l = 1, \dots, k+1). \end{aligned}$$

Hence, the equality (8.38) implies

$$\begin{aligned} P^R(u - u_1)|_{x_1 = l} &= 0 && (l = 0, \dots, k+1), \\ P^R(u - u_1)|_{(l-1,l) \times \partial G} &= 0 && (l = 1, \dots, k+1). \end{aligned}$$

Therefore, $P^R(u - u_1) \in \mathring{W}^1(Q)$. Let $u_0 = P^R(u - u_1) + u_1$. Then $u_0 \in \mathring{W}^1(Q)$ and $R_Q u_0 = R_Q u$.

The proof for $\theta < 1$ is similar. □

The following example shows that Theorem 20.4 cannot be generalized to the case when the difference operator has shifts with respect to several coordinates.

Example 20.2. Let $Q = (0,2) \times (0,2)$. Assume that

$$\begin{aligned} Ru(x_1, x_2) = c_{00} u(x_1, x_2) &+ c_{10} u(x_1 + 1, x_2) \\ &+ c_{10} u(x_1 - 1, x_2) + c_{01} u(x_1, x_2 + 1) + c_{01} u(x_1, x_2 - 1), \end{aligned}$$

where $c_{00}, c_{10}, c_{01} > 0$, $c_{00} = c_{10} + c_{01}$.

Then there exists $u \in L_2(Q)$ such that $R_Q u \in W^1(Q)$, $R_Q u|_{x_2=0} = R_Q u|_{x_2=2} = 0$ and $u \notin W_{A_R}$.

Proof. 1. The decomposition $\mathcal{R}$ of the domain Q consists of four subdomains: $Q_{11} = (0,1) \times (0,1)$, $Q_{12} = (0,1) \times (1,2)$, $Q_{13} = (1,2) \times (0,1)$, $Q_{14} = (1,2) \times (1,2)$. The matrix R_1 has the form

$$R_1 = \begin{pmatrix} c_{00} & c_{01} & c_{10} & 0 \\ c_{01} & c_{00} & 0 & c_{10} \\ c_{10} & 0 & c_{00} & c_{01} \\ 0 & c_{10} & c_{01} & c_{00} \end{pmatrix}.$$

It is easy to show that the principal minors are equal to:

$$\begin{aligned} \Delta_1 &= c_{01} + c_{10} > 0, & \Delta_2 &= c_{10}(2c_{01} + c_{10}) > 0, \\ \Delta_3 &= 2c_{00} c_{01} c_{10} > 0, & \Delta_4 &= 0. \end{aligned}$$

Thus $\operatorname{rank} R_1 = 3$. Hence, by virtue of Lemmas 8.7, 8.12, the operator R_Q is self-adjoint and non-negative, and $0 \in \sigma(R_Q)$.

We define the matrices

$$A_1 = \begin{pmatrix} c_{00} & c_{01} & c_{10} & 0 & 0 & 0 \\ c_{01} & c_{00} & 0 & c_{10} & 0 & 0 \\ c_{10} & 0 & c_{00} & c_{01} & c_{10} & 0 \\ 0 & c_{10} & c_{01} & c_{00} & 0 & c_{10} \\ 0 & 0 & c_{10} & 0 & c_{00} & c_{01} \\ 0 & 0 & 0 & c_{10} & c_{01} & c_{00} \end{pmatrix}, \qquad C_3 = \begin{pmatrix} c_{10} & 0 \\ 0 & c_{10} \\ c_{00} & c_{01} \\ c_{01} & c_{00} \\ c_{10} & 0 \\ 0 & c_{10} \end{pmatrix},$$

$$C_1 = \begin{pmatrix} R_1 & 0 \\ 0 & E \end{pmatrix}, \qquad C_2 = \begin{pmatrix} E & 0 \\ 0 & R_1 \end{pmatrix},$$

where in the matrices C_1, C_2, zeros denote zero matrices of order 2×4 or 4×2, $E = \begin{pmatrix} 1 & 0 \\ 0 & 1 \end{pmatrix}$.

We introduce the subspaces of the space $\mathbb{C}^6$:

$$\mathcal{L} = \mathcal{R}(C_1) \cap \mathcal{R}(C_2), \qquad \mathcal{P} = \{A_1 y : y \in \mathbb{C}^6,\ y_1 = y_2 = y_5 = y_6 = 0\}.$$

Let $z \in \mathcal{P}$, i.e., there exists $y \in \mathbb{C}^6$ such that $z = A_1 y$ and $y_1 = y_2 = y_5 = y_6 = 0$. Then $z = C_1 y_+ = C_2 y_-$, where $y_+ = (0, 0, y_3, y_4, c_{10}y_3, c_{10}y_4)$, $y_- = (c_{10}y_3, c_{10}y_4, y_3, y_4, 0, 0)$. Hence $\mathcal{P} \subset \mathcal{L}$.

2. We prove that $\mathcal{L} \setminus \mathcal{P} \neq \emptyset$.

Consider $\mathcal{N}(C_1)$ and $\mathcal{N}(C_2)$. Clearly each vector from $\mathcal{N}(C_1)$ ($\mathcal{N}(C_2)$) has the following form: the last two (the first two) coordinates are equal to zero and the first four (the last four) put together a vector from $\mathcal{N}(R_1)$. Suppose that $\mathcal{N}(C_1) = \mathcal{N}(C_2)$. Since $\operatorname{rank} R_1 = 3$, then $\dim \mathcal{N}(R_1) = 1$. Therefore, for every non-zero vector $(\alpha_1, \alpha_2, \alpha_3, \alpha_4) \in \mathcal{N}(R_1)$ there is number a such that $(\alpha_1, \alpha_2, \alpha_3, \alpha_4, 0, 0) = a(0, 0, \alpha_1, \alpha_2, \alpha_3, \alpha_4)$. Then we have $\alpha_1 = \alpha_2 = \alpha_3 = \alpha_4 = 0$. This contradiction proves that $\mathcal{N}(C_1) \neq \mathcal{N}(C_2)$. Hence, $\dim(\mathcal{N}(C_1) + \mathcal{N}(C_2)) = 2$. Therefore, since $\mathcal{R}(C_1) \cap \mathcal{R}(C_2) = (\mathcal{N}(C_1) + \mathcal{N}(C_2))^\perp$, we obtain

$$\dim \mathcal{L} = \dim(\mathcal{R}(C_1) \cap \mathcal{R}(C_2)) = 6 - \dim(\mathcal{N}(C_1) + \mathcal{N}(C_2)) = 4.$$

On the other hand, $\dim \mathcal{P} = \operatorname{rank} C_3 = 2$. Thus $\mathcal{P} \subset \mathcal{L}$, $\mathcal{L} \setminus \mathcal{P} \neq \emptyset$.

3. We now construct a function $u \in L_2(Q)$ such that $R_Q u \in W^1(Q)$, $R_Q u|_{x_2=0} = R_Q u|_{x_2=2} = 0$, and $u \notin W_{A_R}$.

Let $\eta_1(x_1) \in \dot{C}^\infty(-1/3, 1/3)$, $\eta_2(x_2) \in \dot{C}^\infty(0, 1)$ and $\eta_1(x_1) = 1$ for $x_1 \in [-1/4, 1/4]$, $\eta_2(x_2) = 1$ for $x_2 \in [1/4, 3/4]$. Assume that $z = (z_1, \ldots, z_6) \in \mathcal{L} \setminus \mathcal{P}$.

We define a function $u_0(x_1,x_2)$ in Q in the following way:

$$u_0(x_1,x_2)=\begin{cases} z_1\eta_1(x_1)\eta_2(x_2) & (0\le x_1\le\frac{1}{3},\ 0\le x_2\le 1),\\ z_2\eta_1(x_1)\eta_2(x_2-1) & (0\le x_1\le\frac{1}{3},\ 1\le x_2\le 2),\\ z_3\eta_1(x_1-1)\eta_2(x_2) & (\frac{2}{3}\le x_1\le\frac{4}{3},\ 0\le x_2\le 1),\\ z_4\eta_1(x_1-1)\eta_2(x_2-1) & (\frac{2}{3}\le x_1\le\frac{4}{3},\ 1\le x_2\le 2),\\ z_5\eta_1(x_1-2)\eta_2(x_2) & (\frac{5}{3}\le x_1\le 2,\ 0\le x_2\le 1),\\ z_6\eta_1(x_1-2)\eta_2(x_2-1) & (\frac{5}{3}\le x_1\le 2,\ 1\le x_2\le 2),\end{cases}$$

$u_0(x_1,x_2)=0$ at the remaining points of Q.

Let us introduce the isomorphism $U_1\colon L_2(Q)\to L_2^4(Q_{11})$ by the formula (8.4). Clearly

$$U_1u_0(x_1,x_2)=\begin{cases}(z_1,z_2,z_3,z_4)\eta_1(x_1)\eta_2(x_2) & (0\le x_1\le\frac{1}{3},\ 0\le x_2\le 1),\\ (z_3,z_4,z_5,z_6)\eta_1(x_1-1)\eta_2(x_2) & (\frac{2}{3}\le x_1\le 1,\ 0\le x_2\le 1),\\ (0,0,0,0) & (\frac{1}{3}\le x_1\le\frac{2}{3},\ 0\le x_2\le 1).\end{cases}$$

Since $z\in\mathcal{L}$, then $U_1u_0\in\mathcal{R}(R_{Q1})$. Hence the proof of Lemma 8.17 implies that

$$\widehat{R}_1(U_1u_0)\in L_2^4(Q_{11})\quad\text{and}\quad R_1(\widehat{R}_1U_1u_0)=U_1u_0.$$

Thus $u=U_1^{-1}\widehat{R}_1U_1u_0\in L_2(Q)$ and $R_Qu=u_0$. By construction, $R_Qu\in C^\infty(\overline{Q})\subset W^1(Q)$ and $R_Qu|_{x_2=0}=R_Qu|_{x_2=2}=0$. Let us prove that $u\notin W_{A_R}$.

Assume to the contrary that $u\in W_{A_R}$. Then, by virtue of Lemma 19.2, there is a sequence $\{w_n\}\subset\dot{C}^\infty(Q)$ such that $\|R_Qu-R_Qw_n\|_{W^1(Q)}\to 0$ as $n\to\infty$. We define vector-valued functions $V,W_n\in L_2^6(0,1)$ assuming $V_i=(U_1R_Qu)_i|_{x_1=0}=(U_1u_0)_i|_{x_1=0}$, $W_{ni}=(U_1R_Qw_n)_i|_{x_1=0}$ $(i=1,2,3,4)$, $V_i=(U_1R_Qu)_{i-2}|_{x_1=1}=(U_1u_0)_{i-2}|_{x_1=1}$, $W_{ni}=(U_1R_Qw_n)_{i-2}|_{x_1=1}$ $(i=5,6)$. A subspace $\mathcal{P}\subset\mathbb{C}^6$ can be represented as a set of all vectors $y=(y_1,\dots,y_6)$ satisfying some system of equations

$$\left.\begin{matrix} b_{11}y_1+ & \cdots & +b_{16}y_6=0\\ \vdots & \ddots & \vdots\\ b_{41}y_1+ & \cdots & +b_{46}y_6=0\end{matrix}\right\}. \tag{20.17}$$

By construction, for every $x_2\in(0,1)$ we have $W(x_2)\in\mathcal{P}$, i.e., $W_n(x_2)$ satisfies the equations (20.17) in the interval (0,1). But Theorem B.6 implies that

$$\|V-W_n\|_{L_2^6(0,1)}\le k_1\|R_Qu-R_Qw_n\|_{W^1(Q)},$$

where $k_1>0$ is a constant. Hence $V(x_2)$ satisfies the equations (20.17) for almost all $x_2\in(0,1)$, i.e., $V(x_2)\in\mathcal{P}$ almost everywhere in the interval $(0,1)$. But $z\in\mathcal{L}\setminus\mathcal{P}$. Therefore $V(x_2)\in\mathcal{L}\setminus\mathcal{P}$ for $x\in[1/4,3/4]$. The resulting contradiction proves that $u\notin W_{A_R}$.

Thus we have constructed a function $u\in L_2(Q)$ such that $R_Qu\in W^1(Q)$, $R_Qu|_{x_2=0}=R_Qu|_{x_2=2}=0$, but $u\notin W_{A_R}$.

Traces of Solutions on the Boundary

It is known that the boundary conditions for the elliptic differential equations with degeneration are sometimes given on certain parts of the boundary only (see O. A. Oleĭnik and E. V. Radkevič [1]). An analogous phenomenon takes place for the differential-difference equations with degeneration. In this subsection we consider the question: when can the traces of functions $u \in \mathcal{D}(\mathcal{A}_R)$ be defined on the manifolds $\Gamma_{rl} \subset \partial Q$?

By Lemma 7.7, for each $r = 1, 2, \dots$ there is a unique $s = s(r)$ such that $N(s) = J(r)$, and we can renumber the subdomains of the sth class so that $\Gamma_{rl} \subset \partial Q_{sl}$ $(l = 1, \dots, N(s))$. Denote by B_r a set of $1 \le l \le N(s)$ such that the lth column of the matrix R_s is not a linear combination of the remaining columns of this matrix $(s = s(r))$.

Theorem 20.5. *Let $Q \subset \mathbb{R}^n$ be a bounded domain with boundary $\partial Q \in C^\infty$ satisfying the condition* 7.1 *or a cylinder $(0,d) \times G$, where $G \subset \mathbb{R}^{n-1}$ is a bounded domain (with boundary $\partial G \in C^\infty$ if $n \ge 3$). Assume that the conditions* 18.1–18.3 *hold. Let $\Gamma_{rl} \subset \partial Q$.*

Then the traces $\mathcal{D}_\nu^{\mu-1} u|_{\Gamma_{rl}}$ $(\mu = 1, \dots, m)$ are defined for all $u \in \mathcal{D}(\mathcal{A}_R)$ if and only if $l \in B_r$. When $l \in B_r$ we have $\mathcal{D}_\nu^{\mu-1} u|_{\Gamma_{rl}} = 0$.

Proof. 1. By virtue of Theorem B.6, we obtain

$$\sum_{j=1}^{N(s)} \|\mathcal{D}_\nu^{\mu-1}(R_Q u)|_{\Gamma_{rj}}\|_{L_2(\Gamma_{rj})} \le k_1 \|R_Q u\|_{W^m(Q)} \qquad (\mu = 1, \dots, m)$$

for all $u \in \dot{C}^\infty(Q)$. From this and from (18.4), (8.38) it follows that

$$\|\mathcal{D}_\nu^{\mu-1}(P^R u)|_{\Gamma_{rl}}\|_{L_2(\Gamma_{rl})} \le k_2 \|u\|_{\mathcal{A}_R} \qquad (\mu = 1, \dots, m). \tag{20.18}$$

Since $\dot{C}^\infty(Q)$ is dense in $W_{\mathcal{A}_R}$, the traces $\mathcal{D}_\nu^{\mu-1}(P^R u)|_{\Gamma_{rl}} \in L_2(\Gamma_{rl})$ are defined for each function $u \in \mathcal{D}(\mathcal{A}_R)$. Hence, by virtue of Lemma 19.1, the traces $\mathcal{D}_\nu^{\mu-1} u|_{\Gamma_{rl}}$ $(\mu = 1, \dots, m)$ are defined in $L_2(\Gamma_{rl})$ if and only if $(P^R u)(x) = u(x)$ for $x \in Q_{sl}$. This is equivalent to the following condition: $\mathcal{N}(R_s) \subset \{(x_1, \dots, x_{l-1}, 0, x_{l+1}, \dots, x_N) \in \mathbb{R}^n\}$, i.e., the lth column of the matrix R_s is not a linear combination of the other columns.

2. Let $l \in B_r$, and let $u \in \mathcal{D}(\mathcal{A}_R)$. Suppose that the sequence $u_p \in \dot{C}^\infty(Q)$ is such that $\|u - u_p\|_{\mathcal{A}_R} \to 0$ as $p \to \infty$. Then, by virtue of (20.18), $\|\mathcal{D}_\nu^{\mu-1}(u - u_p)|_{\Gamma_{rl}}\|_{L_2(\Gamma_{rl})} \to 0$ as $p \to \infty$, i.e., $\mathcal{D}_\nu^{\mu-1} u|_{\Gamma_{rl}} = 0$. □

From Theorem 20.5 and Lemma 6.2 we obtain the following assertion (cf. the proof of Theorem 6.5).

Theorem 20.6. *Let the operators A and R have the form* (20.6) *and* (20.7), *respectively, and let $Q = (0,d) \times G$, where $G \subset \mathbb{R}^{n-1}$ is a bounded domain (with boundary $\partial G \in C^\infty$ if $n \ge 3$). Assume that the conditions* 18.1–18.3 *hold.*

Then for each function $\varphi \in L_2(Q_{11})$ *there exist numbers* $\alpha, \beta \neq 0$ *and a function* $u \in \mathcal{N}(\mathcal{A}_R)$ *such that*

$$\begin{aligned} u(x) &= \alpha\varphi(x) && (x \in Q_{11}), \\ u(x) &= \beta\varphi(x_1 - k, x_2, \dots, x_n) && (x \in Q_{1,k+1}). \end{aligned} \tag{20.19}$$

Example 20.3. Let $A = -\Delta$, $Ru(x) = u(x) + a(u(x_1+1, x_2) + u(x_1-1, x_2)) + u(x_1+2, x_2) + u(x_1-2, x_2)$, $Q = (0,3) \times (0,1) \subset \mathbb{R}^2$, where $|a| < 1$. Then the decomposition $\mathcal{R}$ of the domain Q consists of one class of subdomains $Q_{1l} = (l-1, l) \times (0,1)$ $(l = 1,2,3)$. The matrix

$$R_1 = \begin{pmatrix} 1 & a & 1 \\ a & 1 & a \\ 1 & a & 1 \end{pmatrix}.$$

Evidently the second column of the matrix R_1 is not a linear combination of the first and the third columns, which are equal. Thus, by virtue of Theorem 20.5, the traces $u|_{(1,2)\times\{0\}}$, $u|_{(1,2)\times\{1\}}$ are given for all $u \in \mathcal{D}(\mathcal{A}_R)$, but the traces $u|_{\partial Q_{11}\cap\partial Q}$, $u|_{\partial Q_{13}\cap\partial Q}$ are not defined for some $u \in \mathcal{D}(\mathcal{A}_R)$.

Problem 20.1. *Generalize the results of this chapter to semi-bounded differential-difference operators with degeneration of the form* (9.1).

Applications to Nonlocal Elliptic Problems

Example 20.4. Consider the problem

$$Aw(x) = f_0(x) \qquad (x \in Q = (0,2) \times (0,1)), \tag{20.20}$$

$$w|_{x_2=0} = w|_{x_2=1} = 0 \qquad (0 \le x_1 \le 2), \tag{20.21}$$

$$w(x_1, x_2) = w(x_1+1, x_2) \qquad (0 \le x_1 \le 1; 0 \le x_2 \le 1) \tag{20.22}$$

(cf. (0.5), (0.6)). Here the operator A is strongly elliptic in $\overline{Q}$ and has the form (20.6), $f_0 \in L_2(Q)$.

A function $u \in W^2(Q)$ is called a generalized solution of the problem (20.20)–(20.22) if it satisfies the equation (20.20) and the conditions (20.21), (20.22).

We introduce the difference operator R by the formula $Ru(x) = u(x) + u(x_1+1, x_2) + u(x_1-1, x_2)$ (see Example 8.4). The matrix

$$R_1 = \begin{pmatrix} 1 & 1 \\ 1 & 1 \end{pmatrix}.$$

Thus the operator R_Q satisfies the conditions 18.2, 18.3. For the operators A and R_Q, a constant c_0 in the inequality (18.4) can be taken as equal to zero. Therefore, by virtue of Lemma 19.3, the operator $\mathcal{A}_R^0: \mathcal{R}(R_Q) \to \mathcal{R}(R_Q)$ has a bounded inverse defined on $\mathcal{R}(R_Q)$. Clearly, $\mathcal{R}(R_Q)$ consists of the functions

$u \in L_2(Q)$ satisfying the conditions (20.22). Hence, by virtue of Theorem 20.4, $\mathcal{D}(\mathcal{A}_R^0)$ consists of the functions $u \in L_2(Q)$ such that $u(x)$ satisfies the conditions (20.22), $R_Q u \in W^2(Q)$ and satisfies the conditions (20.21), (20.22). Thus, for every $f_0 \in \mathcal{R}(R_Q)$, there exists a unique generalized solution w of the problem (20.20)–(20.22), and this solution has the form $w = 2(\mathcal{A}_R^0)^{-1} f_0$.

Problem 20.2. *Consider generalization of the elliptic problem with nonlocal conditions of the type* (20.21), (20.22) *and obtain a connection between this problem and the boundary value problem for the elliptic differential-difference equation with degeneration.*

Notes

The results of this chapter were obtained by A. L. Skubachevskiĭ [3] for the case $m = 1$. Theorems concerning the operators of $2m$ order and Theorems 20.5, 20.6 are published here for the first time.

Chapter V

Nonlocal Elliptic Boundary Value Problems

In this chapter we consider mainly elliptic boundary value problems with the support of nonlocal terms inside a domain Q.

Section 21 deals with nonlocal elliptic boundary value problems with a parameter. Here we prove a priori estimates of the solutions and existence of a unique solution for sufficiently large values of a parameter.

In Section 22, we study the elliptic differential equation of the second order with nonlocal terms in a cylinder. In this case the support of the nonlocal terms has a nonempty intersection with the lateral surface of the cylinder. However, the special structure of nonlocal terms allows us to apply the methods of Section 21 and obtain similar results.

In Section 23, we consider the Fredholm property and spectrum of the nonsemibounded elliptic differential-difference operator of the second order in a cylinder. The proofs are based on the reduction to elliptic differential equations with nonlocal boundary conditions studied in Section 22.

Sections 24, 25 are devoted to the theory of multidimensional diffusion processes. Combining the results for nonlocal elliptic boundary value problems and boundary value problems for elliptic differential-difference equations, we consider the existence of Feller semigroups.

21 Nonlocal Elliptic Problems with a Parameter

Formulation of Nonlocal Elliptic Problems with a Parameter

We shall study the equation

$$Au = A^0u + \sum_{j=0}^{2m-1} q^j A_j^1 u = f_0(x) \qquad (x \in Q) \tag{21.1}$$

with *nonlocal boundary conditions*

$$\begin{aligned} B_\mu u &= \Big(B^0_\mu u + \sum_{l=0}^{m_\mu} q^l B^1_{\mu l} u\Big)\Big|_{\partial Q} + \sum_{l=0}^{m_\mu} q^l B^2_{\mu l} u \\ &= f_\mu(x) \qquad (x \in \partial Q,\ \mu = 1, \dots, m). \end{aligned} \tag{21.2}$$

Here

$$A^0 = A^0(x, \mathcal{D}, q) = \sum_{\beta+|\alpha|=2m} a_{\alpha\beta}(x) q^\beta \mathcal{D}^\alpha, \tag{21.3}$$

$$B^0_\mu = B^0_\mu(x, \mathcal{D}, q) = \sum_{\beta+|\alpha|=m_\mu} b_{\mu 0 \alpha\beta}(x) q^\beta \mathcal{D}^\alpha \tag{21.4}$$

are differential operators with complex coefficients $a_{\alpha\beta}, b_{\mu 0\alpha\beta} \in C^\infty(\mathbb{R}^n)$, $n \geq 1$; $Q \subset \mathbb{R}^n$ is a bounded domain with boundary $\partial Q \in C^\infty$ (a bounded, open interval if $n = 1$), q is a complex parameter, $f_0 \in W^k(Q)$, $f_\mu \in W^{k+2m-m_\mu-1/2}(\partial Q)$ are complex-valued functions, and $k \geq \max\{0, m_\mu - 2m + 1\}$ is an integer.

Along with the operators $A^0(x, \mathcal{D}, q)$ and $B^0_\mu(x, \mathcal{D}, q)$, we study the polynomials

$$A^0(x, \xi, q) = \sum_{\beta+|\alpha|=2m} a_{\alpha\beta}(x) q^\beta \xi^\alpha,$$

$$B^0_\mu(x, \xi, q) = \sum_{\beta+|\alpha|=m_\mu} b_{\mu 0\alpha\beta}(x) q^\beta \xi^\alpha,$$

where $\xi = (\xi_1, \dots, \xi_n)$, $\xi^\alpha = \xi_1^{\alpha_1} \cdots \xi_n^{\alpha_n}$.

Denote by θ the closed angle in $\mathbb{C}$

$$\theta = \{q \in \mathbb{C} : \varphi_1 \leq \arg q \leq \varphi_2\}.$$

We assume that the following conditions hold:

21.1. *For $x \in \overline{Q}$ and for all $q \in \theta$, $\nu \neq 0$ and ξ orthogonal to ν in $\mathbb{R}^n$ such that $|q| + |\xi| \neq 0$, the polynomial $A^0(x, \xi + \tau\nu, q)$ in the variable τ has m roots $\tau_1^+(x, \xi, \nu, q), \dots, \tau_m^+(x, \xi, \nu, q)$ with positive imaginary parts and m roots with negative imaginary parts.*

21.2. *The polynomials $B^0_\mu(x, \xi + \tau\nu, q)$ $(\mu = 1, \dots, m)$ in the variable τ are linearly independent modulo the polynomial $\prod_{j=1}^m (\tau - \tau_j^+(x, \xi, \nu, q))$ for all $x \in \partial Q$, $q \in \theta$ and ξ which is orthogonal to ν such that $|\xi| + |q| \neq 0$, where ν is the inner unit normal vector to ∂Q at the point x.*

21.3. $A^1_j : W^{2m-j-r_j}(Q) \to L_2(Q)$, $B^1_{\mu l} : W^{m_\mu - l}(Q) \to L_2(Q)$, $B^2_{\mu l} : W^{m_\mu + 1/2 - l - p_l}(Q) \to L_2(\partial Q)$ *are linear bounded operators such that their restrictions* $A^1_j : W^{k+2m-j-r_j}(Q) \to W^k(Q)$, $B^1_{\mu l} : W^{k+2m-l}(Q) \to W^{k+2m-m_\mu}(Q)$, $B^2_{\mu l} : W^{k+2m-l-p_l}(Q) \to W^{k+2m-m_\mu-1/2}(\partial Q)$ *are also bounded operators, and*

$$\|B^1_{\mu l} u\|_{L_2(Q)} \leq c_1 \|u\|_{W^{m_\mu - l}(Q_\delta)} \qquad (u \in W^{m_\mu - l}(Q)), \tag{21.5}$$

$$\|B^1_{\mu l} u\|_{W^{k+2m-m_\mu}(Q)} \leq c_2 \|u\|_{W^{k+2m-l}(Q_\delta)} \qquad (u \in W^{k+2m-l}(Q)), \tag{21.6}$$

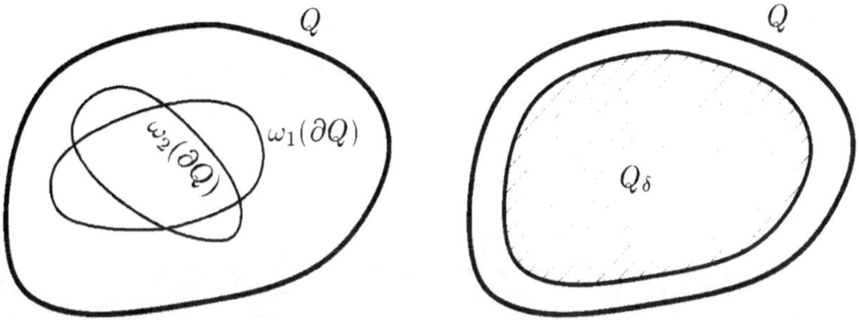

Fig. V.1

where $Q_\delta = \{x \in Q : \rho(x, \partial Q) > \delta > 0\}$, $0 < r_j \le 2m - j$, $j = 0, \dots, 2m-1$, $0 < p_l \le m_\mu + 1/2 - l$, $l = 0, \dots, m_\mu$, $\mu = 1, \dots, m$.

We introduce in the Hilbert spaces $W^l(\Omega)$ $(\Omega = Q, \partial Q)$ and $\mathcal{W}^k(Q, \partial Q) = W^k(Q) \times \prod_{\mu=1}^m W^{k+2m-m_\mu-1/2}(\partial Q)$, the equivalent norms depending on a parameter q,

$$|||u|||_{W^l(\Omega)} = \{\|u\|^2_{W^l(\Omega)} + |q|^{2l}\|u\|^2_{L_2(\Omega)}\}^{1/2},$$

$$|||f|||_{\mathcal{W}^k(Q,\partial Q)} = \left\{ |||f_0|||^2_{W^k(Q)} + \sum_{\mu=1}^m |||f_\mu|||^2_{W^{k+2m-m_\mu-1/2}(\partial Q)} \right\}^{1/2},$$

where $f = (f_0, f_1, \dots, f_m)$.

Evidently, the linear operators $A^0 : W^l(Q) \to W^{l-2m}(Q)$ and $B^0_\mu : W^s(Q) \to W^{s-m_\mu}(Q)$ $(l = 2m, \dots, k+2m,\ s = m_\mu, \dots, k+2m,\ \mu = 1, \dots, m)$ are bounded.

Now we consider an example of this type of nonlocal problem.

Example 21.1. We consider the equation

$$Au = \sum_{\beta + |\alpha| \le 2m} a_{\alpha\beta}(x) q^\beta \mathcal{D}^\alpha u(x) = f_0(x) \qquad (x \in Q) \tag{21.7}$$

with nonlocal boundary conditions

$$\begin{aligned} B_\mu u &= \sum_{s=0}^S \sum_{\beta+|\alpha|\le m_\mu} b_{\mu s\alpha\beta}(x) q^\beta (\mathcal{D}^\alpha u)(\omega_s(x))|_{\partial Q} \\ &= f_\mu(x) \qquad (x \in \partial Q,\ \mu = 1, \dots, m). \end{aligned} \tag{21.8}$$

Here $a_{\alpha\beta}, b_{\mu s\alpha\beta} \in C^\infty(\mathbb{R}^n)$ are complex-valued functions; ω_s are infinitely differentiable nondegenerate transformations mapping some neighborhood γ of the boundary ∂Q onto the set $\omega_s(\gamma)$, $\overline{\omega_s(\gamma)} \subset Q$ if $s > 0$ (see Fig. V.1) $\omega_0(x) \equiv x$.

Assume that the operators A^0, B^0_μ satisfy the conditions 21.1, 21.2. We prove that the operators A and B_μ can be represented in the form

$$Au = A^0u + \sum_{j=0}^{2m-1} q^j A^1_j u, \tag{21.9}$$

$$B_\mu u = \left(B^0_\mu u + \sum_{l=0}^{m_\mu} q^l B^1_{\mu l} u\right)\Big|_{\partial Q} + \sum_{l=0}^{m_\mu} q^l B^2_{\mu l} u, \tag{21.10}$$

where the operators A^1_j, $B^1_{\mu l}$, $B^2_{\mu l}$ satisfy the condition 21.3.
Proof. We denote

$$A^1_j u = \sum_{|\alpha|<2m-j} a_{\alpha j}(x)\mathcal{D}^\alpha u(x) \qquad (j = 0, \dots, 2m-1),$$

$$B^2_{\mu l} u = \left(\sum_{|\alpha|<m_\mu - l} b_{\mu 0 \alpha l}(x)\mathcal{D}^\alpha u(x)\right)\Big|_{\partial Q} \qquad (l = 0, \dots, m_\mu - 1),$$

$B^2_{\mu m_\mu} u = 0$. The operators A^1_j, $B^2_{\mu l}$ satisfy the condition 21.3 for $r_j = 1$, $p_l = 1/2$ $(j = 0, \dots, 2m-1,\ l = 0, \dots, m_\mu)$.

Let us introduce the functions $\eta_s \in \dot{C}^\infty(\mathbb{R}^n)$ such that $\eta_s(x) = 1$ for $x \in \omega_s(\gamma_1)$, $\eta_s(x) = 0$ for $x \notin \omega_s(\gamma_2)$, where γ_1 is some neighborhood of ∂Q and $\overline{\gamma_1} \subset \gamma_2$, $\overline{\gamma_2} \subset \gamma$. Since $\delta = \min_{0<s} \rho(\omega_s(\gamma_2), \partial Q) > 0$, the operators $B^1_{\mu l}$, defined by

$$(B^1_{\mu l}u)(x) = \sum_{0<s} \sum_{|\alpha| \le m_\mu - l} b_{\mu s \alpha l}(x)\eta_s(\omega_s(x))(\mathcal{D}^\alpha u)(\omega_s(x)) \qquad (x \in \gamma \cap Q),$$

$$(B^1_{\mu l}u)(x) = 0 \qquad (x \in Q \setminus \gamma),$$

satisfy the condition 21.3.

By construction, the operators A and B_μ can be written in the form (21.9), (21.10). □

Solvability of Nonlocal Problems with a Parameter

We introduce the bounded operator $\mathcal{L} = \mathcal{L}(q) \colon W^{k+2m}(Q) \to \mathcal{W}^k(Q, \partial Q)$ according to the formula

$$\mathcal{L}u = (Au, B_1 u, \dots, B_m u).$$

Denote by $\mathcal{L}_0$ the operator given by

$$\mathcal{L}_0 u = (A^0 u, (B^0_1 u)|_{\partial Q}, \dots, (B^0_\mu u)|_{\partial Q}).$$

The operator $\mathcal{L}$ corresponds to the nonlocal problem (21.1), (21.2), while the operator $\mathcal{L}_0$ corresponds to the "local" boundary value problem. We introduce the operator $\mathcal{L}_t = \mathcal{L}_0 + t(\mathcal{L} - \mathcal{L}_0)$, where $0 \le t \le 1$. A function $u \in W^{k+2m}(Q)$ is called a *generalized solution* of the boundary value problem (21.1), (21.2) if

$$\mathcal{L}(q)u = f.$$

We now state the main result of this section.

Theorem 21.1. *Let the conditions* 21.1–21.3 *be fulfilled.*

Then we have:

(a) $\mathcal{L}(q)\colon W^{k+2m}(Q) \to \mathcal{W}^k(Q, \partial Q)$ *is a Fredholm operator, and* $\operatorname{ind} \mathcal{L}(q) = 0$ *for all* $q \in \mathbb{C}$.

(b) *There exists* $q_1 > 0$ *such that for* $q \in \{q \in \theta : |q| \ge q_1\}$ *the operator* $\mathcal{L}(q)$ *has a bounded inverse* $\mathcal{L}^{-1}(q)\colon \mathcal{W}^k(Q, \partial Q) \to W^{k+2m}(Q)$.

(c) *The operator function* $\mathcal{L}^{-1}(q)\colon \mathcal{W}^k(Q, \partial Q) \to W^{k+2m}(Q)$ *is a finitely meromorphic Fredholm operator function in* $\mathbb{C}$.

The proof of this theorem is based on the following statement:

Lemma 21.1. *Let conditions* 21.1–21.3 *be fulfilled.*

Then there exists $q_1 > 0$ *such that for* $q \in \{q \in \theta : |q| \ge q_1\}$ *and* $0 \le t \le 1$ *for all* $u \in W^{k+2m}(Q)$ *we have the estimate*

$$c_3 |||\mathcal{L}_t u|||_{\mathcal{W}^k(Q,\partial Q)} \le |||u|||_{W^{k+2m}(Q)} \le c_4 |||\mathcal{L}_t u|||_{\mathcal{W}^k(Q,\partial Q)}, \tag{21.11}$$

where $c_3, c_4 > 0$ *do not depend on* q, t, *and* u.

Proof. The left part of the inequality (21.11) follows from condition 21.3 and inequalities (B.21), (B.20).

Let us prove the right part of inequality (21.11). Denote $\mathcal{L}_t u = f$. Then

$$\mathcal{L}_0 u = f + \Phi, \tag{21.12}$$

where

$$\Phi = \Big(-t \sum_{j=0}^{2m-1} q^j A_j^1 u, -t \sum_{l=0}^{m_\mu} q^l ((B_{1l}^1 u)|_{\partial Q} + B_{1l}^2 u), \dots, \\ -t \sum_{l=0}^{m_\mu} q^l ((B_{ml}^1 u)|_{\partial Q} + B_{ml}^2 u) \Big),$$

$$f = (f_0, f_1, \dots, f_m).$$

By virtue of Theorem C.7, there exists $q_0 > 0$ such that for $q \in \{q \in \theta : |q| \ge q_0\}$ a solution of the "local" equation (21.12) is estimated as

$$|||u|||_{W^{k+2m}(Q)} \le k_1 |||f + \Phi|||_{\mathcal{W}^k(Q,\partial Q)}. \tag{21.13}$$

From the inequality (B.20) and condition 21.3 it follows that

$$\begin{aligned}
&|||tq^j A_j^1 u|||_{W^k(Q)} \\
&\quad \le k_2 |q|^j (\|u\|_{W^{k+2m-j-r_j}(Q)} + |q|^k \|u\|_{W^{2m-j-r_j}(Q)}) \\
&\quad \le k_3 |q|^{-r_j} |||u|||_{W^{k+2m}(Q)}, \qquad (21.14) \\
&|||tq^l B_{\mu l}^2 u|||_{W^{k+2m-m_\mu-1/2}(\partial Q)} \\
&\quad \le k_4 |q|^l (\|u\|_{W^{k+2m-l-p_l}(Q)} + |q|^{k+2m-m_\mu-1/2} \|u\|_{W^{m_\mu+1/2-l-p_l}(Q)}) \\
&\quad \le k_5 |q|^{-p_l} |||u|||_{W^{k+2m}(Q)}. \qquad (21.15)
\end{aligned}$$

Without loss of generality, we can assume that a boundary $\partial Q_\delta \in C^\infty$. Using the condition 21.3 and the inequalities (B.20), (B.21), we obtain

$$\begin{aligned}
&|||tq^l (B_{\mu l}^1 u)|_{\partial Q}|||_{W^{k+2m-m_\mu-1/2}(\partial Q)} \\
&\quad \le k_6 |q|^l (\|u\|_{W^{k+2m-l}(Q_\delta)} + |q|^{k+2m-m_\mu-1} (\|tB_{\mu l}^1 u\|_{W^1(Q)} + |q| \|tB_{\mu l}^1 u\|_{L_2(Q)})) \\
&\quad \le k_7 |q|^l (\|u\|_{W^{k+2m-l}(Q_\delta)} + \|tB_{\mu l}^1 u\|_{W^{k+2m-m_\mu}(Q)} + |q|^{k+2m-m_\mu} \|tB_{\mu l}^1 u\|_{L_2(Q)}) \\
&\quad \le k_8 |q|^l (\|u\|_{W^{k+2m-l}(Q_\delta)} + |q|^{k+2m-m_\mu} \|u\|_{W^{m_\mu-l}(Q_\delta)}) \\
&\quad \le k_9 |||u|||_{W^{k+2m}(Q_\delta)}. \qquad (21.16)
\end{aligned}$$

We introduce a function

$$\xi \in \dot{C}^\infty(\mathbb{R}^n), \quad \xi(x) = 1 \ (x \in Q_{\delta/2}), \quad \xi(x) = 0 \ (x \notin Q_{\delta/4}), \qquad (21.17)$$

where $\overline{Q}_\delta \subset Q_{\delta/2}$, $\overline{Q}_{\delta/2} \subset Q_{\delta/4}$. From the inequality (21.16), Theorem C.7, the inequality (B.20), Leibniz' formula, the condition 21.3 and the inequality (21.14), it follows that

$$\begin{aligned}
&|||tq^l (B_{\mu l}^1) u|_{\partial Q}|||_{W^{k+2m-m_\mu-1/2}(\partial Q)} \le k_9 |||\xi u|||_{W^{k+2m}(Q)} \le k_{10} |||A^0(\xi u)|||_{W^k(Q)} \\
&\quad \le k_{11} \Big\{ |||\xi A^0 u|||_{W^k(Q)} + \sum_{j=0}^{2m-1} |q|^j \sum_{|\alpha|=1}^{2m-j} \sum_{|\beta| \le 2m-|\alpha|-j} (\|\mathcal{D}^\alpha \xi \mathcal{D}^\beta u\|_{W^k(Q)} \\
&\qquad + |q|^k \|\mathcal{D}^\alpha \xi \mathcal{D}^\beta u\|_{L_2(Q)}) \Big\} \\
&\quad \le k_{12} \Big\{ \|A^0 u\|_{W^k(Q)} + \sum_{j=0}^{2m-1} |q|^j (\|u\|_{W^{k+2m-1-j}(Q)} + |q|^k \|u\|_{W^{2m-1-j}(Q)}) \Big\} \\
&\quad \le k_{13} \Big\{ \Big|\Big|\Big| \Big(A^0 + t \sum_{j=0}^{2m-1} q^j A_j^1 \Big) u \Big|\Big|\Big|_{W^k(Q)} \\
&\qquad + \Big(\sum_{j=0}^{2m-1} |q|^{-r_j} + |q|^{-1} \Big) |||u|||_{W^{k+2m}(Q)} \Big\} \qquad (21.18)
\end{aligned}$$

for $q \in \{q \in \theta : |q| \ge q_0\}$. Here $k_1, \dots, k_{13} > 0$ do not depend on t, q, and u.

Choosing $q_1 > q_0$ such that

$$k_1\left\{(k_3+mk_{13})\sum_{j=0}^{2m-1}|q_1|^{-r_j}+mk_5\sum_{l=0}^{m_\mu}|q_1|^{-p_l}+mk_{13}|q_1|^{-1}\right\}<\frac{1}{2},$$

by virtue of the inequalities (21.13)–(21.15), (21.18), we obtain the right part of the estimate (21.11) with constant $c_4 = 2k_1(1+mk_{13})$ for $q \in \{q \in \theta : |q| \geq q_1\}$.
□

Proof of Theorem 21.1. Combining Theorem C.8 and Lemmas 21.1, 1.2, we obtain the statement (b). For $\mu \in \{q \in \theta : |q| \geq q_1\}$ and $q \in \mathbb{C}$, we have $\mathcal{L}(q)\mathcal{L}^{-1}(\mu) = I + (\mathcal{L}(q) - \mathcal{L}(\mu))\mathcal{L}^{-1}(\mu)$, where I is the identity operator in $\mathcal{W}^k(Q, \partial Q)$. By virtue of condition 21.3, the operator $\mathcal{L}(q) - \mathcal{L}(\mu)\colon W^{k+2m-1}(Q) \to \mathcal{W}^k(Q, \partial Q)$ is bounded. From the compactness of the imbedding of $W^{k+2m}(Q)$ into $W^{k+2m-1}(Q)$ (see Theorem B.8) it follows that the operator $(\mathcal{L}(q) - \mathcal{L}(\mu))\mathcal{L}^{-1}(\mu)\colon \mathcal{W}^k(Q, \partial Q) \to \mathcal{W}^k(Q, \partial Q)$ is compact. Thus, by Theorem A.1, the operator $\mathcal{L}(q)$ is Fredholm, and $\operatorname{ind} \mathcal{L}(q) = 0$ for all $q \in \mathbb{C}$. Now the statement (c) follows from Theorem A.9.
□

Nonlocal Perturbations of the Dirichlet problem

We assume that the operators A^0, B^0_μ satisfy the following conditions:

21.4. $A^0 u = \sum_{|\alpha|=2m} a_\alpha(x)\mathcal{D}^\alpha u(x) + q^{2m}u(x)$ *is the differential operator with real coefficients* $a_\alpha \in C^\infty(\mathbb{R}^n)$; $\sum_{|\alpha|=2m} a_\alpha(x)\xi^\alpha > 0$ *for* $0 \neq \xi \in \mathbb{R}^n$ *and* $x \in \overline{Q}$.

21.5. $B^0_\mu u = (-i\partial/\partial\nu)^{\mu-1}$ $(\mu = 1, \ldots, m)$ *are boundary operators of the Dirichlet problem.*

Evidently, if the conditions 21.4, 21.5 are fulfilled, then the operators A^0, B^0_μ satisfy the conditions 21.1, 21.2 for $|\arg q| \leq (\pi - \varepsilon)/2m$, where $0 < \varepsilon < \pi$ is arbitrary. Therefore, by virtue of Theorem 21.1, the following statement is valid:

Theorem 21.2. *Let conditions 21.3–21.5 hold.*
Then the operator $\mathcal{L}(q)\colon W^{k+2m}(Q) \to \mathcal{W}^k(Q, \partial Q)$ *is Fredholm, and* $\operatorname{ind} \mathcal{L}(q) = 0$ *for all* $q \in \mathbb{C}$.

For every $0 < \varepsilon < \pi$, *there exists* $q_\varepsilon > 0$ *such that for* $|q| \geq q_\varepsilon$, $|\arg q| \leq (\pi - \varepsilon)/2m$ *the operator* $\mathcal{L}(q)$ *has a bounded inverse* $\mathcal{L}^{-1}(q)\colon \mathcal{W}^k(Q, \partial Q) \to W^{k+2m}(Q)$.

We introduce the unbounded operator $\mathcal{A}_\gamma\colon W^k(Q) \to W^k(Q)$ given by

$$\mathcal{A}_\gamma u = Au \qquad (u \in \mathcal{D}(\mathcal{A}_\gamma)),$$
$$\mathcal{D}(\mathcal{A}_\gamma) = \{u \in W^{k+2m}(Q) : B_\mu u = 0,\ \mu = 1, \ldots, m\}.$$

Here, in the operators A, B_μ we set $q = 0$.

Theorem 21.3. *Let conditions* 21.3–21.5 *be fulfilled.*

Then we have:

(a) *The operator* $\mathcal{A}_\gamma\colon W^k(Q) \to W^k(Q)$ *is Fredholm, and* $\operatorname{ind}\mathcal{A}_\gamma = 0$.

(b) *The spectrum* $\sigma(\mathcal{A}_\gamma)$ *is discrete; for* $\lambda \notin \sigma(\mathcal{A}_\gamma)$, *the resolvent* $R(\lambda, \mathcal{A}_\gamma)\colon W^k(Q) \to W^k(Q)$ *is a compact operator.*

(c) *For every* $0 < \varepsilon < \pi$, *all points of the spectrum* $\sigma(\mathcal{A}_\gamma)$, *except possibly a finite number of them, belong to the angle of the complex plane* $|\arg\lambda| < \varepsilon$.

The proof follows from the substitution $\lambda = -q^{2m}$ and Theorems 21.2, A.8, B.8.

Example 21.2. We consider the boundary value problem

$$-\Delta u + q^2 u = f_0(x) \qquad (x \in Q = S_1(0) \subset \mathbb{R}^2), \tag{21.19}$$

$$u|_{r=1} - u|_{r=1/2} + \int_{1/4}^{1} u(r,\varphi)\,dr = f_1(\varphi) \qquad (0 \le \varphi \le 2\pi), \tag{21.20}$$

where r, φ are polar coordinates of the point $x \in \mathbb{R}^2$. Denote $B_{10}^1 u = -\eta(r - 1/2)u(r-1/2,\varphi)$, $B_{10}^2 u = \int_{1/4}^{1} u(r,\varphi)\,dr$, where $\eta \in \dot{C}^\infty(1/4, 3/4)$, $\eta(r) = 1$ for $r \in (3/8, 5/8)$. From Example 21.1 it follows that the operator B_{10}^1 satisfies the condition 21.3 for $\delta = 1/4$. The operator $B_{10}^2\colon W^s(Q) \to W^s(\partial Q)$ is bounded for the integer $s \ge 0$. By virtue of interpolation Theorem B.15, this statement is valid also if $s = l - 1/2$ $(l = 1, 2, \ldots)$. The operator B_{10}^2 thus satisfies the condition 21.3, where $p_0 = 1/2$. Hence Theorem 21.2 remains valid for the operator-valued function $\mathcal{L}(q)$ corresponding to the boundary value problem (21.19), (21.20).

Similarly, we can see that the following example satisfies the conditions 21.3–21.5.

Example 21.3. We consider the boundary value problem

$$\Delta\Delta u + q^4 u = f_0(x) \qquad (x \in Q = S_1(0) \subset \mathbb{R}^2), \tag{21.21}$$

$$\begin{aligned} u(r,\varphi)|_{r=1} - 2u(r,\varphi)|_{r=1/2} + u(r,\varphi-1)|_{r=1/3} &= f_1(\varphi), \\ \left.\frac{\partial u}{\partial r}\right|_{r=1} + Gu = f_2(\varphi) \qquad (0 \le \varphi \le 2\pi). & \end{aligned} \tag{21.22}$$

Here $Gu = \{F^{-1}[(1+|\xi|^2)^{1/2}F(\eta u)]\}|_{r=1}$, $\eta \in \dot{C}^\infty(\mathbb{R}^2)$, $\eta(x) = 1$ for $r < 1/2$, $\eta(x) = 0$ for $r > 3/4$; $F(v) = (Fv)(\xi)$ is the Fourier transform with respect to x, $F^{-1}(w) = (F^{-1}w)(x)$ is the inverse Fourier transform with respect to ξ, $\xi \in \mathbb{R}^2$.

22 Elliptic Equations with Nonlocal Boundary Conditions in a Cylinder

Formulation of the Nonlocal Elliptic Problem in a Cylinder

In this section we consider the nonlocal elliptic problem in a cylinder, which is a generalization of the problem (0.3), (0.4). In this case there are nonlocal terms near the boundary. Nevertheless, we obtain theorems on solvability in Sobolev spaces analogous to Theorems 21.2, 21.3.

We consider the equation

$$Au = A^0 u + A^1 u = f_0(x) \qquad (x \in Q) \tag{22.1}$$

with *nonlocal boundary conditions*

$$\begin{gathered} B_\mu u = (u + B_\mu^1 u)|_{x_1 = t_\mu} + B_\mu^2 u = f_\mu(x') \qquad (x' \in G,\ \mu = 1, 2), \\ u|_{[0,d]\times\partial G} = 0. \end{gathered} \tag{22.2}$$

Here $Q = (0, d) \times G$, $G \subset \mathbb{R}^{n-1}$ is a bounded domain (with boundary $\partial G \in C^\infty$ if $n \geq 3$), $x = (x_1, \dots, x_n) \in \mathbb{R}^n$, $x' = (x_2, \dots, x_n) \in \mathbb{R}^{n-1}$,

$$A^0 = -\sum_{i,j=1}^{n} \frac{\partial}{\partial x_i} a_{ij} \frac{\partial}{\partial x_j}, \tag{22.3}$$

$a_{ij} = a_{ji} \in C^\infty(\mathbb{R}^n)$ are real-valued functions; $f_0 \in L_2(Q)$, $f_\mu \in W^{3/2}(G)$ are complex-valued functions, $t_1 = 0$, $t_2 = d$.

We shall assume that the following conditions are fulfilled:

22.1. $\sum_{i,j=1}^{n} a_{ij}(x)\xi_i\xi_j > 0$ *for* $x \in \overline{Q}$ *and* $0 \neq \xi \in \mathbb{R}^n$.

22.2. $A^1\colon W^{2-r}(Q) \to L_2(Q)$, $B_\mu^1\colon L_2(Q) \to L_2(Q)$, $B_\mu^2\colon W^{1/2-p}(Q) \to L_2(G)$ *are linear bounded operators such that their restrictions* $B_\mu^1\colon W_0^2(Q) \to W_0^2(Q)$, $B_\mu^2\colon W_0^{2-p}(Q) \to W_0^{3/2}(G)$ *are also bounded operators, and*

$$\|B_\mu^1 u\|_{L_2(Q)} \leq c_1 \|u\|_{L_2(\widetilde{Q}_\delta)} \qquad (u \in L_2(Q)), \tag{22.4}$$

$$\|B_\mu^1 u\|_{W^2(Q)} \leq c_2 \|u\|_{W^2(\widetilde{Q}_\delta)} \qquad (u \in W_0^2(Q)), \tag{22.5}$$

where $0 < r \leq 2$, $0 < p \leq 1/2$, $\delta > 0$.

Here $W_0^k(Q)$, $W_0^k(\widetilde{Q}_\delta)$, and $W_0^k(G)$ are the subspaces of functions in $W^k(Q)$, $W^k(\widetilde{Q}_\delta)$, and $W^k(G)$, respectively, whose traces vanish on $[0, d] \times \partial G$, $[\delta, d - \delta] \times \partial G$, and ∂G, respectively; $k \geq 1$, $\widetilde{Q}_\delta = (\delta, d - \delta) \times G$, $\delta > 0$ (see Fig. V.2).

Example 22.1.

$$Au = -\sum_{i,j=1}^{n} \frac{\partial}{\partial x_i} a_{ij} \frac{\partial u}{\partial x_j} + \sum_{i=1}^{n} a_i \frac{\partial u}{\partial x_i} + a_0 u = f_0(x) \qquad (x \in Q) \tag{22.6}$$

with nonlocal boundary conditions

$$\begin{aligned} & u|_{[0,d]\times\partial G} = 0, \\ B_\mu u &= u|_{x_1 = t_\mu} + \sum_{i=1}^{m} b_{\mu i}(x') u|_{x_1 = d_i} + \int_0^d b_\mu(x) u(x)\, dx_1 \\ &= f_\mu(x') \qquad (x' \in G,\ \mu = 1, 2). \end{aligned} \tag{22.7}$$

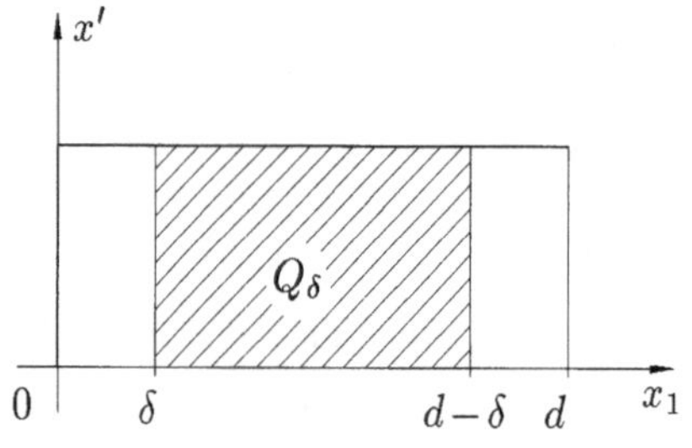

Fig. V.2

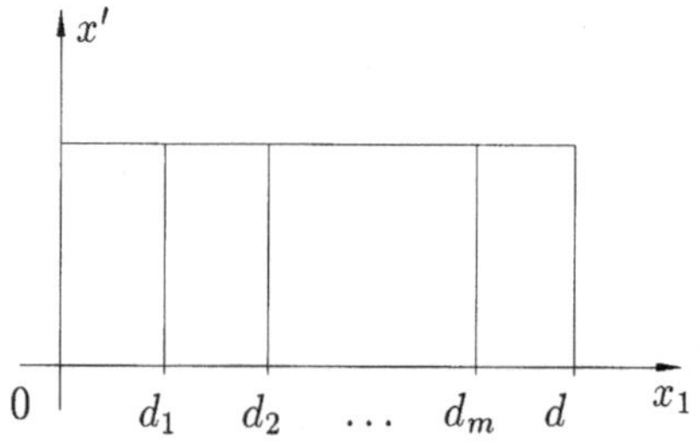

Fig. V.3

Here $Q = (0,d) \times G$, $G \subset \mathbb{R}^{n-1}$ is a bounded domain (with boundary $\partial G \in C^\infty$ if $n \geq 3$), $x = (x_1, \dots, x_n) \in \mathbb{R}^n$, $x' = (x_2, \dots, x_n) \in \mathbb{R}^{n-1}$; $a_{ij}, a_i, a_0 \in C^\infty(\mathbb{R}^n)$ are real-valued functions, $a_{ij} = a_{ji}$; $b_\mu \in C^\infty(\mathbb{R}^n)$ and $b_{\mu i} \in C^\infty(\mathbb{R}^{n-1})$ are complex-valued functions, $0 < d_i < d$, $t_1 = 0$, $t_2 = d$ (see Fig. V.3).

Assume that the operator A^0 satisfies the condition 22.1. We prove that the operators A and B_μ can be represented in the form

$$Au = A^0 u + A^1 u, \tag{22.8}$$

$$B_\mu u = (u + B_\mu^1 u)|_{x_1 = t_\mu} + B_\mu^2 u \qquad (\mu = 1,2), \tag{22.9}$$

where the operators A^1, B_μ^1, B_μ^2 satisfy the condition 22.2.

Proof. We denote

$$A^1 u = \sum_{i=1}^n a_i \frac{\partial u}{\partial x_i} + a_0 u.$$

Clearly the operator A^1 satisfies the condition 22.2 for $r = 1$. We introduce the function $\eta(x_1)$ such that $\eta \in \dot{C}^\infty(\mathbb{R})$, $\eta(x_1) = 1$ for $x_1 \in (-\delta, \delta)$, $\eta(x_1) = 0$ for $x_1 \notin (-2\delta, 2\delta)$, where $4\delta = \min_{\mu,i} |t_\mu - d_i|$.

Let

$$\begin{aligned}
(B_1^1u)(x) &= \eta(x_1)\sum_{i=1}^{m} b_{1i}(x')u(x_1+d_i,x') \qquad (x\in(0,4\delta)\times G),\\
(B_1^1u)(x) &= 0 \qquad (x\in Q\setminus(0,4\delta)\times G),\\
(B_2^1u)(x) &= \eta(x_1-d)\sum_{i=1}^{m} b_{2i}(x')u(x_1-d+d_i,x')\\
&\qquad\qquad\qquad\qquad\qquad\qquad\qquad (x\in(d-4\delta,d)\times G),\\
(B_2^1u)(x) &= 0 \qquad (x\in Q\setminus(d-4\delta,d)\times G).
\end{aligned}$$

By definition, the operators B_μ^1 satisfy the condition 22.2 (see Fig. V.3). We denote

$$B_\mu^2 u = \int_0^d b_\mu(x)u(x)\,dx_1 \qquad (\mu=1,2).$$

By virtue of Theorem B.5 concerning an extension of functions and interpolation Theorem B.14, the operators B_μ^2 satisfy the condition 22.2 for $p=1/2$. By construction, the operators A and B_μ can be written in the form (22.8), (22.9). □

Solvability and Spectrum

We define the operators $\mathcal{L}=\mathcal{L}(\lambda)$, $\mathcal{L}_0=\mathcal{L}_0(\lambda)\colon W_0^2(Q)\to\mathcal{W}^0(Q,G)$ by the formulas

$$\begin{aligned}
\mathcal{L}u &= (Au-\lambda u, B_1u, B_2u),\\
\mathcal{L}_0u &= (A^0u-\lambda u, u|_{x_1=0}, u|_{x_1=d}),
\end{aligned}$$

where $\mathcal{W}^0(Q,G)=L_2(Q)\times W_0^{3/2}(G)\times W_0^{3/2}(G)$.

A function $u\in W_0^2(Q)$ is called a *generalized solution* of the boundary value problem (22.1), (22.2) if

$$\mathcal{L}u=f,$$

where $f=(f_0,f_1,f_2)\in\mathcal{W}^0(Q,G)$.

Now we formulate the main result of this section:

Theorem 22.1. *Let conditions* 22.1, 22.2 *be fulfilled.*

Then we have:

(a) *The operator* $\mathcal{L}(\lambda)\colon W_0^2(Q)\to\mathcal{W}^0(Q,G)$ *is Fredholm, and* $\operatorname{ind}\mathcal{L}(\lambda)=0$ *for all* $\lambda\in\mathbb{C}$.

(b) *For every* $\varepsilon>0$, *there exists* $\lambda_0=\lambda_0(\varepsilon)>0$ *such that for* $\lambda\in\Omega_{\varepsilon,\lambda_0}=\{\lambda\in\mathbb{C}:|\arg\lambda|\ge\varepsilon,\ |\lambda|\ge\lambda_0\}$ *the operator* $\mathcal{L}(\lambda)$ *has a bounded inverse* $\mathcal{L}^{-1}(\lambda)\colon\mathcal{W}^0(Q,G)\to W_0^2(Q)$.

(c) *The operator function* $\mathcal{L}^{-1}(\lambda)\colon\mathcal{W}^0(Q,G)\to W_0^2(Q)$ *is a finitely meromorphic Fredholm operator function in* $\mathbb{C}$.

We introduce the unbounded operator $\mathcal{A}_\gamma: L_2(Q) \to L_2(Q)$ with domain $\mathcal{D}(\mathcal{A}_\gamma) = \{u \in \overset{\circ}{W}{}^2(Q) : B_\mu u = 0,\ \mu = 1, 2\}$ by the formula $\mathcal{A}_\gamma u = Au$.

Theorem 22.2 follows from Theorem 22.1.

Theorem 22.2. *Let conditions 22.1, 22.2 hold.*

Then we have:

(a) *The operator $\mathcal{A}_\gamma: L_2(Q) \to L_2(Q)$ is Fredholm, and* $\operatorname{ind} \mathcal{A}_\gamma = 0$.

(b) *The spectrum $\sigma(\mathcal{A}_\gamma)$ is discrete; for $\lambda \notin \sigma(\mathcal{A}_\gamma)$, the resolvent $R(\lambda, \mathcal{A}_\gamma)$: $L_2(Q) \to L_2(Q)$ is a compact operator.*

(c) *For every $0 < \varepsilon < \pi$, all points of the spectrum $\sigma(\mathcal{A}_\gamma)$, except possibly for a finite number of them, belong to the angle of the complex plane $|\arg \lambda| < \varepsilon$.*

Example 22.2 (cf. (0.3), (0.4)).We consider the boundary value problem

$$A^0 u = f_0(x) \qquad (x \in Q = (0,2) \times (0,1)), \tag{22.10}$$

$$\begin{gathered} u|_{x_1=0} = \gamma_1 u|_{x_1=1}, \quad u|_{x_1=2} = \gamma_2 u|_{x_1=1}, \\ u|_{x_2=0} = u|_{x_2=1} = 0, \end{gathered} \tag{22.11}$$

where $\gamma_1, \gamma_2 \in \mathbb{R}$, $f_0 \in L_2(Q)$, $a_{ij} = a_{ji} \in C^\infty(\mathbb{R}^2)$, and the operator A^0 of the form (22.3) is strongly elliptic in $\overline{Q}$.

The existence and uniqueness of a generalized solution $u \in W^2(Q)$ of the boundary value problem (22.10), (22.11) for $|\gamma_1|, |\gamma_2| \le 1$ follow from Theorem 22.2. In fact, let $u_0 \in W^2(Q)$ be a generalized solution of the boundary value problem (22.10), (22.11) for $f_0 = 0$. Then $u_0 \in W^k_{\text{loc}}(Q)$ ($k = 1, 2, \ldots$). From this and from the imbedding theorem for $n = 2$ (see Theorem B.7) we obtain $u_0 \in C(\overline{Q}) \cap C^\infty(Q)$. Therefore, by virtue of the maximum principle, $u_0 = 0$. Thus, by Theorem 22.2, $0 \notin \sigma(\mathcal{A}_\gamma)$.

Proof of Theorem 22.1

In the spaces $W^k(\Omega)$ $(\Omega = Q, G)$ and $\mathcal{W}^0(Q, G) = L_2(Q) \times \overset{\circ}{W}{}^{3/2}(G) \times \overset{\circ}{W}{}^{3/2}(G)$ we introduce the equivalent norms depending on a parameter q

$$|||u|||_{W^k(\Omega)} = \{\|u\|^2_{W^k(\Omega)} + |q|^{2k} \|u\|^2_{L_2(\Omega)}\}^{1/2},$$

$$|||f|||_{\mathcal{W}^0(Q,G)} = \left\{ \|f_0\|^2_{L_2(Q)} + \sum_{\mu=1}^{2} |||f_\mu|||^2_{W^{3/2}(G)} \right\}^{1/2},$$

where $u \in W^k(\Omega)$, $f = (f_0, f_1, f_2) \in \mathcal{W}^0(Q, G)$.

Lemma 22.1. *There exists a linear operator $S: \overset{\circ}{W}{}^{3/2}(G) \times \overset{\circ}{W}{}^{3/2}(G) \to \overset{\circ}{W}{}^2(Q)$ such that $(Sw)|_{x_1 = t_\mu} = w_\mu$ and*

$$|||Sw|||_{W^2(Q)} \le c_3 \sum_{\mu=1}^{2} |||w_\mu|||_{W^{3/2}(G)} \tag{22.12}$$

for any $w = (w_1, w_2) \in W_0^{3/2}(G) \times W_0^{3/2}(G)$ *and* $|q| \geq 1 > 0$, *where* $c_3 > 0$ *does not depend on* w *and* q.

Proof. Since $\partial G \in C^\infty$, we can use a partition of unity and a rectification of the boundary ∂G. So to prove Lemma 22.1 it is sufficient to construct a linear operator $\widehat{S}: W_0^{3/2}(\mathbb{R}_+^{n-1}) \to W_0^2(\mathbb{R}_+^n)$ bounded in the norm $|||\cdot|||$ such that $(\widehat{S}u)|_{x_1=0, x_2>0} = u$ for all $u \in W_0^{3/2}(\mathbb{R}_+^{n-1})$. Here $\mathbb{R}_+^n = \{x = (x_1, \dots, x_n) : x_1 > 0\}$, $\mathbb{R}_+^{n-1} = \{x' = (x_2, \dots, x_n) : x_2 > 0\}$, $W_0^k(\cdot) = \{u \in W^k(\cdot) : u|_{x_2=0} = 0\}$.

We define the operator $S_1: L_2(\mathbb{R}_+^{n-1}) \to L_2(\mathbb{R}^{n-1})$ by the formula

$$(S_1 u)(x') = \begin{cases} u(x'), & x' \in \mathbb{R}_+^{n-1}, \\ -u(-x_2, x_3, \dots, x_{n-1}), & x' \in \mathbb{R}^{n-1} \setminus \mathbb{R}_+^{n-1}. \end{cases}$$

Evidently, the restriction of the operator $S_1: W_0^k(\mathbb{R}_+^{n-1}) \to W_0^k(\mathbb{R}^{n-1})$ $(k = 1, 2)$ is a bounded operator. Using this property and interpolation Theorem B.14, we can easily prove that the restriction of the operator $S_1: W_0^{3/2}(\mathbb{R}_+^{n-1}) \to W_0^{3/2}(\mathbb{R}^{n-1})$ is a bounded operator in the norm $|||\cdot|||$.

Let us consider the operator $S_2: W^{3/2}(\mathbb{R}^{n-1}) \to W^2(\mathbb{R}_+^n)$ defined by $S_2\varphi = F'^{-1}(vF'\varphi)$. Here $\varphi \in W^{3/2}(\mathbb{R}^{n-1})$; $F'\varphi = (F'\varphi)(\xi)$ is the Fourier transform of a function $\varphi(x')$ with respect to x'; $F'^{-1}\psi = (F'^{-1}\psi)(x')$ is the inverse Fourier transform of a function $\psi(\xi')$ with respect to ξ'; $v = \exp(-\omega x_1)$. $\omega = \{\sum_{j=2}^n \xi_i^2 + |q|^2\}^{1/2}$, $\xi' = (\xi_2, \dots, \xi_n) \in \mathbb{R}^{n-1}$. It is easy to see that

$$|||S_2\varphi|||_{W^2(\mathbb{R}_+^n)} \leq k_1 |||\varphi|||_{W^{3/2}(\mathbb{R}^{n-1})}, \qquad (S_2\varphi)|_{x_1=0} = \varphi.$$

Therefore, the operator $\widehat{S} = S_2 S_1: W_0^{3/2}(\mathbb{R}_+^{n-1}) \to W^2(\mathbb{R}_+^n)$ is bounded in the norm $|||\cdot|||$. The Fourier transform (or the inverse Fourier transform) with respect to x' (or ξ') maps a function odd in x_2 (or ξ_2) onto a function odd in ξ_2 (or x_2). Therefore, the function $S_2 S_1 u \in W^2(\mathbb{R}_+^n)$ is odd in x_2. Hence $(S_2 S_1 u)|_{x_2=0} = 0$. □

Lemma 22.2. *Let condition* 22.1 *be fulfilled.*

Then, for every $\varepsilon > 0$ *and* $\lambda \in \Omega_{\varepsilon,1}$, *there exists a unique solution of the equation* $\mathcal{L}_0 u = f$ *for each* $f \in \mathcal{W}^0(Q, G)$, *and*

$$c_4 |||f|||_{\mathcal{W}^0(Q,G)} \leq |||u|||_{W^2(Q)} \leq c_5 |||f|||_{\mathcal{W}^0(Q,G)}, \tag{22.13}$$

where $c_4 = c_4(\varepsilon) > 0$, $c_5 = c_5(\varepsilon) > 0$ *do not depend on* λ *and* u. *We put* $q = |\lambda|^{1/2}$ *in the norm* $|||\cdot|||$.

Proof. 1. First we prove this statement in the case $f_1 = f_2 = 0$. We introduce the unbounded operator $\mathcal{A}_0: L_2(Q) \to L_2(Q)$ with domain $\mathcal{D}(\mathcal{A}_0) = W^2(Q) \cap \mathring{W}^1(Q)$ by the formula $\mathcal{A}_0 u = A^0 u$.

By virtue of Remarks C.1, C.2 and Banach's inverse operator theorem, the operator $\mathcal{A}_0$ is self-adjoint, $\sigma(\mathcal{A}_0) \subset (0, \infty)$, and

$$\|\mathcal{A}_0^{-1} f_0\|_{W^2(Q)} \le k_1 \|f_0\|_{L_2(Q)} \tag{22.14}$$

for all $f_0 \in L_2(Q)$, where $k_1 > 0$ does not depend on f_0.

Let us consider the equation

$$\mathcal{A}_0 u - \lambda u = f_0. \tag{22.15}$$

From (22.14) it follows that

$$\|u\|_{W^2(Q)} \le k_1(|\lambda| \cdot \|u\|_{L_2(Q)} + \|f_0\|_{L_2(Q)}). \tag{22.16}$$

On the other hand, by virtue of Theorem C.4, the set of eigenfunctions $\{v_s\}$ of the operator $\mathcal{A}_0$ is an orthonormal basis in the space $L_2(Q)$, while the set of functions $\{v_s/\sqrt{\lambda_s}\}$ is an orthonormal basis in the space $\mathring{W}^1(Q)$ with inner product

$$(v, w)'_{\mathring{W}^1(Q)} = \sum_{i,j=1}^{n} \int_Q a_{ij} v_{x_j} \overline{w}_{x_i}\, dx. \tag{22.17}$$

Here λ_s is an eigenvalue of the operator $\mathcal{A}_0$ corresponding to an eigenfunction v_s. Hence, for all $f_0 \in L_2(Q)$ and $\lambda \in \Omega_{\varepsilon,1}$,

$$u = \sum_{s=1}^{\infty} \frac{(f_0, v_s)_{L_2(Q)}}{\lambda_s - \lambda} v_s, \tag{22.18}$$

where the series (22.18) converges in the space $\mathring{W}^1(Q)$. From (22.18) it follows that for all $f_0 \in L_2(Q)$ and $\lambda \in \Omega_{\varepsilon,1}$

$$|\lambda| \cdot \|u\|_{L_2(Q)} \le k_2 \|f_0\|_{L_2(Q)}, \tag{22.19}$$

where $k_2 = k_2(\varepsilon) > 0$ does not depend on λ and f_0.

Using the inequalities (22.16), (22.19), we obtain

$$|||u|||_{W^2(Q)} \le k_3 \|f_0\|_{L_2(Q)} \tag{22.20}$$

for all $f_0 \in L_2(Q)$ and $\lambda \in \Omega_{\varepsilon,1}$, where $k_3 = k_3(\varepsilon) > 0$ does not depend on λ and f_0.

2. From the above-mentioned properties of the operator $\mathcal{A}_0$ and Lemma 22.1 it follows that for all $f \in \mathcal{W}^0(Q, G)$ and $\lambda \in \Omega_{\varepsilon,1}$ there exists a unique solution of the equation $\mathcal{L}_0 u = f$. Moreover, this solution has the form

$$u = R(\lambda, \mathcal{A}_0)[f_0 - (A^0 - \lambda I)Sf'] + Sf',$$

where $f' = (f_1, f_2)$. Hence, by virtue of (22.12), (22.20), we have

$$\begin{aligned} |||u|||_{W^2(Q)} &\le k_4(\|f_0 - (A^0 - \lambda I)Sf'\|_{L_2(Q)} + |||Sf'|||_{W^2(Q)}) \\ &\le k_5(\|f_0\|_{L_2(Q)} + |||Sf'|||_{W^2(Q)}) \le k_6 |||f|||_{\mathcal{W}^0(Q,G)}, \end{aligned}$$

where $k_i = k_i(\varepsilon) > 0$ $(i = 4, 5, 6)$ do not depend on λ and f_0. The left part of (22.13) follows from Theorem B.6. □

Lemma 22.3. *Let conditions* 22.1, 22.2 *hold.*

Then, for every $\varepsilon > 0$, *there exists* $\lambda_0 = \lambda_0(\varepsilon) > 0$ *such that, if* $\lambda \in \Omega_{\varepsilon,\lambda_0}$ *and* $0 \le t \le 1$, *for* $u \in \overset{\circ}{W}{}^2(Q)$ *we have the estimate*

$$c_6 |||\mathcal{L}_t u|||_{\mathcal{W}^0(Q,G)} \le |||u|||_{W^2(Q)} \le c_7 |||\mathcal{L}_t u|||_{\mathcal{W}^0(Q,G)}, \tag{22.21}$$

where $q = |\lambda|^{1/2}$, $c_6 = c_6(\varepsilon) > 0$, $c_7 = c_7(\varepsilon) > 0$ *do not depend on* λ, t, *and* u, $\mathcal{L}_t = \mathcal{L}_0 + t(\mathcal{L} - \mathcal{L}_0)$.

Proof. Denote $\mathcal{L}_t u = f$. By virtue of Lemma 22.2, for every $\varepsilon > 0$ and $\lambda \in \Omega_{\varepsilon,1}$

$$|||u|||_{W^2(Q)} \le c_5 \Big\{ \|f_0 - tA^1 u\|_{L_2(Q)} + \sum_{\mu=1}^{2} |||f_\mu - t(B^1_\mu u)|_{x_1=t_\mu} - tB^2_\mu u|||_{W^{3/2}(G)} \Big\}.$$

By virtue of condition 22.2 and inequality (B.20), we obtain

$$\|tA^1 u\|_{L_2(Q)} \le k_1 q^{-r} |||u|||_{W^2(Q)},$$
$$|||tB^2_\mu u|||_{W^{3/2}(G)} \le k_2 q^{-p} |||u|||_{W^2(Q)}.$$

We introduce a function $\xi \in \dot{C}^\infty(\mathbb{R})$ such that $\xi(x_1) = 1$ for $x_1 \in (\delta/2, d - \delta/2)$, $\xi(x_1) = 0$ for $x_1 \notin (\delta/4, d - \delta/4)$. From condition 22.2, inequalities (B.20), (B.21), Lemma 22.2 and the Leibniz' formula we have

$$\begin{aligned} |||t(B^1_\mu u)|_{x_1=t_\mu}|||_{W^{3/2}(G)} &\le k_3 |||u|||_{W^2(\widetilde{Q}_\delta)} \le k_3 |||u\xi|||_{W^2(Q)} \\ &\le k_4 \|(A^0 - \lambda I)(\xi u)\|_{L_2(Q)} \\ &\le k_5 (\|(A - \lambda I)u\|_{L_2(Q)} + (q^{-1} + q^{-r})|||u|||_{W^2(Q)}). \end{aligned}$$

Choosing $\lambda_0 > 1$ so that $c_5((k_1 + 2k_5)|\lambda_0|^{-r/2} + 2k_2|\lambda_0|^{-p/2} + 2k_5|\lambda_0|^{-1/2}) < 1/2$, we obtain the right part of (22.21) for $\lambda \in \Omega_{\varepsilon,\lambda_0}$. The left part of (22.21) is obvious. □

Proof of Theorem 22.1. Combining Lemmas 22.2, 22.3, and 1.2, we obtain the statement (b). Using Theorems B.8, A.1, A.9, as in the proof of Theorem 21.1 we prove the statements (a) and (c). □

23 Elliptic Differential-Difference Equations in a Cylinder

Smoothness of Generalized Solutions

In this section we consider the differential-difference equation

$$AR_Q u = f_0(x) \quad (x \in Q) \tag{23.1}$$

with the boundary condition

$$u|_{\partial Q} = 0. \tag{23.2}$$

Here

$$Aw = -\sum_{i,j=1}^{n} \frac{\partial}{\partial x_i} a_{ij} \frac{\partial w}{\partial x_j} + \sum_{i=1}^{n} a_i \frac{\partial w}{\partial x_i} + a_0 w,$$

$a_{ij} = a_{ji}$, $a_i, a_0 \in C^\infty(\mathbb{R}^n)$ are real-valued functions; $Q = (0,d) \times G \subset \mathbb{R}^n$, $G \subset \mathbb{R}^{n-1}$ is a bounded domain (with boundary $\partial G \in C^\infty$ if $n \geq 3$),

$$\begin{aligned} R_Q &= P_Q R I_Q, \\ Ru(x) &= \sum_{i=-k}^{k} b_i u(x_1 + i, x_2, \dots, x_n), \end{aligned} \tag{23.3}$$

$d = k + \theta$, k is a natural number, $0 < \theta \leq 1$, $b_i \in \mathbb{C}$, $f_0 \in L_2(Q)$.

We assume that the operator A is elliptic in $\overline{Q}$, i.e., it satisfies the following condition:

23.1. $\sum_{i,j=1}^{n} a_{ij}(x)\xi_i\xi_j \neq 0$ $(x \in \overline{Q},\ 0 \neq \xi \in \mathbb{R}^n)$.

We define the matrix R_1 of order $(k+1)\times(k+1)$ with the elements $r^1_{ij} = b_{j-i}$. Denote by R_2 the matrix of order $k \times k$ obtained from the matrix R_1 by deleting the last column and the last row (see Example 8.2).

Definition 23.1. The equation (23.1) and the corresponding differential-difference operator AR_Q are said to be *elliptic* in $\overline{Q}$ if the operator A is elliptic in $\overline{Q}$, and the operator R_Q satisfies the condition

23.2. $\det R_s \neq 0$ $(s = 1, 2)$.

Theorem 23.1. *Let the equation* (23.1) *be strongly elliptic in* $\overline{Q}$.
Then it is elliptic in $\overline{Q}$.

Proof. We consider the matrices $R_{ijs}(x)$ $(x \in \overline{Q}_{s1})$ given by (9.4). Here subdomains Q_{sl} are defined in Example 8.2. By virtue of Theorem 9.1, for $x \in \overline{Q}_{s1}$ and $0 \neq \xi \in \mathbb{R}^n$ the matrices $B_s(x,\xi) + B_s^*(x,\xi)$ are positive definite, where

$$B_s(x,\xi) = \sum_{i,j=1}^{n} R_{ijs}(x)\xi_i\xi_j.$$

We denote by $e_l^s(g_l^s)$ the lth row of the matrix $B_s(x,\xi)$ (R_1). Clearly

$$e_l^s = \left(\sum_{i,j=1}^{n} a_{ij}(x_1 + l - 1, x_2, \dots, x_n)\xi_i\xi_j \right) g_l^s. \quad \text{Hence,}$$

$$\det B_s(x,\xi) = \prod_{l=1}^{N(s)} \left(\sum_{i,j=1}^{n} a_{ij}(x_1 + l - 1, x_2, \dots, x_n)\xi_i\xi_j \right) \det R_s. \tag{23.4}$$

Since $B_s(x,\xi) + B_s^*(x,\xi)$ is positive definite, then $\det B_s(x,\xi) \neq 0$ $(x \in \overline{Q}_{s1}$, $0 \neq \xi \in \mathbb{R}^n)$. Therefore, (23.4) implies that

$$\sum_{i,j=1}^{n} a_{ij}(x)\xi_i\xi_j \neq 0 \quad \left(x \in \bigcup_l \overline{Q}_{sl},\ 0 \neq \xi \in \mathbb{R}^n \right), \quad \det R_s \neq 0.$$

Hence, we have $\sum_{i,j=1}^{n} a_{ij}(x)\xi_i\xi_j \neq 0$ $(x \in \overline{Q},\ 0 \neq \xi \in \mathbb{R}^n)$, $\det R_s \neq 0$ $(s = 1, 2$ if $\theta < 1$ and $s = 1$ if $\theta = 1)$. In order to prove that $\det R_2 \neq 0$ for $\theta = 1$, we denote by $B_2(x,\xi)$ $(x \in \overline{Q}_{11},\ 0 \neq \xi \in \mathbb{R}^n)$ the matrix of order $k \times k$ obtained from $B_1(x,\xi)$ by deleting the last column and the last row. Since the matrix $B_2(x,\xi) + B_2^*(x,\xi)$ is positive definite, then, by the preceding, $\det R_2 \neq 0$. □

We introduce the unbounded operator $\mathcal{A}_R\colon L_2(Q) \to L_2(Q)$ acting in the space of distributions $\mathcal{D}'(Q)$ by the formulas

$$\mathcal{A}_R u = AR_Q u,$$
$$\mathcal{D}(\mathcal{A}_R) = \{u \in \mathring{W}^1(Q) : \mathcal{A}_R u \in L_2(Q)\}.$$

Definition 23.2. A function u is called a *generalized solution* of the boundary value problem (23.1), (23.2) if $u \in \mathcal{D}(\mathcal{A}_R)$ and

$$\mathcal{A}_R u = f_0. \tag{23.5}$$

Theorem 23.2. *Let the equation* (23.1) *be elliptic in* $\overline{Q}$, *and let* u *be a generalized solution of the boundary value problem* (23.1), (23.2).

Then $R_Q u \in W^2(Q)$ *and* $u \in W^2(Q_{sl})$, *where* $Q_{1l} = (l-1, l-1+\theta) \times G$ $(l = 1, \dots, k+1)$, $Q_{2l} = (l-1+\theta, l) \times G$ $(l = 1, \dots, k)$ *if* $\theta < 1$; $Q_{1l} = (l-1, l) \times G$ $(l = 1, \dots, k+1)$ *if* $\theta = 1$.

Proof. By virtue of the condition 23.2 and Theorem 8.1, there are numbers $\gamma_i^1, \gamma_i^2 \in \mathbb{C}$ $(i = 1, \dots, k)$ such that the operator R_Q maps $\mathring{W}^1(Q)$ onto $W_\gamma^1(Q)$ continuously and in a one-to-one manner, where $W_\gamma^1(Q)$ is a subspace of functions in $W^1(Q)$ satisfying the conditions (13.31). Hence from Lemma 13.5 it follows that $R_Q u \in W^2(Q)$. Thus, by Lemma 8.15, $u \in W^2(Q_{sl})$. □

Problem 23.1. *Let* $u \in \mathring{W}^1(Q)$ *be a generalized solution of the boundary value problem*

$$-\sum_{i,j=1}^{n} \frac{\partial}{\partial x_i} R_{ijQ} \frac{\partial u}{\partial x_j} = f_0(x) \qquad (x \in Q), \tag{23.6}$$

$$u|_{\partial Q} = 0. \tag{23.7}$$

Here $Q = (0, d) \times G \subset \mathbb{R}^n$, $G \subset \mathbb{R}^{n-1}$ *is a bounded domain (with boundary* $\partial G \in C^\infty$ *if* $n \geq 3$),

$$R_{ijQ} = P_Q R_{ij} I_Q,$$

$$R_{ij}u(x) = \sum_{l=-k}^{k} b_{ijl}(x) u(x_1 + l, x_2, \dots, x_n), \tag{23.8}$$

$d = k + \theta$, k *is a natural number,* $0 < \theta \leq 1$; $b_{ijl} \in C^\infty(\mathbb{R}^n)$ *are complex-valued functions; the equation* (23.6) *is strongly elliptic in* $\overline{Q}$.
Is it true that $u \in W^2(Q_{sl})$?

Solvability and Spectrum

Suppose that the conditions 23.1, 23.2 are fulfilled. We introduce the unbounded operator $\mathcal{A}_\gamma: L_2(Q) \to L_2(Q)$ with domain $\mathcal{D}(\mathcal{A}_\gamma) = W^1_\gamma(Q) \cap W^2(Q)$ according to the formula $\mathcal{A}_\gamma u = Au$. Here the subspace $W^1_\gamma(Q)$ was defined in the proof of Theorem 23.2.

From Theorems 8.1, 22.2, 23.2 and Theorems A.1, A.8 we have

Theorem 23.3. *Let the equation* (23.1) *be elliptic in* $\overline{Q}$.

Then the operator $\mathcal{A}_R: L_2(Q) \to L_2(Q)$ *is Fredholm, and* $\operatorname{ind}\mathcal{A}_R = 0$. *If* $0 \notin \sigma(\mathcal{A}_\gamma)$, *then the spectrum* $\sigma(\mathcal{A}_R)$ *is discrete, and for* $\lambda \notin \sigma(\mathcal{A}_R)$ *the resolvent* $R(\lambda, \mathcal{A}_R): L_2(Q) \to L_2(Q)$ *is a compact operator.*

Theorem 23.4. *Let the equation* (23.1) *be elliptic in* $\overline{Q}$, *and let* $a_i(x) \equiv 0$ $(i = 1, \dots, n)$, $b_j = \overline{b_{-j}}$ $(j = 0, \dots, k)$. *Suppose that the functions* $a_{ij}(x)$, $a_0(x)$ *are* 1*-periodic in* x_1.

Then the operator $\mathcal{A}_R: L_2(Q) \to L_2(Q)$ *is self-adjoint. The spectrum* $\sigma(\mathcal{A}_R)$ *consists of real isolated eigenvalues of finite multiplicity.*

Proof. 1. By virtue of condition 23.2 and Theorem 8.1, there exist numbers $\gamma_i^1, \gamma_i^2 \in \mathbb{C}$ $(i = 1, \dots, k)$ such that the operator R_Q maps $\mathring{W}^1(Q)$ onto $W^1_\gamma(Q)$ continuously and in a one-to-one manner.

From Theorem 22.2 it follows that the spectrum $\sigma(\mathcal{A}_\gamma)$ consists of isolated eigenvalues of finite multiplicity. By the Banach inverse operator theorem, for $\lambda_0 \in \mathbb{R}\backslash\sigma(\mathcal{A}_\gamma)$ the resolvent $R(\lambda_0, \mathcal{A}_\gamma)$ maps $L_2(Q)$ onto $W^1_\gamma(Q) \cap W^2(Q)$ continuously. Hence the operator $\mathcal{A}_R - \lambda_0 R_Q$ has an inverse operator $(\mathcal{A}_R - \lambda_0 R_Q)^{-1} = R_Q^{-1} R(\lambda_0, \mathcal{A}_\gamma)$, continuous from $L_2(Q)$ into $\mathring{W}^1(Q)$. Thus, by Theorem A.8, the spectrum $\sigma(\mathcal{A}_R - \lambda_0 R_Q)$ is discrete.

2. We now prove that the operator $\mathcal{A}_R - \lambda_0 R_Q$ is self-adjoint. By virtue of Lemmas 8.10, 8.11, 8.14, for all $u \in \mathcal{D}(\mathcal{A}_R)$, $v \in \dot{C}^\infty(Q)$

$$((\mathcal{A}_R - \lambda_0 R_Q)u, v)_{L_2(Q)} = \sum_{i,j=1}^{n} (a_{ij} R_Q u_{x_j}, v_{x_i})_{L_2(Q)} + ((a_0 - \lambda_0) R_Q u, v)_{L_2(Q)}$$

and for all $u \in \dot{C}^\infty(Q)$, $v \in \mathcal{D}(\mathcal{A}_R)$

$$(u, (\mathcal{A}_R - \lambda_0 R_Q)v)_{L_2(Q)} = \sum_{i,j=1}^{n} (a_{ij} R_Q u_{x_j}, v_{x_i})_{L_2(Q)} + ((a_0 - \lambda_0) R_Q u, v)_{L_2(Q)}.$$

Since $\dot{C}^\infty(Q)$ is dense in $\mathring{W}^1(Q)$, then these identities are valid for all $u \in \mathcal{D}(\mathcal{A}_R)$ and $v \in \mathcal{D}(\mathcal{A}_R)$. Hence $\mathcal{A}_R - \lambda_0 R_Q \subset (\mathcal{A}_R - \lambda_0 R_Q)^*$. Thus, since the spectrum $\sigma(\mathcal{A}_R - \lambda_0 R_Q)$ is discrete, by Theorem A.6, $\mathcal{A}_R - \lambda_0 R_Q = (\mathcal{A}_R - \lambda_0 R_Q)^*$.

3. By virtue of Lemma 8.11, $\mathcal{A}_R = \mathcal{A}_R^*$. Let $\lambda \in \mathbb{C}\setminus\mathbb{R}$. Then there exists the bounded operator $(\mathcal{A}_R - \lambda I)^{-1}: L_2(Q) \to L_2(Q)$. From this it follows that

$$(\mathcal{A}_R - \lambda I)^{-1} = (\mathcal{A}_R - \lambda_0 R_Q)^{-1}(I + (\lambda_0 R_Q - \lambda I)(\mathcal{A}_R - \lambda_0 R_Q)^{-1})^{-1}.$$

Hence $(\mathcal{A}_R - \lambda I)^{-1}: L_2(Q) \to L_2(Q)$ is a compact operator. Thus, by Theorem A.8, the spectrum $\sigma(\mathcal{A}_R)$ is discrete, and for $\lambda \notin \sigma(\mathcal{A}_R)$ the resolvent $R(\lambda, \mathcal{A}_R): L_2(Q) \to L_2(Q)$ is a compact operator. □

Example 23.1. Let $\mathcal{A}_R = -\Delta R_Q$, where $Q = (0,2) \times (0,1)$, $Ru(x) = u(x) + \gamma_1 u(x_1+1, x_2) + \gamma_2 u(x_1 - 1, x_2)$, $\gamma_1, \gamma_2 \in \mathbb{R}$. Clearly, $Q_{1l} = (l-1, l) \times (0,1)$ $(l = 1, 2)$,

$$R_1 = \begin{pmatrix} 1 & \gamma_1 \\ \gamma_2 & 1 \end{pmatrix}.$$

If $\gamma_1\gamma_2 \neq 1$, then the condition 23.2 is fulfilled. Hence, by virtue of Theorem 23.3, the operator $\mathcal{A}_R: L_2(Q) \to L_2(Q)$ is Fredholm, and $\operatorname{ind} \mathcal{A}_R = 0$. Furthermore, if $\gamma_1 = \gamma_2$, $\gamma_1 \neq \pm 1$, then, by Theorem 23.4, the spectrum $\sigma(\mathcal{A}_R)$ consists of real isolated eigenvalues of finite multiplicity.

As Example 23.2 shows, the spectrum $\sigma(\mathcal{A}_R)$ is not semi-bounded in general. This is related to the question of whether the operator R_Q has a constant sign (see Example 9.3 and Theorem 10.2).

Example 23.2. Let $\mathcal{A}_R = -\Delta R_Q$, where $Q = (0,2) \times (0,1)$, $Ru(x) = u(x) + 2u(x_1+1, x_2) + 2u(x_1 - 1, x_2)$. Evidently, $Q_{1l} = (l-1, l) \times (0,1)$ $(l = 1, 2)$, $R_1 = \begin{pmatrix} 1 & 2 \\ 2 & 1 \end{pmatrix}$. Denote

$$u_n^{\pm}(x) = \begin{cases} \eta_n(x_1, x_2) & (x \in Q_{11}), \\ \pm\eta_n(x_1 - 1, x_2) & (x \in Q_{12}), \end{cases}$$

where $\eta_n \in \dot{C}^\infty(Q_{11})$, $\|\eta_n\|_{L_2(Q_{11})} = 1$, $\|\nabla\eta_n\|_{L_2^2(Q_{11})} \to \infty$. Then

$$(\mathcal{A}_R u_n^{\pm}, u_n^{\pm})_{L_2(Q)} = 2(\nabla(\eta_n \pm 2\eta_n), \nabla\eta_n)_{L_2^2(Q_{11})} \to \pm\infty \quad \text{as } n \to \infty.$$

From Theorem 23.4 it follows that $\mathcal{A}_R = \mathcal{A}_R^*$. Therefore, by virtue of the spectral resolution of a self-adjoint operator, the spectrum $\sigma(\mathcal{A}_R)$ is not semi-bounded.

Condition 23.2

If the condition 23.2 does not hold, then the boundary value problem (23.1), (23.2) may be equivalent to an elliptic differential equation with nonlocal conditions relating the values of the solution on the different pieces of the boundary. Thus, if the image of the operator corresponding to this problem is not closed in $L_2(Q)$, then the image of the original problem is not closed in $L_2(Q)$.

Example 23.3. Let us consider the operator $\mathcal{A}_R = -\Delta R_Q : L_2(Q) \to L_2(Q)$ with domain $\mathcal{D}(\mathcal{A}_R) = \{u \in \mathring{W}^1(Q) : \mathcal{A}_R u \in L_2(Q)\}$ acting in the space of distributions $\mathcal{D}'(Q)$. Here $Q = (0,2) \times (0,1)$, $Ru(x) = u(x_1+1, x_2) + u(x_1-1, x_2)$. It is obvious that

$$R_1 = \begin{pmatrix} 0 & 1 \\ 1 & 0 \end{pmatrix},$$

i.e., the condition 23.2 is not fulfilled, although $\det R_1 \neq 0$. The operator R_Q maps $\mathring{W}^1(Q)$ onto $W^1_{\gamma'}(Q)$ continuously and in a one-to-one manner. Here $W^1_{\gamma'}(Q)$ is the subspace of functions in $W^1(Q)$ satisfying the conditions

$$u|_{x_2=0} = u|_{x_2=1} = 0, \quad u|_{x_1=1} = 0, \quad u|_{x_1=0} = u|_{x_1=2}. \tag{23.9}$$

Let us consider the unbounded operator $\mathcal{A}_{\gamma'} = -\Delta : L_2(Q) \to L_2(Q)$ with domain $\mathcal{D}(\mathcal{A}_{\gamma'}) = \{u \in W^1_{\gamma'}(Q) : \mathcal{A}_{\gamma'} u \in L_2(Q)\}$ acting in the space of distributions $\mathcal{D}'(Q)$. By construction, $\mathcal{A}_R = \mathcal{A}_{\gamma'} R_Q$. Hence, if $0 \in \sigma_c(\mathcal{A}_{\gamma'})$, then $0 \in \sigma_c(\mathcal{A}_R)$, where $\sigma_c(\cdot)$ is the continuous spectrum of the operator.

Example 23.4. We consider the spectrum $\sigma(\mathcal{A}_{\gamma'})$. Clearly the numbers $\pi^2(m^2+l^2)$ are the eigenvalues of $\mathcal{A}_{\gamma'}$ corresponding to eigenfunctions $\sin \pi m x_1 \cdot \sin \pi l x_2$ $(m, l = 1, 2, \ldots)$.

Suppose $\lambda \neq \pi^2(m^2+l^2)$. It can be shown by the method of separation of variables that the equation $(\mathcal{A}_{\gamma'} - \lambda I)u = f$ has a unique solution for all $f \in \mathcal{P}$, where $\mathcal{P}$ is the set of linear combinations of the functions $e^{i\pi k x_1} \sin \pi s x_2$ $(k = 0, \pm 1, \pm 2, \ldots,\ s = 1, 2, \ldots)$. For sufficiently large s, the solution of the equation $(\mathcal{A}_{\gamma'} - \lambda I)u_s = f_s$ $(f_s = \sin \pi s x_2)$ has the form

$$u_s = 2^{-1}\omega_s^{-2}(2 - e^{\omega_s(x_1-1)} - e^{-\omega_s(x_1-1)}) \sin \pi s x_2,$$

where $\omega_s = \sqrt{\pi^2 s^2 - \lambda}$, $r_s = \operatorname{Re} \omega_s > 0$. It is easy to see that $\|u_s\|^2_{L_2(Q)} = 2^{-2}|\omega_s|^{-4} r_s^{-1}(e^{2r_s} + o(e^{2r_s})) \to +\infty$ as $s \to +\infty$, $\|f_s\|^2_{L_2(Q)} = 1$. Hence $\lambda \in \sigma_c(\mathcal{A}_{\gamma'})$.

Thus $\sigma(\mathcal{A}_{\gamma'}) = \sigma_p(\mathcal{A}_{\gamma'}) \cup \sigma_c(\mathcal{A}_{\gamma'}) = \mathbb{C}$, where $\sigma_p(\mathcal{A}_{\gamma'})$ is the point spectrum of $\mathcal{A}_{\gamma'}$. From this it follows that $0 \in \sigma_c(\mathcal{A}_{\gamma'})$, i.e., $0 \in \sigma_c(\mathcal{A}_R)$.

Problem 23.2. *Let the equation* (23.1) *be elliptic, i.e.,* $\det R_1 \neq 0$, $\det R_2 \neq 0$. *Is the spectrum* $\sigma(\mathcal{A}_R)$ *discrete?*

24 Applications to the Multidimensional Diffusion Processes

The Hille–Yosida Theorem

First we restate a version of the Hille–Yosida theorem (see K. Taira [1]) for our present use.

Let X be a closed linear subspace in $C(\overline{Q})$, where $Q \subset \mathbb{R}^n$ is a bounded domain with boundary $\partial Q \in C^\infty$, $n \geq 2$.

Definition 24.1. A family $\{T_t\}_{t\geq 0}$ of bounded linear operators acting on X is called a *Feller semigroup* on X if it satisfies the following conditions:

24.1. $T_{t+s} = T_t \cdot T_s$ $(t, s \geq 0)$, $T_0 = I$.

24.2. $\{T_t\}$ *is strongly continuous in* 0:

$$\lim_{t\to 0} \|T_t f - f\|_{C(\overline{Q})} = 0 \qquad (f \in X).$$

24.3. $\{T_t\}$ *is contractive and non-negative on* X:

$$\|T_t\| \leq 1 \ (t \geq 0), \text{ and } T_t f \geq 0 \text{ for any } f \in X \text{ such that } f \geq 0.$$

Definition 24.2. A linear operator $\mathcal{A}: X \to X$ is called the *infinitesimal generator* of $\{T_t\}$ if

$$\mathcal{A}u = \lim_{t\to 0} \frac{T_t u - u}{t} \qquad (u \in \mathcal{D}(\mathcal{A})), \tag{24.1}$$

$$\mathcal{D}(\mathcal{A}) = \{u \in X : \text{the limit (24.1) exists in } X\}. \tag{24.2}$$

Theorem 24.1 (Hille–Yosida). 1. *Let* $\{T_t\}_{t\geq 0}$ *be a Feller semigroup on* X, *and let* $\mathcal{A}: X \to X$ *be its infinitesimal generator. Then we have:*

(a) *The domain* $\mathcal{D}(\mathcal{A})$ *is everywhere dense in* X.

(b) *For each* $\lambda > 0$, *the operator* $\lambda I - \mathcal{A}$ *has a bounded inverse* $(\lambda I - \mathcal{A})^{-1}$: $X \to X$ *with norm* $\|(\lambda I - \mathcal{A})^{-1}\| \leq 1/\lambda$.

(c) *For each* $\lambda > 0$, *the operator* $(\lambda I - \mathcal{A})^{-1}: X \to X$ *is non-negative.*

2. *Conversely, if* $\mathcal{A}$ *is a linear operator from* X *into itself satisfying condition* (a) *and if there is a constant* $\lambda_0 \geq 0$ *such that for all* $\lambda > \lambda_0$ *conditions* (b), (c) *are satisfied, then* $\mathcal{A}$ *is the infinitesimal generator of some Feller semigroup* $\{T_t\}_{t\geq 0}$ *on* X, *which is uniquely determined by* $\mathcal{A}$.

Feller Semigroups and Boundary Value Problems

In [1] and [2], W. Feller considered the general form of boundary conditions for one-dimensional diffusion processes. This problem was reduced to the study of nonlocal boundary value problems. He proved that if a differential operator $\mathcal{A}$ is an infinitesimal generator of some Feller semigroup, then $\mathcal{D}(\mathcal{A})$ consists of functions satisfying nonlocal boundary conditions. Conversely, if $\mathcal{D}(\mathcal{A})$ consists of functions with such nonlocal conditions, then $\mathcal{A}$ is an infinitesimal generator of a Feller semigroup. These conditions, in contrast to the classical conditions, related the values of a function and its derivatives at the end points of an interval to values within the interval.

An analogous problem for multidimensional diffusion processes in the domain $Q \subset \mathbb{R}^n$ was studied by A. D. Ventsel′ [1]. He obtained a general form for boundary conditions for an infinitesimal generator of a Feller semigroup. In that paper, a particular case, when Q is a ball and these nonlocal conditions are invariant with respect to rotation, is also considered. It was proved that a corresponding elliptic differential operator $\mathcal{A}$ is an infinitesimal generator of a Feller semigroup.

In the general case, the problem of constructing of a Feller semigroup whose infinitesimal generator is characterized by nonlocal boundary conditions is unsolved.

This problem arises in biophysics (see W. Feller [2]). Nonlocal terms in the boundary conditions correspond to diffusion in a cell, in which a particle arriving at the membrane can later jump at a point $x \in \overline{Q}$. The "local" terms in the boundary conditions correspond to the absorption, reflection, viscosity phenomena, and diffusion along the boundary.

We consider a differential operator A of the form

$$Au = \sum_{i,j=1}^{n} a_{ij}(x)u_{x_i x_j}(x) + \sum_{i=1}^{n} a_i(x)u_{x_i}(x) + a(x)u(x) \qquad (x \in Q) \tag{24.3}$$

with domain $\mathcal{D}(A) = C^2(\overline{Q})$. Here $a_{ij}, a_i, a \in C^\infty(\mathbb{R}^n)$. We assume also that the following condition holds:

24.4. $\sum_{i,j} a_{ij}(x)\xi_i\xi_j > 0$, $a(x) \le 0$ *for all* $x \in \overline{Q}$ *and* $0 \ne \xi \in \mathbb{R}^n$.

Suppose in some neighborhood U of each point $x^0 \in \partial Q$ there is defined an infinitely differentiable, nondegenerate coordinate transformation $x \to t = t(x)$ such that:

(a) $U \cap Q = \{x \in U : t_n(x) > 0\}$;
(b) $U \cap \partial Q = \{x \in U : t_n(x) = 0\}$;
(c) $t_i(x^0) = 0$ $(i = 1, \dots, n)$;
(d) the functions $t_1, \dots, t_n$ can be extended to C^∞ functions on $\mathbb{R}^n$.

We assume that $t_n(x) = \rho(x, U \cap \partial Q)$. Thus

$$\frac{\partial u(x^0)}{\partial t_n} = \frac{\partial u(x^0)}{\partial \nu} \qquad (x^0 \in \partial Q),$$

where ν is the inner unit normal to ∂Q at the point $x^0 \in \partial Q$.

Lemma 24.1. *Let the condition* 24.4 *be fulfilled.*

Then the operator A *has a closure* $\overline{A}$. *If* $u \in \mathcal{D}(\overline{A})$ *and* u *has a positive maximum at the point* $x^0 \in Q$, *then* $\overline{A}u(x^0) \le 0$.

Proof. 1. We first show that the operator A is closable. It suffices to prove that if $\|u_m\|_{C(\overline{Q})} \to 0$ $(u_m \in C^2(\overline{Q}))$, $\|Au_m - f\|_{C(\overline{Q})} \to 0$, then $f = 0$.

Assume to the contrary that $f(y) > 0$ $(y \in Q)$. Then for some $\varepsilon > 0$ and $r > 0$ for sufficiently large m we have $Au_m(x) > \varepsilon$ for $|x - y| \le r$. We consider the function

$$v_m(x) = u_m(x) - \frac{\varepsilon|x-y|^2}{\max\limits_{|x-y|\le r} A|x-y|^2}.$$

For $|x-y| \le r$ and sufficiently large m we obtain $Av_m(x) > 0$. Therefore $v_m(x)$ does not have a maximum for $|x-y| < r$. Thus $v_m(y) < \max_{|x-y|=r} v_m(x)$. Hence we obtain

$$u_m(y) < \max_{|x-y|=r} u_m(x) - \frac{\varepsilon r^2}{\max\limits_{|x-y|\le r} A|x-y|^2}. \tag{24.4}$$

But such a sequence cannot converge uniformly to zero.

2. For the proof of the second statement it suffices to prove that if $\|u_m - u\|_{C(\overline{Q})} \to 0$ $(u_m \in C^2(\overline{Q}))$, $\|Au_m - f\|_{C(\overline{Q})} \to 0$, $f(x^0) > 0$ $(x^0 \in Q)$, then a function u does not have a positive maximum at the point x^0. In fact, assuming that $f(x^0) > 0$, we obtain (24.4) for $y = x^0$. Hence, the sequence u_m cannot converge to a function having a positive maximum at x^0. □

Remark 24.1. Clearly every restriction of $\overline{A}$ is also closable.

Theorem 24.2. *Assume the condition 24.4 holds. Let $\{T_t\}_{t\ge 0}$ be a Feller semigroup on X, and let $\mathcal{A}$ be its infinitesimal generator, which is a restriction of $\overline{A}$.*

Then every function $u \in \mathcal{D}(\mathcal{A}) \cap C^2(\overline{Q})$ satisfies at each point $x^0 \in \partial Q$ a nonlocal condition of the form

$$\begin{aligned} Bu \; = \; & \gamma(x^0)u(x^0) + \int_{\overline{Q}} \left[u(x^0) - u(y) + \sum_{i=1}^{n-1} \frac{\partial u(x^0)}{\partial t_i} t_i(y) \right] m(x^0, dy) \\ & - \mu(x^0) \frac{\partial u(x^0)}{\partial \nu} + \sum_{i=1}^{n-1} \beta_i(x^0) \frac{\partial u(x^0)}{\partial t_i} + \sigma(x^0) Au(x^0) \\ & - \sum_{i,j=1}^{n-1} \alpha_{ij}(x^0) \frac{\partial^2 u(x^0)}{\partial t_i \partial t_j} = 0 \qquad (x^0 \in \partial Q), \end{aligned} \tag{24.5}$$

where:

1. $\gamma(x^0) \ge 0$;
2. $\mu(x^0) \ge 0$;
3. $\sigma(x^0) \ge 0$;
4. *the matrix $\|\alpha_{ij}\|$ is symmetric and non-negative;*
5. *$m(x^0, \cdot)$ is a non-negative Borel measure on $\overline{Q}$ such that*

$$m(x^0, \overline{Q} \setminus U) < \infty,$$

$$\int_{\overline{Q}\cap U} \left[t_n(y) + \sum_{j=1}^{n-1} t_j^2(y) \right] m(x^0, dy) < \infty;$$

6. *if $\gamma(x^0) = \mu(x^0) = \sigma(x^0) = \beta_i(x^0) = \alpha_{ij}(x^0) = 0$, then $m(x^0, \overline{Q} \setminus x^0) > 0$.*

Theorem 24.2 was proved by A. D. Ventsel′ [1].

Lemma 24.2. *Let a linear operator $G: X \to Y$ with domain $\mathcal{D}(G) \subset X$ have a closure $\overline{G}$, where $Y = X$ or $Y = C(\overline{Q})$. Assume that for some λ the range $\mathcal{R}(\lambda I - G)$ is dense in Y, and that there exists a bounded operator $(\lambda I - G)^{-1}$ from $\mathcal{R}(\lambda I - G)$ into X.*

Then $\mathcal{R}(\lambda I - \overline{G}) = Y$, the operator $(\lambda I - \overline{G})$ has a bounded inverse $(\lambda I - \overline{G})^{-1}: Y \to X$ and $\|(\lambda I - \overline{G})^{-1}\| = \|(\lambda I - G)^{-1}\|$.

Proof. Denote by $\overline{(\lambda I - G)^{-1}}$ a continuous extension of the operator $(\lambda I - G)^{-1}$. It suffices to remark that $\overline{(\lambda I - G)^{-1}}$ is the inverse operator of $(\lambda I - \overline{G})$. □

Theorem 24.3. *Let a linear operator $\mathcal{G}: X \to C(\overline{Q})$ with domain $\mathcal{D}(\mathcal{G}) \subset X$ is closable. Let $G: X \to X$ be a linear operator with domain $\mathcal{D}(G) \subset X$ such that $G \subset \mathcal{G}$. Suppose that G, $\mathcal{G}$ satisfy the following conditions:*

(a) *$\mathcal{D}(G)$ is dense in X.*

(b) *If $u \in \mathcal{D}(\mathcal{G})$ takes a positive maximum at $x^0 \in \overline{Q}$, then there is a point $x^1 \in \overline{Q}$ such that $u(x^1) = u(x^0)$ and $\mathcal{G}u(x^1) \le 0$.*

(c) *The range $\mathcal{R}(\lambda I - \mathcal{G})$ is dense in $C(\overline{Q})$ and the range $\mathcal{R}(\lambda I - G)$ is dense in X for $\lambda > 0$.*

Then $\overline{G}$ is the infinitesimal generator of a Feller semigroup on X, which is uniquely determined by G.

Proof. 1. First we prove that for each $\lambda > 0$ the operator $\lambda I - \mathcal{G}$ has a bounded inverse $(\lambda I - \mathcal{G})^{-1}$ and

$$\|(\lambda I - \mathcal{G})^{-1}\| \le 1/\lambda. \tag{24.6}$$

Let us consider a solution u of the equation

$$(\lambda I - \mathcal{G})u = f. \tag{24.7}$$

Without loss of generality, we assume that $\max_{x \in \overline{Q}} |u(x)| = u(x^0) > 0$. Then, by virtue of the condition (b), there exists a point $x^1 \in \overline{Q}$ such that $u(x^1) = u(x^0)$ and $\mathcal{G}u(x^1) \le 0$. Therefore (24.7) implies that

$$\|u\|_{C(\overline{Q})} = u(x^0) = u(x^1) = (\mathcal{G}u(x^1) + f(x^1))/\lambda \le \|f\|_{C(\overline{Q})}/\lambda.$$

The operator $\lambda I - \mathcal{G}$ thus has a bounded inverse and $\|(\lambda I - \mathcal{G})^{-1}\| \le 1/\lambda$.

2. Now we prove that for each $\lambda > 0$ the operator $(\lambda I - \mathcal{G})^{-1}$ is non-negative. Assume to the contrary that, for some $f \ge 0$ a solution u of the equation (24.7) takes negative values, i.e., $\min_{x \in \overline{Q}} u(x) = u(x^0) < 0$. Denote $v(x) = -u(x)$. By virtue of the condition (b), there exists a point $x^1 \in \overline{Q}$ such that $v(x^1) = v(x^0)$ and $\mathcal{G}v(x^1) \le 0$. From the equality $(\lambda I - \mathcal{G})v = -f$ we obtain

$$0 < v(x^0) = v(x^1) = (\mathcal{G}v(x^1) - f(x^1))/\lambda \le 0.$$

3. The operators $\lambda I - \overline{G}$ and $\lambda I - \overline{\mathcal{G}}$ have bounded inverse operators $(\lambda I - \overline{G})^{-1}: X \to X$ and $(\lambda I - \overline{\mathcal{G}})^{-1}: C(\overline{Q}) \to X$ for $\lambda > 0$, while

$$\|(\lambda I - \overline{G})^{-1}\| \le \|(\lambda I - \overline{\mathcal{G}})^{-1}\| \le 1/\lambda.$$

This follows from the existence of inverse operators $(\lambda I - \mathcal{G})^{-1}$, $(\lambda I - G)^{-1}$, condition (c), inequality (24.6) and Lemma 24.2.

4. Let us now prove that the operator $(\lambda I - \overline{\mathcal{G}})^{-1}$ is non-negative. First let $f \in C(\overline{Q})$ and $f(x) > 0$ $(x \in \overline{Q})$. We can find a sequence $\{f_m\} \subset \mathcal{D}((\lambda I - \mathcal{G})^{-1})$ such that $\lim_{m\to\infty} \|f_m - f\|_{C(\overline{Q})} = 0$. Hence, there exists $M > 0$ such that $f_m(x) > 0$ $(x \in \overline{Q})$ for $m \geq M$. Thus,

$$(\lambda I - \overline{\mathcal{G}})^{-1} f = \lim_{m\to\infty} (\lambda I - \mathcal{G})^{-1} f_m \geq 0.$$

If $f \in C(\overline{Q})$, $f(x) \geq 0$ $(x \in \overline{Q})$, there is a sequence $\{F_k\} \subset C(\overline{Q})$ such that $F_k(x) > 0$ $(x \in \overline{Q})$ and $\lim_{k\to\infty} \|F_k - f\|_{C(\overline{Q})} = 0$. From this it follows that $(\lambda I - \overline{\mathcal{G}})^{-1} f \geq 0$ for every function $f \in C(\overline{Q})$ such that $f(x) \geq 0$ $(x \in \overline{Q})$.

5. We have thus proved that for each $\lambda > 0$ the operator $\lambda I - \overline{G}$ has a bounded non-negative inverse $(\lambda I - \overline{G})^{-1}: X \to X$ with norm $\|(\lambda I - \overline{G})^{-1}\| \leq 1/\lambda$. Hence, by virtue of Theorem 24.1, $\overline{G}$ is the generator of a Feller semigroup on X, which is uniquely determined by G. □

Theorem 24.4. *Suppose the condition $\gamma(x^0) = \mu(x^0) = \sigma(x^0) = 0$ implies that $m(x^0, Q) > 0$. Let $u \in \{u \in C^2(\overline{Q}) : Bu = 0\}$, and let $u(x)$ take a positive maximum at $x^0 \in \overline{Q}$.*

Then there is a point $x^1 \in \overline{Q}$ such that $u(x^1) = u(x^0)$ and $Au(x^1) \leq 0$.

Proof. If $x^0 \in Q$, then the statement of Theorem 24.4 follows from Lemma 24.1. Let $\max_{x\in\overline{Q}} u(x) = u(x^0) > 0$ and $u(x^0) > u(x)$ $(x \in Q)$, where $x^0 \in \partial Q$. Then

$$\frac{\partial u(x^0)}{\partial \nu} \leq 0, \qquad \frac{\partial u(x^0)}{\partial t_i} = 0 \; (i = 1, \ldots, n-1), \qquad \sum_{i,j=1}^{n-1} \alpha_{ij}(x^0) \frac{\partial^2 u(x^0)}{\partial t_i \partial t_j} \leq 0.$$

If $\sigma(x^0) > 0$, then the condition (24.5) implies that $Au(x^0) \leq 0$.

Let $\sigma(x^0) = 0$. Then the left part of (24.5) contains only non-negative terms. Hence each term equals zero. Thus

$$\gamma(x^0)u(x^0) = \mu(x^0)\frac{\partial u(x^0)}{\partial \nu} = \int_Q (u(x^0) - u(x)) m(x^0, dx) = 0.$$

From this it follows that $\gamma(x^0) = m(x^0, Q) = 0$. Then from the condition of Theorem 24.4 we obtain $\mu(x^0) > 0$. Hence $\partial u(x^0)/\partial \nu = 0$.

We can define a linear nondegenerate transformation of coordinates $x \to y = y(x)$ such that at the point x^0 the operator A will have the form:

$$Au(x^0) = \sum_{i=1}^{n} b_{ii}(x^0) u_{y_i y_i}(x^0) + \sum_{i=1}^{n} b_i(x^0) u_{y_i}(x^0) + a(x^0)u(x^0),$$

where $b_{ii}(x^0) > 0$ $(i = 1, \ldots, n)$. The equalities $\partial u(x^0)/\partial t_i = 0$ $(i = 1, \ldots, n-1)$, $\partial u(x^0)/\partial \nu = 0$ imply that $\partial u(x^0)/\partial y_i = 0$ $(i = 1, \ldots, n)$. Hence, since $u(x)$ has a maximum at x^0, $u_{y_i y_i}(x^0) \leq 0$ $(i = 1, \ldots, n)$. Thus $Au(x^0) \leq 0$. □

Existence of Feller Semigroups

Definition 24.3. The nonlocal boundary condition (24.5) is said to be *transversal* on ∂Q if

$$\mu(x)+\sigma(x)>0 \qquad (x\in\partial Q). \tag{24.8}$$

The condition (24.8) implies that one of the reflection or viscosity phenomena occurs at each point of ∂Q. Nonlocal terms in the boundary condition (24.5) correspond to the jump phenomenon on the boundary and the inward jump phenomenon from the boundary (see Fig. 0.4).

Analytically, the condition (24.8) implies that the nonlocal perturbation in the boundary condition (24.5) has lowest order with respect to dominant terms (compare with the operators $B^2_{\mu l}$ in (21.2)).

The problem of construction of Feller semigroups has been studied only in the transversal case and under some additional assumptions on a measure $m(x^0,\cdot)$. Further we apply the methods of this chapter to investigation of both transversal and non-transversal cases.

We consider the following boundary condition

$$\begin{aligned} Bu \;=\; & \sum_{s=0}^{N} b_s(x^0)u(\omega_s(x^0))+\int_\Omega c(x^0,y)u(y)\,dy-\mu(x^0)\,\frac{\partial u(x^0)}{\partial\nu}\\ & +\sum_{i=1}^{n-1}\beta_i(x^0)\,\frac{\partial u(x^0)}{\partial t_i}+\sigma(x^0)Au(x^0)\\ & -\sum_{i,j=1}^{n-1}\alpha_{ij}(x^0)\,\frac{\partial^2 u(x^0)}{\partial t_i\partial t_j}=0 \qquad (x^0\in\partial Q). \end{aligned} \tag{24.9}$$

Here $\Omega\subset Q$ is a bounded domain, $x\to t=t(x)$ is a nondegenerate, infinitely differentiable coordinate transformation satisfying the conditions (a)–(d) of the previous subsection, $b_s,\mu,\beta_i,\sigma,\alpha_{ij}\in C^\infty(\mathbb{R}^n)$, $c\in C^\infty(\mathbb{R}^n\times\mathbb{R}^n)$, $\mu(x),\sigma(x)\ge 0$; the matrix $\|\alpha_{ij}\|$ is symmetric and non-negative, $\omega_0(x)\equiv x$, ω_s $(s\ge 1)$ are infinitely differentiable nondegenerate transformations mapping some neighborhood γ of the boundary ∂Q onto $\omega_s(\gamma)$ so that $\omega_s(\partial Q)\subset\overline{Q}$; ν is the inner unit normal to ∂Q at the point $x\in\partial Q$.

Moreover, let the following condition hold:

24.5. *For all* $x\in\partial Q$ *and* $y\in\overline{\Omega}$,

$$b_s(x),c(x,y)\le 0 \quad (s\ge 1), \qquad b_0(x),\mu(x),\sigma(x)\ge 0,$$

$$\sum_{s\ge 1}|b_s(x)|+\int_\Omega |c(x,y)|\,dy\le b_0(x).$$

We denote by X^k_p $(k\ge p-1,\ p=1,2,3)$ a linear manifold of functions in $C^k(\overline{Q})$ satisfying the condition (24.9). Here we assume that the form of (24.9) depends on p.

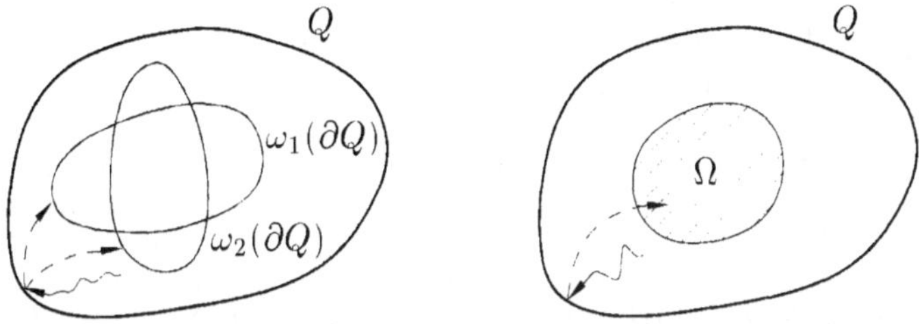

Fig. V.4

Suppose one of the following conditions is fulfilled:

24.6. *If* $p = 1$, *then* $b_0(x) > 0$ *and* $\mu(x) = \sigma(x) = \beta_i(x) = \alpha_{ij}(x) \equiv 0$ $(x \in \partial Q)$, $\overline{\omega_s(\gamma)} \subset Q$ *for* $s \geq 1$, *and* $\overline{\Omega} \subset Q$ (see Fig. V.4).

24.7. *If* $p = 2$, *then* $\mu(x) > 0$ *and* $\sigma(x) = \alpha_{ij}(x) \equiv 0$ $(x \in \partial Q)$.

24.8. *If* $p = 3$, *then* $\sigma(x) > 0$ *and* $\sum_{i,j} \alpha_{ij}\eta_i\eta_j > 0$ *for* $x \in \partial Q$ *and* $0 \neq \eta \in \mathbb{R}^{n-1}$.

Lemma 24.3. *Let the conditions* 24.4, 24.5 *be fulfilled. Suppose also that one of the conditions*: 24.6, 24.7, *or* 24.8 *holds.*

Then the boundary condition (24.9) *can be represented in the form* (24.5).

Proof. In order to prove the lemma, it suffices to show that there are $\gamma(x^0) \geq 0$ and a non-negative Borel measure $m(x^0, \cdot)$ such that

$$\sum_{s=0}^{N} b_s(x^0)u(\omega_s(x^0)) + \int_\Omega c(x^0, y)u(y)dy$$
$$= \gamma(x^0)u(x^0) + \int_{\overline{Q}} [u(x^0) - u(y)]m(x^0, dy). \quad (24.10)$$

Let

$$\gamma(x^0) = b_0(x^0) - \sum_{s \geq 1} |b_s(x^0)| - \int_\Omega |c(x^0, y)|\, dy,$$
$$m(x^0, A) = \sum_{s \in S} |b_s(x^0)| + \int_{\Omega \cap \mathcal{B}} |c(x^0, y)|\, dy$$

for any Borel set $\mathcal{B} \subset \overline{Q}$, where $S = S(x^0) = \{1 \leq s : \mathcal{B} \cap \omega_s(x^0) \neq \emptyset\}$. By virtue of the condition 24.5, $\gamma(x^0) \geq 0$ and $m(x^0, \cdot)$ is a non-negative Borel measure on $\overline{Q}$. Thus every function $u \in C(\overline{Q})$ satisfies (24.10). □

We introduce the unbounded operators $A_1 \colon X_1^0 \to X_1^0$ and $A_p \colon C(\overline{Q}) \to C(\overline{Q})$ $(p = 2, 3)$ by the formulas $A_1 u_1 = A u_1$ for $u_1 \in \mathcal{D}(A_1) = \{u \in X_1^2 : A_1 u \in X_1^0\}$ and $A_p u_p = A u_p$ for $u \in \mathcal{D}(A_p) = X_p^2$ $(p = 2, 3)$. We denote by $\overline{A_1}$ and $\overline{A_p}$ $(p = 2, 3)$ the closures of the operators A_1 in X_1^0 and A_p in $C(\overline{Q})$.

Lemma 24.4. *Let the conditions* 24.4–24.6 *be fulfilled.*
Then $\overline{(C^\infty(\overline{Q}) \cap \mathcal{D}(A_1))} = X_1^0$.

Proof. Let $u \in X_1^0$. For each $\varepsilon > 0$, there is $u_1 \in C^\infty(\overline{Q})$ such that $\|u-u_1\|_{C(\overline{Q})} < \varepsilon$. We introduce the functions

$$\varphi(x) = -\frac{1}{b_0(x)}\Big\{\sum_{s\geq 1} b_s(x)u_1(\omega_s(x)) + \int_\Omega c(x,y)u_1(y)\,dy\Big\} \qquad (x \in \partial Q),$$

$$f(x) = -\frac{\eta(x)}{b_0(x)}\Big\{\sum_{s\geq 1} b_s(x)(Au_1)(\omega_s(x)) + \int_\Omega c(x,y)(Au_1)(y)\,dy\Big\} \quad (x \in \overline{Q}),$$

where $\eta \in \dot{C}^\infty(\mathbb{R}^n)$, $\eta(x) = 1$ $(x \in \partial Q)$, $\operatorname{supp}\eta \subset \gamma$. Let u_2 be a solution of the boundary value problem

$$(Au_2)(x) = f(x) \quad (x \in Q), \qquad u_2(x) = \varphi(x) \quad (x \in \partial Q). \tag{24.11}$$

By definition, $\varphi \in C^\infty(\partial Q)$, $f \in C^\infty(\overline{Q})$. From the conditions 24.4–24.6, Theorems C.3, C.6, B.7, and the maximum principle it follows that there exists a unique solution of the boundary value problem (24.11) $u_2 \in C^\infty(\overline{Q})$. By virtue of (24.9) and conditions 24.5, 24.6, we obtain $|u_2(x) - u(x)| < \varepsilon$ $(x \in \partial Q)$. Clearly this inequality is also valid for $x \in \Gamma^\delta = \{x \in \mathbb{R}^n : \rho(x, \partial Q) < \delta\}$, where $\delta > 0$ is sufficiently small. Suppose that $0 < \sigma < \delta$ is such that $\Gamma^\sigma \cap \{\overline{\Omega} \cup (\bigcup_{s\geq 1} \omega_s(\partial Q))\} = \emptyset$ (see condition 24.6). We introduce the function

$$w(x) = \xi(x)u_2(x) + (1 - \xi(x))u_1(x),$$

where $\xi \in C^\infty(\mathbb{R}^n)$, $\xi(x) = 1$ $(x \in \Gamma^{\sigma/2})$, $\operatorname{supp}\xi \subset \Gamma^\sigma$, $0 \leq \xi(x) \leq 1$. By definition, $w \in C^\infty(\overline{Q}) \cap X_1^0$, $Aw \in X_1^0$ and $\|w - u\|_{C(\overline{Q})} < \varepsilon$. □

Theorem 24.5. *Let the conditions* 24.4–24.6 *hold.*
Then the operator $\overline{A}_1 \colon X_1^0 \to X_1^0$ *is the infinitesimal generator of a Feller semigroup on* X_1^0, *which is uniquely determined by* A_1.

Proof. By Lemma 24.4, $\overline{\mathcal{D}(A_1)} = X_1^0$.

Now we define the unbounded operator $\mathcal{A}_1 \colon X_1^0 \to C(\overline{Q})$ with domain $\mathcal{D}(\mathcal{A}_1) = X_1^2$. By virtue of Lemma 24.3, we can rewrite (24.9) in the form (24.5). Then the conditions of Theorem 24.4 are fulfilled. Therefore, by Theorem 24.4, if $u \in \mathcal{D}(\mathcal{A}_1)$ takes a positive maximum at $x^0 \in \overline{Q}$, then there is a point $x^1 \in \overline{Q}$ such that $u(x^1) = u(x^0)$ and $\mathcal{A}_1 u(x^1) \leq 0$.

Thus, by virtue of Theorem 24.3, it suffices to show that for $\lambda > 0$ the range $\mathcal{R}(\lambda I - \mathcal{A}_1)$ is dense in $C(\overline{Q})$ and the range $\mathcal{R}(\lambda I - A_1)$ is dense in X_1^0. By Theorem B.7, $W^{k+2}(Q) \subset C^2(\overline{Q})$ and $W^k(Q) \subset C(\overline{Q})$ for $k > n/2$. We introduce the unbounded operator $A_1' \colon W^k(Q) \to W^k(Q)$ by the formula $A_1' u = A_1 u$ with domain $\mathcal{D}(A_1') = W^{k+2}(Q) \cap X_1^0$. By virtue of Theorem 21.3, the operator $\lambda I -$

$A_1'\colon W^k(Q) \to W^k(Q)$ is Fredholm and $\operatorname{ind}(\lambda I - A_1') = 0$. If $u_0 \in \mathcal{N}(\lambda I - A_1')$, then $u_0 \in C^2(\overline{Q})$. Hence, by Theorem 24.4, $u_0(x) \equiv 0$. Therefore, since the operator $\lambda I - A_1'$ is Fredholm and $\operatorname{ind}(\lambda I - A_1') = 0$, it follows that the equation $(\lambda I - A_1')u = f$ has a unique solution $u \in W^{k+2}(Q) \cap X_1^0 \subset X_1^2$ for all $f \in W^k(Q)$. From this, the density of $W^k(Q)$ in $C(\overline{Q})$ and Lemma 24.4 it follows that $\overline{\mathcal{R}(\lambda I - A_1')} = C(\overline{Q})$ and $\overline{\mathcal{R}(\lambda I - A_1)} = X_1^0$ for $\lambda > 0$. Since $A_1' \subset \mathcal{A}_1$, we obtain $\mathcal{R}(\lambda I - \mathcal{A}_1) = C(\overline{Q})$. □

Lemma 24.5. *Suppose that the conditions* 24.4, 24.5, 24.7 *hold if* $p = 2$, *and that the conditions* 24.4, 24.5, 24.8 *hold if* $p = 3$. *We also assume that the following condition is fulfilled:*

24.9. *either* $\overline{\omega_s(\gamma)} \subset Q$, *or* $\omega_s(\partial Q) = \partial Q$ *for* $s \geq 1$; $\overline{\Omega} \subset Q$.

Then $\overline{(C^\infty(\overline{Q}) \cap \mathcal{D}(A_p))} = C(\overline{Q})$, *where* $p = 2, 3$.

Proof. Let $u \in C(\overline{Q})$. For each $\varepsilon > 0$, there is $u_1 \in C^\infty(\overline{Q})$ such that $\|u - u_1\|_{C(\overline{Q})} < \varepsilon$. We introduce the function

$$\psi(x) = -\Big\{\sum_{s\geq 0} b_s(x)u_1(\omega_s(x)) + \int_\Omega c(x,y)u_1(y)dy\Big\} \qquad (x \in \partial Q).$$

There exists a function $u_2 \in C^\infty(\overline{Q})$ such that

$$u_2(x^0) = u_1(x^0) \qquad (x^0 \in \partial Q), \tag{24.12}$$

$$-\mu(x^0)\frac{\partial u_2(x^0)}{\partial \nu} + \sum_{i=1}^{n-1} \beta_i(x^0)\frac{\partial u_2(x^0)}{\partial t_i} + \sigma(x^0)Au_2(x^0)$$
$$-\sum_{i,j=1}^{n-1} \alpha_{ij}(x^0)\frac{\partial^2 u_2(x^0)}{\partial t_i \partial t_j} = \psi(x^0) \qquad (x^0 \in \partial Q). \tag{24.13}$$

Then, for sufficiently small $\delta > 0$,

$$|u_2(x) - u(x)| < \varepsilon \qquad (x \in \Gamma^\delta).$$

Now we define the function

$$w(x) = \xi(x)u_2(x) + (1 - \xi(x))u_1(x),$$

where $\xi \in C^\infty(\mathbb{R}^n)$, $\xi(x) = 1$ $(x \in \Gamma^{\sigma/2})$, $\operatorname{supp}\xi \subset \Gamma^\sigma$, $0 \leq \xi(x) \leq 1$. A number σ is such that $0 < \sigma < \delta$, $\Gamma^\sigma \cap \{\overline{\Omega} \cup (\bigcup_s \omega_s(\partial Q))\} = \emptyset$, where $s \in \{s : \omega_s(\partial Q) \subset Q\}$ (see condition 24.9). By definition, $w \in C^\infty(\overline{Q}) \cap X_p^2$ and $\|w - u\|_{C(\overline{Q})} < \varepsilon$. □

Theorem 24.6. *Let the conditions* 24.4, 24.5, 24.7 *hold if* $p = 2$, *and let the conditions* 24.4, 24.5, 24.8 *hold if* $p = 3$. *Assume that the condition* 24.9 *is fulfilled.*

Then the operator $\overline{A}_p\colon C(\overline{Q}) \to C(\overline{Q})$ *is the infinitesimal generator of a Feller semigroup on* $C(\overline{Q})$, *which is uniquely determined by* A_p $(p = 2, 3)$.

Proof. By Lemma 24.5, $\overline{\mathcal{D}(A_p)} = C(\overline{Q})$.

By virtue of Lemma 24.3, we can rewrite (24.9) in the form (24.5). Hence, from Theorem 24.4 it follows that, if $u \in \mathcal{D}(A_p)$ takes a positive maximum at $x^0 \in \overline{Q}$, then there is a point $x^1 \in \overline{Q}$ such that $u(x^1) = u(x^0)$ and $A_p u(x^1) \leq 0$.

Thus, by virtue of Theorem 24.3, it suffices to show that for $\lambda > 0$ the range $\mathcal{R}(\lambda I - A_p)$ is dense in $C(\overline{Q})$. By Theorem B.7, $W^{k+2}(Q) \subset C^2(\overline{Q})$ and $W^k(Q) \subset C(\overline{Q})$ for $k > n/2$. We introduce the unbounded operator $A_p' \colon W^k(Q) \to W^k(Q)$ by the formula $A_p' u = Au$ with domain $\mathcal{D}(A_p') = W^{k+2}(Q) \cap X_p^2$. By virtue of the conditions 24.4, 24.7, 24.8 and Theorem 21.1, the operator $\lambda I - A_p' \colon W^k(Q) \to W^k(Q)$ is Fredholm, and $\operatorname{ind}(\lambda I - A_p') = 0$. If $u_0 \in \mathcal{N}(\lambda I - A_p')$, then $u_0 \in C^2(\overline{Q})$. Hence, by Theorem 24.4, $u_0(x) \equiv 0$. Therefore the equation $(\lambda I - A_p')u = f$ has a unique solution $u \in W^{k+2}(Q) \cap X_p^2 \subset X_p^2$ for all $f \in W^k(Q)$. Since $W^k(Q)$ is dense in $C(\overline{Q})$, we obtain $\overline{\mathcal{R}(\lambda I - A_p')} = C(\overline{Q})$. Hence $\overline{\mathcal{R}(\lambda I - A_p)} = C(\overline{Q})$. □

Remark 24.2. From Theorem 5.2 of K. Sato and T. Ueno [1] it follows that Lemma 24.4 remains true without condition 24.9. Thus the condition 24.9 in Theorem 24.6 can be omitted.

A similar result is not valid in the case of nonlocal perturbation of the Dirichlet problem. If a support of nonlocal terms has non-empty intersection with the boundary ∂Q, then there exists the operator $\overline{A}_1$, which is not the infinitesimal generator of a Feller semigroup. Moreover, in Section 25, we shall show that such operator $\overline{A}_1$ can be constructed for arbitrary small coefficients in nonlocal terms.

25 Elliptic Problems with Nonlocal Conditions near the Boundary and Feller Semigroups

Formulation of Problem

Let $Q \subset \mathbb{R}^2$ be a bounded domain with boundary $\partial Q \in C^\infty$, which outside the disks $S_{1/8}((i4/3, j4/3))$ $(i,j = 0,1)$ coincides with the boundary of the square $(0,4/3) \times (0,4/3)$. We denote $\Gamma_1 = \{x \in \partial Q : x_1 < 1/3, x_2 < 1/3\}$, $\Gamma_2 = \{x \in \partial Q : 1 < x_1,\ 1 < x_2\}$, $\Gamma_3 = \partial Q \setminus (\overline{\Gamma}_1 \cup \overline{\Gamma}_2)$ (cf. Examples 11.2, 13.1).

Let X_γ^k be a linear manifold of functions in $C^k(\overline{Q})$ satisfying the conditions

$$\begin{aligned} u(x)|_{\Gamma_j} - \gamma u(x + h_j)|_{\Gamma_j} &= 0 \qquad (x \in \Gamma_j,\ j = 1,2), \\ u(x)|_{\overline{\Gamma}_3} &= 0 \qquad (x \in \overline{\Gamma}_3), \end{aligned} \tag{25.1}$$

where $0 < \gamma < 1$, $h_j = (-1)^{j+1}(1,1)$.

We introduce the unbounded operator $G_\gamma \colon X_\gamma^0 \to X_\gamma^0$ by the formula $G_\gamma u = \Delta u$ for $u \in \mathcal{D}(G_\gamma) = \{u \in X_\gamma^2 : G_\gamma u \in X_\gamma^0\}$. Denote by $\overline{G}_\gamma$ the closure of the operator G_γ in X_γ^0.

As in the proof of Lemma 24.3, it is easy to show that the nonlocal conditions (25.1) can be represented in the form (24.5).

However, we have the following:

Theorem 25.1. *For any* $0 < \gamma < 1$, $\overline{G}_\gamma$ *is not the infinitesimal generator of a Feller semigroup on* X_γ^0.

For the proof, in the next subsection we consider the Poisson equation with nonlocal conditions of the type (25.1) in weighted spaces. We prove that a corresponding operator A_γ is Fredholm. Using a relation between nonlocal elliptic boundary value problem and boundary value problem for elliptic differential-difference equation, we obtain that $\operatorname{ind} A_\gamma < 0$. On the other hand, we prove that $G_\gamma \subset A_\gamma$. Therefore $\mathcal{R}(\lambda I - \overline{G}_\gamma) \neq X_\gamma^0$. Hence, by virtue of the Hille–Yosida theorem, $\overline{G}_\gamma$ is not the infinitesimal generator of a Feller semigroup on X_γ^0.

Nonlocal Elliptic Problems in Weighted Spaces

We consider the nonlocal elliptic boundary value problem

$$\Delta u(x) = f_0(x) \qquad (x \in Q), \tag{25.2}$$

$$\left.\begin{aligned} B_j u = u(x)|_{\Gamma_j} - \gamma u(x+h_j)|_{\Gamma_j} = 0 \qquad (x \in \Gamma_j,\ j=1,2),\\ B_3 u = u(x)|_{\Gamma_3} = 0 \qquad (x \in \Gamma_3). \end{aligned}\right. \tag{25.3}$$

Let $\Omega = Q$ or $\Omega = \theta_m = \{(r,\varphi) : a_m < \varphi < b_m\}$ $(m = 1,\dots,4)$ be a half-plane, where r, φ are polar coordinates with pole at the point 0, $a_1 = 0$, $b_1 = \pi$, $a_2 = \pi/2$, $b_2 = 3\pi/2$, $a_3 = -\pi/2$, $b_3 = \pi/2$, $a_4 = -\pi$, $b_4 = 0$. We denote $g^1 = (1/3, 0)$, $g^2 = (4/3, 1)$, $g^3 = (0, 1/3)$, $g^4 = (1, 4/3)$. Let $\mathcal{K} = \bigcup_{m=1}^4 g^m$ if $\Omega = Q$, and $\mathcal{K} = \{0\}$ if $\Omega = \theta_m$. We define the weighted space $H_0^k(\Omega)$ as the completion of the set $C_0^\infty(\overline{\Omega} \setminus \mathcal{K})$ with respect to the norm

$$\|u\|_{H_0^k(\Omega)} = \left\{ \sum_{|\alpha| \le k} \int_\Omega \rho^{2(|\alpha|-k)} |\mathcal{D}^\alpha u(x)|^2 dx \right\}^{1/2},$$

where $\rho \in C^\infty(\mathbb{R}^2 \setminus \mathcal{K})$ is a real-valued function; for $x \in S_{1/8}(g^m)$, the function $\rho(x)$ coincides with the distance to g^m and $\rho(x) \ge c > 0$ for $x \notin \bigcup_m S_{1/8}(g^m)$ if $\Omega = Q$; $\rho(x) = r$ if $\Omega = \theta$; $C_0^\infty(\overline{\Omega} \setminus \mathcal{K})$ is the set of infinitely differentiable functions in $\overline{\Omega}$ with compact supports belonging to $\overline{\Omega} \setminus \mathcal{K}$.

Clearly $H_0^0(\Omega) = L_2(\Omega)$.

Let $H_{0,\gamma}^2(Q) = \{u \in H_0^2(Q) : B_j u = 0\ (j = 1,2,3)\}$. We introduce a bounded linear operator $A_\gamma \colon H_{0,\gamma}^2(Q) \to L_2(Q)$ by the formula $A_j u = \Delta u$ $(u \in H_{0,\gamma}^2(Q))$. In order to prove the Fredholm property of operator A_γ, we consider the auxiliary boundary value problem

$$\Delta U_m = f_{0m}(x) \qquad (x \in \theta_m,\ m = 1,\dots,4), \tag{25.4}$$

$$\left.\begin{aligned} &U_1|_{\varphi=0} = 0,\ U_2|_{\varphi=3\pi/2} = 0,\ U_3|_{\varphi=\pi/2} = 0,\ U_4|_{\varphi=-\pi} = 0,\\ &U_1|_{\varphi=\pi} = \gamma U_2|_{\varphi=\pi},\ U_2|_{\varphi=\pi/2} = \gamma U_1|_{\varphi=\pi/2},\\ &U_3|_{\varphi=-\pi/2} = \gamma U_4|_{\varphi=-\pi/2},\ U_4|_{\varphi=0} = \gamma U_3|_{\varphi=0}. \end{aligned}\right\} \tag{25.5}$$

Remark 25.1. Let $0 < 4\delta < 1/8$. Denote $U_m(x) = u(x+g^m)$, $f_{0m}(x) = f_0(x+g^m)$ $(x \in S_{4\delta}(0),\ m = 1,\dots,4)$. If $\operatorname{supp} u \subset \mathcal{K}^{4\delta} = \{x \in \mathbb{R}^2 : \rho(x,\mathcal{K}) < 4\delta\}$, then the problem (25.2), (25.3) is equivalent to the problem (25.4), (25.5).

We denote $H_0^{k,4}(\theta) = \prod_{m=1}^4 H_0^k(\theta_m)$, $H_{0,\gamma}^{k,4}(\theta) = \{U \in H_0^{k,4}(\theta) \colon U$ satisfies the conditions (25.5)$\}$, where $U = (U_1,\dots,U_4)$. We define a bounded linear operator $L_\gamma \colon H_{0,\gamma}^{2,4}(\theta) \to L_2^4(\theta)$ by the formula $L_\gamma U = (\Delta U_1,\dots,\Delta U_4)$ $(U \in H_{0,\gamma}^{2,4}(\theta))$.

Lemma 25.1. *For any $0 < \gamma < 1$, the operator $L_\gamma \colon H_{0,\gamma}^{2,4}(\theta) \to L_2^4(\theta)$ has a bounded inverse $L_\gamma^{-1} \colon L_2^4(\theta) \to H_{0,\gamma}^{2,4}(\theta)$.*

Proof. 1. We pass to polar coordinates r, φ and denote $V_1(r,\varphi) = U_1(r,\varphi)$, $V_2(r,\varphi) = U_2(r, 3\pi/2 - \varphi)$, $V_3(r,\varphi) = U_3(r,\pi/2 - \varphi)$, $V_4(r,\varphi) = U_4(r,\pi + \varphi)$, $f_1(r,\varphi) = f_{01}(r,\varphi)$, $f_2(r,\varphi) = f_{02}(r,3\pi/2 - \varphi)$, $f_3(r,\varphi) = f_{03}(r,\pi/2 - \varphi)$, $f_4 = f_{04}(r,\pi + \varphi)$. Then the nonlocal boundary value problem (25.4), (25.5) will take the form

$$\begin{gathered}\frac{1}{r}\frac{\partial}{\partial r}\left(r\frac{\partial V_m}{\partial r}\right) + \frac{1}{r^2}\frac{\partial V_m}{\partial \varphi^2} = f_m(r,\varphi),\\ (0 < r < \infty,\ 0 < \varphi < \pi,\ m = 1,\dots,4)\end{gathered} \tag{25.6}$$

$$\left.\begin{aligned} &V_m|_{\varphi=0} = 0 && (m = 1,\dots,4),\\ &V_m|_{\varphi=\pi} = \gamma V_{m+1}|_{\varphi=\pi/2}, && \\ &V_{m+1}|_{\varphi=\pi} = \gamma V_m|_{\varphi=\pi/2} && (m = 1,3).\end{aligned}\right\} \tag{25.7}$$

Making the transformation $\tau = -\ln r$, we have

$$\begin{gathered}\frac{\partial^2 V_m}{\partial \tau^2} + \frac{\partial^2 V_m}{\partial \varphi^2} = F_m(\tau,\varphi)\\ (-\infty < \tau < \infty,\ 0 < \varphi < \pi,\ m = 1,\dots,4),\end{gathered} \tag{25.8}$$

$$\left.\begin{aligned} &V_m|_{\varphi=0} = 0 && (-\infty < \tau < \infty,\ m = 1,\dots,4),\\ &V_m|_{\varphi=\pi} = \gamma V_{m+1}|_{\varphi=\pi/2}, && \\ &V_{m+1}|_{\varphi=\pi} = \gamma V_m|_{\varphi=\pi/2} && (-\infty < \tau < \infty,\ m = 1,3),\end{aligned}\right\} \tag{25.9}$$

where $F_m(\tau,\varphi) = e^{-2\tau} f_m(\tau,\varphi)$.

The nonlocal problem (25.8), (25.9) becomes the following, after the Fourier transform with respect to τ:

$$\frac{d^2\widehat{V}_m}{d\varphi^2} - \lambda^2 \widehat{V}_m = \widehat{F}_m(\lambda,\varphi) \qquad (0 < \varphi < \pi,\ m = 1,\dots,4), \tag{25.10}$$

$$\left.\begin{aligned} &\widehat{V}_m|_{\varphi=0} = 0 && (m = 1,\dots,4),\\ &\widehat{V}_m|_{\varphi=\pi} = \gamma \widehat{V}_{m+1}|_{\varphi=\pi/2}, && \\ &\widehat{V}_{m+1}|_{\varphi=\pi} = \gamma \widehat{V}_m|_{\varphi=\pi/2} && (m = 1,3),\end{aligned}\right\} \tag{25.11}$$

where

$$\widehat{V}_m(\lambda,\varphi) = (2\pi)^{-1/2}\int_{-\infty}^{\infty} e^{-i\lambda\tau} V_m(\tau,\varphi)\,d\tau.$$

2. Now we estimate the norms. Clearly

$$k_1\|v\|^2_{H^2_0(\theta_1)} \le \sum_{\alpha_1+\alpha_2\le 2}\int_{-\infty}^{+\infty} d\tau \int_0^\pi \left|\frac{\partial^{\alpha_1+\alpha_2}v}{\partial\tau^{\alpha_1}\partial\varphi^{\alpha_2}}\right|^2 e^{2\tau}\,d\varphi$$
$$\le k_2\|v\|^2_{H^2_0(\theta_1)} \qquad (v\in H^2_0(\theta_1)). \tag{25.12}$$

Using the complex analog of Parseval's equality

$$\int_{-\infty}^{+\infty}|w(\tau)|^2e^{2\tau}\,d\tau = \int_{-\infty+i}^{+\infty+i}|\widehat{w}(\lambda)|^2d\lambda \qquad (we^\tau\in L_2(-\infty.\infty)) \tag{25.13}$$

and the interpolation inequality (B.20), from (25.12) we have

$$k_3\|v\|^2_{H^2_0(\theta_1)} \le \int_{-\infty+i}^{+\infty+i}\left(\|\hat{v}\|^2_{W^2(0,\pi)}+|\lambda|^4\|\hat{v}\|^2_{L_2(0,\pi)}\right)d\lambda$$
$$\le k_4\|v\|^2_{H^2_0(\theta_1)} \qquad (v\in H^2_0(\theta_1)). \tag{25.14}$$

3. We now find the eigenvalues of the problem

$$\frac{d^2W_m}{d\varphi^2}-\lambda^2W_m=0 \qquad (0<\varphi<\pi,\ m=1,2), \tag{25.15}$$

$$\left.\begin{aligned} &W_m|_{\varphi=0}=0 \qquad (m=1,2),\\ &W_1|_{\varphi=\pi}=\gamma W_2|_{\varphi=\pi/2},\ W_2|_{\varphi=\pi}=\gamma W_1|_{\varphi=\pi/2}.\end{aligned}\right\} \tag{25.16}$$

Substituting a general solution of the homogeneous equations (25.15) $W_1=c_1e^{\lambda\varphi}+c_2e^{-\lambda\varphi}$, $W_2=c_3e^{\lambda\varphi}+c_4e^{-\lambda\varphi}$ into the nonlocal conditions (25.16), we obtain

$$\left.\begin{array}{llllr} c_1 & +\,c_2 & & & =0,\\ & & c_3 & +\,c_4 & =0,\\ c_1e^{\lambda\pi} & +\,c_2e^{-\lambda\pi} & -\,c_3\gamma e^{\lambda\pi/2} & -\,c_4\gamma e^{-\lambda\pi/2} & =0,\\ c_1\gamma e^{\lambda\pi/2} & +\,c_2\gamma e^{-\lambda\pi/2} & -\,c_3e^{\lambda\pi} & -\,c_4e^{-\lambda\pi} & =0.\end{array}\right\} \tag{25.17}$$

The eigenvalues of the problem (25.15), (25.16) are zeros of the determinant $\Delta(\lambda)$ of the system of linear algebraic equations (25.17). It is easy to see that

$$\Delta(\lambda)=(e^{\lambda\pi/2}-e^{-\lambda\pi/2})^2(e^{\lambda\pi}+e^{-\lambda\pi}+2-\gamma^2). \tag{25.18}$$

Therefore, since $0<\gamma<1$, then $\lambda_s=i2s$ $(s=\pm1,\pm2,\dots)$, $\lambda_p^\pm=i(\pm(2/\pi)\arctan\sqrt{4\gamma^{-2}-1}+2p)$ $(p=0,\pm1,\pm2,\dots)$. Thus the line $\operatorname{Im}\lambda=1$ does not contain any eigenvalue of the problem (25.15), (25.16).

4. We note that the nonlocal problem with a parameter (25.10), (25.11) consists of two independent nonlocal problems having the same form. Therefore, by virtue of Lemma 1.4 and Theorem 1.3, for any $\widehat{F}=(\widehat{F}_1,\dots,\widehat{F}_4)\in L_2^4(0,\pi)$ there

is a unique solution $\widehat{V} = (\widehat{V}_1, \dots, \widehat{V}_4) \in W^{2,4}(0, \pi)$ of the problem (25.10), (25.11) for $\operatorname{Im} \lambda = 1$, and

$$\sum_{m=1}^{4} |||\widehat{V}_m|||^2_{W^2(0,\pi)} \le k_5 \sum_{m=1}^{4} \|\hat{F}_m\|^2_{L_2(0,\pi)} \qquad (\operatorname{Im} \lambda = 1), \tag{25.19}$$

where $k_5 > 0$ does not depend on λ and $\widehat{F}$. Integrating (25.19) over the line $\operatorname{Im} \lambda = 1$ and using (25.13), we have

$$\int_{-\infty+i}^{+\infty+i} \sum_{m=1}^{4} |||\widehat{V}_m|||^2_{W^2(0,\pi)} \, d\lambda \le k_6 \sum_{m=1}^{4} \|f_{0m}\|^2_{L_2(\theta_m)}.$$

Hence, by virtue of (25.14), there is a unique solution $V = (V_1, \dots, V_4) \in \prod_{m=1}^{4} H_0^2(\theta_1)$ of the problem (25.6), (25.7). Furthermore,

$$V_m(\tau, \varphi) = (2\pi)^{-1/2} \int_{-\infty+i}^{+\infty+i} \widehat{V}_m(\lambda, \varphi) e^{i\lambda\tau} d\lambda,$$

$$\sum_{m=1}^{4} \|V_m\|^2_{H_0^2(\theta_1)} \le k_7 \sum_{m=1}^{4} \|f_{0m}\|^2_{L_2(\theta_m)}.$$

Thus, for any $f = (f_{01}, \dots, f_{04}) \in L_2^4(\theta)$, there exists a unique solution $U \in H_0^{k,4}(\theta)$ of the problem (25.4), (25.5), and

$$\|U\|^2_{H_0^{2,4}(\theta)} \le k_7 \|f\|^2_{L_2^4(\theta)}. \quad \square$$

Theorem 25.2. *Let* $0 < \gamma < 1$.
Then the operator $A_\gamma \colon H^2_{0,\gamma}(Q) \to L_2(Q)$ *is Fredholm.*

Proof. By virtue of Theorem A.5, it suffices to show that the operator A_γ has both a left regularizer and a right regularizer.

1. We introduce functions $\eta, \eta_1 \in \dot{C}^\infty(\mathbb{R}^2)$ such that $0 \le \eta(x), \eta_1(x) \le 1$, $\eta(x) = 1$ $(x \in S_{2\delta}(0))$, $\operatorname{supp} \eta \subset S_{3\delta}(0)$, $\eta_1(x) = 1$ $(x \in S_{3\delta}(0))$, $\operatorname{supp} \eta_1 \subset S_{4\delta}(0)$. Let

$$\xi(x) = \sum_{m=1}^{4} \eta(x - g^m), \qquad \psi(x) = 1 - \xi(x).$$

We also introduce a function $\psi_1 \in C^\infty(\mathbb{R}^2)$ such that $0 \le \psi_1(x) \le 1$, $\psi_1(x) = 1$ $(x \notin \mathcal{K}^{2\delta})$, $\psi_1(x) = 0$ $(x \in \mathcal{K}^\delta)$.

We define an operator $W \colon L_2(Q) \to L_2^4(\theta)$ with domain $\mathcal{D}(W) = \{v \in L_2(Q) : \operatorname{supp} v \subset \mathcal{K}^{4\delta}\}$ by the formulas

$$\begin{aligned} (Wv)_m(x) &= v(x + g^m) && (x \in \theta_m,\ |x| < 4\delta,\ m = 1, \dots, 4), \\ (Wv)_m(x) &= 0 && (x \in \theta_m,\ |x| \ge 4\delta,\ m = 1, \dots, 4). \end{aligned}$$

The equality

$$R_1 f_0 = W^{-1}\eta_1 L_\gamma^{-1} W\xi f_0$$

defines a linear bounded operator $R_1: L_2(Q) \to H^2_{0,\gamma}(Q)$. The Leibniz formula implies that

$$\Delta R_1 f_0 = \xi f_0 + T_1 f_0, \tag{25.20}$$

where $T_1 f_0 = \sum_{|\alpha|\le 1} W^{-1}\eta_\alpha \mathcal{D}^\alpha L_\gamma^{-1} W\xi f_0$, $\eta_\alpha \in C^\infty(\mathbb{R}^2)$, $\operatorname{supp}\eta_\alpha \subset S_{4\delta}(0)$. Clearly $T_1: L_2(Q) \to H^1_0(Q)$ is a bounded operator. By virtue of the compactness of the imbedding operator of $H^1_0(Q)$ into $L_2(Q)$ (see V. A. Kondrat′ev [1], Lemma 3.5), the operator $T_1: L_2(Q) \to L_2(Q)$ is compact.

2. We define functions $\mu_1, \mu_2 \in \dot{C}^\infty(\mathbb{R}^2)$ such that $0 \le \mu_j(x) \le 1$, $\mu_j(x) = 1$ $(x \in (\Gamma_j + h_j) \setminus \mathcal{K}^{\delta/2})$, and $\mu_j(x) = 0$ $(x \notin ((\Gamma_j + h_j) \setminus \mathcal{K}^{\delta/2})^{\delta/4})$.

Let $W^2_{\gamma,\delta}(Q)$ be a subspace of $W^2(Q)$ consisting of functions satisfying the nonlocal conditions

$$u(x)|_{\partial Q} - \gamma \sum_{j=1}^{2} \mu_j(x + h_j) u(x + h_j)|_{\partial Q} = 0. \tag{25.21}$$

We define a bounded operator $A_{\gamma,\delta}: W^2_{\gamma,\delta}(Q) \to L_2(Q)$ given by $A_{\gamma,\delta} u = \Delta u$ $(u \in W^2_{\gamma,\delta}(Q))$. By virtue of Theorem 21.3, the operator $A_{\gamma,\delta}$ is Fredholm. Hence, by Theorem A.5, there is a bounded operator $R_0: L_2(Q) \to W^2_{\gamma,\delta}(Q)$, which is the left and the right regularizer of $A_{\gamma,\delta}$. The equality

$$R_2 f_0 = \psi_1 R_0 \psi f_0$$

defines a linear bounded operator $R_2: L_2(Q) \to W^2_{\gamma,\delta}(Q)$. Since $\psi_1(x) = 0$ $(x \in \mathcal{K}^\delta)$, then $R_2: L_2(Q) \to H^2_{0,\gamma}(Q)$ is a bounded operator. By virtue of the Leibniz formula, we have

$$\Delta R_2 f_0 = \psi f_0 + T_2 f_0,$$

where $T_2 f_0 = \sum_{|\alpha|\le 1} \psi_\alpha \mathcal{D}^\alpha R_0 \psi f_0 + \psi_1 T_0 \psi f_0$, $\psi_\alpha \in C^\infty(\mathbb{R}^2)$, $\operatorname{supp}\psi_\alpha \subset \overline{\mathcal{K}^{2\delta}} \setminus \mathcal{K}^\delta$, and $T_0: L_2(Q) \to L_2(Q)$ is a compact operator. Thus $T_2: L_2(Q) \to L_2(Q)$ is a compact operator.

3. We denote $R = R_1 + R_2$. Clearly $R: L_2(Q) \to H^2_{0,\gamma}(Q)$ is a bounded operator, and

$$R f_0 = f_0 + (T_1 + T_2) f_0,$$

where $T_1, T_2: L_2(Q) \to L_2(Q)$ are compact operators. Hence the operator R is a right regularizer of A_γ.

Similarly it is easy to prove that R is a left regularizer of A_γ. □

Proof of Theorem 25.1

Lemma 25.2. $G_\gamma \subset A_\gamma$.

Proof. In order to prove that $X_\gamma^2 \subset H_{0,\gamma}^2(Q)$, it suffices to show that if $u \in X_\gamma^2$, then

$$I_m = \int_{Q_m} \rho^{2(|\alpha|-2)} |\mathcal{D}^\alpha u|^2 dx < \infty \qquad (m = 1, \dots, 4,\ |\alpha| \le 1),$$

where $Q_m = S_{1/8}(g^m) \cap Q$. Suppose, for example, that $m = 1, 2$. We set $Q_m^j = \{x \in Q_m : 2^{-j-1} < |x - g^m| < 2^{-j}\}$ $(j = 3, 4, \dots)$. Then, passing to the new variables $x' = (x - g^m)2^{j-3} + g^m$, we obtain

$$I_m = \sum_{j=3}^{\infty} \int_{Q_m^j} |x - g^m|^{2(|\alpha|-2)} |\mathcal{D}^\alpha u|^2\, dx \le k_1 \sum_j 2^{2j} \int_{Q_m^3} |\mathcal{D}_{x'}^\alpha u|^2\, dx'. \qquad (25.22)$$

It is easy to see, that two polynomials of first order defined in Q_1^3 and Q_2^3 and satisfying conditions (25.3) are identically zero. Hence in the subspace of functions $W^2(\Lambda)$ satisfying conditions (25.3) it is possible to introduce the equivalent norm

$$\|u\|_{W^2(\Lambda)} = \left\{ \sum_{|\beta|=2} \int_\Lambda |\mathcal{D}^\beta u|^2 dx \right\}^{1/2},$$

where $\Lambda = Q_1^3 \cup Q_2^3$. Thus from (25.22) we obtain

$$\begin{aligned} I_m &\le k_2 \sum_j \sum_{m=1,2} \sum_{|\beta|=2} 2^{2j} \int_{Q_m^3} |\mathcal{D}_{x'}^\beta u|^2\, dx' \\ &\le k_3 \sum_j \sum_{m=1,2} \sum_{|\beta|=2} \int_{Q_m^j} |\mathcal{D}^\beta u|^2\, dx. \end{aligned}$$

The lemma is proved. □

We introduce the unbounded operator $\mathcal{A}_\gamma\colon L_2(Q) \to L_2(Q)$ acting in the space of distributions $\mathcal{D}'(Q)$ according to the formula $\mathcal{A}_\gamma u = \Delta u$ with domain $\mathcal{D}(\mathcal{A}_\gamma) = \{u \in W_\gamma^1(Q) : \mathcal{A}_\gamma u \in L_2(Q)\}$, where $W_\gamma^1(Q)$ is the space of functions in $W^1(Q)$ satisfying (25.3).

Lemma 25.3. *Let* $0 < \gamma < 1$.

Then the equation $\mathcal{A}_\gamma u = f_0$ *has a unique solution for any* $f \in L_2(Q)$.

The proof follows from Example 13.1.

Theorem 25.3. *For any* $0 < \gamma < 1$, $\operatorname{ind} A_\gamma < 0$.

Proof. From Theorem 8.1 and Example 13.1 it follows that the operator R_Q maps $\mathring{W}^1(Q)$ onto $W^1_\gamma(Q)$ continuously and in a one-to-one manner, where

$$Ru(x) = u(x) + \gamma u(x_1+1, x_2+1) + \gamma(x_1-1, x_2-1).$$

Therefore the equation

$$\mathcal{A}_\gamma u = f_0 \tag{25.23}$$

is equivalent to the equation

$$\mathcal{A}_R v = f_0, \tag{25.24}$$

where $\mathcal{A}_R\colon L_2(Q) \to L_2(Q)$ is the unbounded operator acting in $\mathcal{D}'(Q)$ by the formula $\mathcal{A}_R v = \Delta R_Q v$ with domain $\mathcal{D}(\mathcal{A}_R) = \{u \in \mathring{W}^1(Q) : \mathcal{A}_R u \in L_2(Q)\}$.

By virtue of Example 11.2, there is $f_0 \in L_2(Q)$ such that $R_Q v \notin W^2(Q)$. Hence the solution of the equation (25.23) $u = R_Q v \notin W^2(Q)$. Since $H^2_{0,\gamma}(Q) \subset W^2(Q)$, we have $u \notin H^2_{0,\gamma}(Q)$. On the other hand, the imbedding $H^2_{0,\gamma}(Q) \subset W^1_\gamma(Q)$ implies that A_γ is a restriction of $\mathcal{A}_\gamma$ to $H^2_{0,\gamma}(Q)$. Therefore, by virtue of Lemma 25.3 and Theorem 25.2, $\dim \mathcal{N}(A_\gamma) = 0$, $0 < \operatorname{codim} \mathcal{R}(A_\gamma) < \infty$. Thus $\operatorname{ind} A_\gamma < 0$. □

Proof of Theorem 25.1. Assume to the contrary that: $\overline{G}_\gamma$ is the infinitesimal generator of a Feller semigroup on X^0_γ. Then, by virtue of the Hille–Yosida theorem, $\overline{\mathcal{R}(\lambda I - G_\gamma)} = X^0_\gamma$. Hence $\mathcal{R}(\lambda I - G_\gamma)$ is also dense in $L_2(Q)$. Therefore, by Lemma 25.2, since $\mathcal{R}(\lambda I - A_\gamma)$ is closed in $L_2(Q)$, we have $\mathcal{R}(\lambda I - A_\gamma) = L_2(Q)$. On the other hand, $\operatorname{ind} A_\gamma = \operatorname{ind}(\lambda I - A_\gamma) < 0$. Thus $\operatorname{codim} \mathcal{R}(\lambda I - A_\gamma) > 0$. This contradiction shows that $\mathcal{R}(\lambda I - G_\gamma)$ is not dense in X^0_γ. □

Remark 25.2. Clearly, if $u \in \mathcal{D}(G_\gamma)$ takes a positive maximum at $x^0 \in \overline{Q}$, then $G_\gamma u(x^0) \le 0$. Similarly to the proof of Lemma 24.4 it is easy to show that $\overline{C^\infty(\overline{Q}) \cap \mathcal{D}(G_\gamma)} = X^0_\gamma$. Hence the conditions (a) and (b) of Theorem 24.3 are fulfilled. From the proof of Theorem 25.1 it follows that $\mathcal{R}(\lambda I - G_\gamma)$ is not dense in X^0_γ, i.e., the condition (c) does not hold. Thus, if an elliptic operator with nonlocal conditions (24.5) has zero index, then we can hope that it gives an infinitesimal generator of a Feller semigroup. On the other hand, an elliptic operator with nonzero index does not correspond to an infinitesimal generator of a Feller semigroup.

Notes

An elliptic differential equation of the second order with nonlocal boundary conditions connecting the values of the unknown function on a boundary with its values on some manifold inside domain (see (0.1), (0.2)) was first studied by A. V. Bitsadze and A. A. Samarskiĭ [1]. In that paper they proved the existence and uniqueness of a classical solution of the boundary value problem (0.3), (0.4).

Section 21. In the case when $S = 1$ Theorem 21.1 for the problem (21.7), (21.8) was proved by Ya. A. Roitberg and Z. G. Sheftel′ [2]. Unlike that paper, our approach allows to study nonlocal elliptic problems for a finite or infinite number of manifolds, which can have non-empty intersections with each other. In the case $S \geq 1$. Theorems 21.1–21.3 and Lemma 21.1 were obtained by A. L. Skubachevskiĭ [4, 20]. All these assertions were proved for the problem (21.1), (21.2) with abstract nonlocal terms.

Solvability and stability of index of elliptic problems with general nonlocal conditions were studied by A. L. Skubachevskiĭ [7, 16].

Necessary and sufficient conditions for the existence of classical solutions of the Laplace equation or the biharmonic equation with nonlocal conditions were obtained by A. V. Bitsadze [1,2].

Section 22. The problem (22.6), (22.7) and the problem (22.1), (22.2) with abstract nonlocal terms are generalizations of the problem (0.3), (0.4). Theorems 22.1, 22.2 and Lemma 22.3 are due to A. L. Skubachevskiĭ [4].

In the general case ($\Gamma_2 \neq \emptyset$, $\overline{\omega(\Gamma_1)} \cap \overline{\Gamma_1} \neq \emptyset$) the theory of nonlocal elliptic boundary value problems in weighted spaces was studied by A. L. Skubachevskiĭ [9, 11, 14, 15, 16].

The results of Section 23 are adapted from A. L. Skubachevskiĭ [1].

Section 24. W. Feller [1, 2] determined all the diffusion processes in one dimension. A. D. Ventsel′ [1] studied a general form of boundary conditions in multidimensional case. Further results concerning multidimensional diffusion processes with nonlocal boundary conditions are due to K. Sato and T. Ueno [1], J. M. Bony, P. Courrege, and P. Priouret [1], C. Cancelier [1], T. Ueno [1, 2], S. Watanabe [1], K. Taira [2], Y. Ishikawa [1].

Theorems 24.2, 24.4 and in the transversal case Theorem 24.3 are due to A. D. Ventsel′ [1].

The time non-transversal case was first studied by A. L. Skubachevskiĭ [13]. Theorems 24.5, 24.6 are adapted from this paper. There an operator with the boundary conditions of the type (24.5) was also constructed, which is not the infinitesimal generator of a Feller semigroup (see Section 25). For further results on the existence of Feller semigroups in the non-transversal case, see A. L. Skubachevskiĭ [20]. The weighted spaces in the theory of elliptic equations in domains with corners were studied by V. A. Kondrat′ev [1].

Appendix

This appendix is a summary of the basic definitions and results concerning linear operators, functional spaces and elliptic differential equations. The material in this appendix is given for completeness to minimize the necessity for consulting too many outside references.

A Linear Operators

Fredholm Operators

Let H_1, H_2 be Hilbert spaces. For simplicity, in this book we consider separable Hilbert spaces. A closed linear operator $A: H_1 \to H_2$ is called a *Fredholm* operator if $\mathcal{R}(A)$ is closed in H_2 and $\dim \mathcal{N}(A) < \infty$, $\operatorname{codim} \mathcal{R}(A) < \infty$, where $\mathcal{N}(A)$ and $\mathcal{R}(A)$ are the null space (or kernel) and the range of the operator A, respectively. The *index* of the Fredholm operator A is defined by $\operatorname{ind} A = \dim \mathcal{N}(A) - \operatorname{codim} \mathcal{R}(A)$.

The next statement follows from the definition of a Fredholm operator.

Theorem A.1. *Let $A: H_1 \to H_2$ be a closed linear operator with domain $\mathcal{D}(A)$. Let $B: H_2 \to H_1$ be a bounded linear operator mapping H_2 onto $\mathcal{R}(B)$ in a one-to-one manner, and let $\mathcal{D}(A) \subset \mathcal{R}(B)$. Suppose that the operator $AB: H_2 \to H_2$ with domain $\mathcal{D}(AB) = \{x \in H_2 : Bx \in \mathcal{D}(A)\}$ is Fredholm.*

Then the operator $A: H_1 \to H_2$ is Fredholm, and $\operatorname{ind} A = \operatorname{ind} AB$.

Theorem A.2. *Let $A: H_1 \to H_2$ be a bounded Fredholm operator, and let $T: H_1 \to H_2$ be a compact operator.*

Then the operator $A + T: H_1 \to H_2$ is Fredholm, and $\operatorname{ind}(A + T) = \operatorname{ind} A$.

For a proof, see S. G. Krein [1], Theorem 16.4.

Let $A: H_1 \to H_2$ be a bounded linear operator. A bounded linear operator $\mathcal{R}_1: H_2 \to H_1$ is called a *left regularizer* of A if $\mathcal{R}_1 A = I_1 + T_1$, where $T_1: H_1 \to H_1$ is a compact operator, and $I_1: H_1 \to H_1$ is the identity operator. A bounded linear operator $\mathcal{R}_2: H_2 \to H_1$ is called a *right regularizer* of A if $A\mathcal{R}_2 = I_2 + T_2$, where $T_2: H_2 \to H_2$ is a compact operator, and $I_2: H_2 \to H_2$ is the identity operator.

Theorem A.3. *Let H_1, H_2, H be Hilbert spaces. Let $A: H_1 \to H_2$ be a bounded linear operator. Suppose that $H_1 \subset H$, and that the imbedding operator from H_1 into H is compact.*

Then the following three conditions are equivalent:

(a) *The range $\mathcal{R}(A)$ is closed in H_2, and $\dim \mathcal{N}(A) < \infty$.*

(b) *There exists a constant $c > 0$ such that*

$$\|u\|_{H_1} \leq c(\|Au\|_{H_2} + \|u\|_H) \tag{A.1}$$

for all $u \in H_2$.

(c) *The operator A has a left regularizer.*

Theorem A.3 follows from Theorems 7.1, 14.3 in S. G. Krein [1].

Theorem A.4. *Let H_1, H_2 be Hilbert spaces. Let $A: H_1 \to H_2$ be a bounded linear operator.*

Then the range $\mathcal{R}(A)$ is closed in H_2, and $\operatorname{codim} \mathcal{R}(A) < \infty$ *if and only if the operator A has a right regularizer.*

For a proof, see S. G. Krein [1], Theorem 15.2.

From Theorems A.3, A.4 we obtain:

Theorem A.5. *Let H_1, H_2, H be Hilbert spaces. Let $A: H_1 \to H_2$ be a bounded linear operator. Suppose that $H_1 \subset H$, and that the imbedding operator from H_1 into H is compact.*

Then the following three conditions are equivalent:

(a) *The operator $A: H_1 \to H_2$ is Fredholm.*

(b) *The operator A has both a right and a left regularizer.*

(c) *A priori estimate* (A.1) *is fulfilled, and the operator A has a right regularizer.*

Resolvents and Spectra

Let $A: H \to H$ be a closed linear operator in a complex Hilbert space H. The *resolvent set* of A is defined as the set of scalars $\lambda \in \mathbb{C}$ such that the operator $A - \lambda I$ has a bounded inverse $(A - \lambda I)^{-1}: H \to H$. We denote by $\rho(A)$ the resolvent set of A. The operator $(A - \lambda I)^{-1} = R(\lambda, A)$ $(\lambda \in \rho(A))$ is called the *resolvent* of A. The complement of $\rho(A)$ is called the *spectrum* of A and is denoted by $\sigma(A)$. A number $\lambda \in \mathbb{C}$ is called an *eigenvalue* of A and $x \neq 0$ an *eigenvector* (*eigenfunction*) of A corresponding to λ if $(A - \lambda I)x = 0$. The null space $\mathcal{N}(A - \lambda I)$ is called the eigenspace of A corresponding to λ, and the dimension of $\mathcal{N}(A - \lambda I)$ is called the multiplicity of λ.

The spectrum $\sigma(A)$ may be divided into three disjoint subsets:

- A *point spectrum* $\sigma_p(A)$ consisting of all eigenvalues of A;
- A *continuous spectrum* $\sigma_c(A) \subset \sigma(A) \setminus \sigma_p(A)$ consisting of all $\lambda \in \mathbb{C}$ such that the range $\mathcal{R}(A - \lambda I)$ is dense in H and $\mathcal{R}(A - \lambda I) \neq H$;
- A *remainder spectrum* $\sigma_r(A) = \sigma(A) \setminus (\sigma_p(A) \cup \sigma_c(A))$.

Theorem A.6. *Let* $A_1, A_2: H \to H$ *be densely defined linear operators, and let* $A_1 \subset A_2^*$, $A_2 \subset A_1^*$. *Suppose that there exists* $\lambda \notin \sigma(A_1)$ *such that* $\overline{\lambda} \notin \sigma(A_2)$.

Then $A_1 = A_2^*$, $A_2 = A_1^*$.

For a proof, see Corollary 13 in N. Dunford and J. T. Schwartz [2], Chapter XIV, Section 6.

Denote by $P(A)$ a set of all isolated eigenvalues of finite multiplicity, where $A: H \to H$ is a closed linear operator. Let $C(A) = \sigma(A) \setminus P(A)$. The spectrum $\sigma(A)$ is said to be *discrete* if $\sigma(A) = P(A)$.

The next theorem follows from Lemma 2 in N. I. Akhiezer and I. M. Glazman [1], Chapter VII, Subsec. 94.

Theorem A.7. *Let* $A: H \to H$ *be a self-adjoint operator with domain* $\mathcal{D}(A)$, *and let* $B: H \to H$ *be a bounded self-adjoint operator.*

Then the operator $A + B: H \to H$ *with domain* $\mathcal{D}(A+B) = \mathcal{D}(A)$ *is self-adjoint. If* $C(A) \subset [a, b]$, *then* $C(A+B) \subset [a - \|B\|, b + \|B\|]$.

Theorem A.8. *Let* $A: H \to H$ *be a closed linear operator. Suppose that the resolvent* $R(\lambda, A): H \to H$ *exists and is compact for some* $\lambda \in \mathbb{C}$.

Then the spectrum $\sigma(A)$ *is discrete, and* $R(\lambda, A): H \to H$ *is compact for each* $\lambda \in \rho(A)$.

For a proof, see Theorem 6.29 in T. Kato [1], Chapter III, Section 6.

Let H_1, H_2 be Hilbert spaces. We consider an operator function $A(\lambda): H_1 \to H_2$ for $\lambda \in \Lambda$, where $\Lambda \in \mathbb{C}$ is an open connected set, $A(\lambda): H_1 \to H_2$ is a linear bounded operator for each $\lambda \in \Lambda$.

The operator function $A(\lambda): H_1 \to H_2$ is said to be *analytic* in Λ if in some neighborhood of an arbitrary point $\xi \in \Lambda$ the operator function $A(\lambda): H_1 \to H_2$ can be expanded in a Taylor series which converges in a norm.

We can similarly introduce the concept of a *meromorphic* operator function.

An operator function which is meromorphic in Λ is said to be *finitely meromorphic*, if the operators in the negative powers $(\lambda - \xi)$ of the expansion in the Laurent series in the neighborhood of any pole $\xi \in \Lambda$ are finite-dimensional.

We shall say that a meromorphic operator function is *Fredholm* in Λ if for any point $\xi \in \Lambda$ the operator in the zero power $(\lambda - \xi)$ of the expansion in a Laurent series in the neighborhood of the point ξ is a Fredholm operator.

Theorem A.9. *Let* $\Lambda \subset \mathbb{C}$ *be an open connected set. Suppose that* $A(\lambda): H_1 \to H_2$ *is a finitely meromorphic Fredholm operator function in* Λ, *and at some point* $\lambda_0 \in \Lambda$, *which is not a pole,* $A(\lambda_0)$ *has a bounded inverse* $A(\lambda_0)^{-1}: H_2 \to H_1$.

Then the operator function $A(\lambda)^{-1}: H_2 \to H_1$ *is a finitely meromorphic Fredholm operator function in* Λ.

This statement is adapted from P. M. Blekher [1], Theorem 1. The proof is based on the results of I. C. Gohberg and M. G. Krein [1].

Sectorial Operators

The properties of sectorial operators in this subsection are adapted from T. Kato [1], Chapter V, Section 3 and Chapter VI, Section 2.

Let H be a Hilbert space. A linear operator $B: H \to H$ is said to be *m-accretive* if, for each $\operatorname{Re}\lambda > 0$, there is a bounded inverse operator $(B + \lambda I)^{-1}: H \to H$, and

$$\|(B + \lambda I)^{-1}\| \le (\operatorname{Re}\lambda)^{-1}, \tag{A.2}$$

where I is the identity operator in H. Denote $\Theta(B) = \{(Bu, u) : u \in \mathcal{D}(B),\ \|u\| = 1\}$. If an operator $B: H \to H$ is m-accretive, then B is closed, $\mathcal{D}(B)$ is dense in H, and $\Theta(B) \subset \{\lambda \in \mathbb{C} : \operatorname{Re}\lambda \ge 0\}$.

We say that a linear operator $B: H \to H$ is *quasi-m-accretive* if $B + \alpha I$ is m-accretive for some $\alpha \in \mathbb{R}$. An operator B is said to be *sectorial* if there are $\theta < \pi/2$ and $\gamma \in \mathbb{R}$ such that $\Theta(B) \subset \{\lambda \in \mathbb{C} : |\arg(\lambda - \gamma)| \le \theta\}$. A number γ is called a *vertex of sectorial operator* B. An operator $B: H \to H$ is said to be *m-sectorial* if it is sectorial and quasi-m-accretive.

We now consider a sesquilinear form $b[u, v]$ with domain $\mathcal{D}(b) \subset H$. A form b is said to be *symmetric* if $b[u, v] = \overline{b[v, u]}$. We define the *adjoint form* b^* by the formulas

$$b^*[u, v] = \overline{b[v, u]}, \qquad \mathcal{D}(b^*) = \mathcal{D}(b). \tag{A.3}$$

Clearly the forms

$$p = \frac{1}{2}(b + b^*), \qquad q = \frac{1}{2i}(b - b^*) \tag{A.4}$$

are symmetric, and

$$a = p + iq. \tag{A.5}$$

Denote $\Theta(b) = \{b[u] : u \in \mathcal{D}(b),\ \|u\| = 1\}$, where $b[u] = b[u, u]$. A form b is said to be *sectorial* if there are $\theta < \pi/2$ and $\gamma \in \mathbb{R}$ such that $\Theta(b) \subset \{\lambda \in \mathbb{C} : |\arg(\lambda - \gamma)| \le \theta\}$. A number γ is called a *vertex of sectorial form* b. A sectorial form b is said to be *closed* if the conditions $u_n \in \mathcal{D}(b)$, $\|u_n - u\| \to 0$, and $b[u_n - u_m] \to 0$ as $n, m \to \infty$ imply that $u \in \mathcal{D}(b)$ and $b[u_n - u] \to 0$ as $n \to \infty$.

Let b be a sectorial form. We introduce the inner product in $H_p = \mathcal{D}(b)$ by the formula

$$(u, v)_p = p[u, v] + (1 - \gamma)(u, v) \qquad ((u, v) \in \mathcal{D}(b)), \tag{A.6}$$

where γ is a vertex of b. A sectorial form b is closed in H if and only if a pre-Hilbert space H_p is complete (see Theorem 1.11 in T. Kato [1], Chapter VI, Section 1).

A linear subspace $\mathcal{D}' \subset \mathcal{D}(b)$ is called a *core* of form b if a restriction of b to $\mathcal{D}'$ has a closure which equals b.

Theorem A.10. *Let $b[u, v]$ $(u, v \in \mathcal{D}(b))$ be a densely defined, closed sectorial sesquilinear form in H, and let γ be a vertex of b.*

Then there is an m-sectorial operator $B: H \to H$ *such that*

(a) $\mathcal{D}(B) \subset \mathcal{D}(b)$, *and*

$$b[u,v] = (Bu, v) \qquad (u \in \mathcal{D}(B),\ v \in \mathcal{D}(b)); \tag{A.7}$$

(b) *the operator* $B + (1-\gamma)I: H \to H$ *has a bounded inverse* $(B+(1-\gamma)I)^{-1}$: $H \to H_p$;

(c) $\mathcal{D}(B)$ *is a core of* b;

(d) *if* $u \in \mathcal{D}(b)$, $f \in H$, *and the equality*

$$b[u,v] = (f, v) \tag{A.8}$$

holds for every v *belonging to a core of* b, *then* $u \in \mathcal{D}(B)$ *and* $Bu = f$.

The operator B *is uniquely determined by the condition* (a).

For a proof, see Theorem 2.1 in T. Kato [1], Chapter VI, Section 2.

We say that B is the m-sectorial operator associated with the form b. Denote $B = B_b$.

The next two statements follow from Theorem A.10 (see Theorems 2.5, 2.6 in T. Kato [1], Chapter VI, Section 2).

Theorem A.11. *Let the conditions of Theorem* 1.10 *be fulfilled, and let* $B = B_b$. *Then* $B_b^* = B_{b^*}$.

Theorem A.12. *Let* b *be a densely defined, symmetric closed form bounded from below.*

Then the operator $B = B_b$ *associated with* b *is self-adjoint and bounded from below. Furthermore, the operator* B *and the form* b *have the same lower bound.*

We denote by G a densely defined, sectorial operator. We consider the form $g[u,v] = (Gu, v)$ with $\mathcal{D}(g) = \mathcal{D}(G)$. Let b be a closure of the form g, and let $B = B_b$ be an m-sectorial operator associated with b. Since $\mathcal{D}(G)$ is a core of b, by Theorem A.10, $G \subset B$. The operator B is called a *Friedrichs extension* of G.

Using Theorem $A.12$, we obtain:

Theorem A.13. *Let* G *be a densely defined, symmetric operator bounded from below.*

Then the Friedrichs extension $B: H \to H$ *of* G *is self-adjoint. Furthermore, the operators* B *and* G *have the same lower bound.*

For a proof of Theorem A.13, see also N. Dunford and J. T. Schwartz [2], Chapter XII, Section 5.

We now state some auxiliary results concerning m-sectorial operators.

Theorem A.14. *Let a Hilbert space H_1 be dense in H, and let the imbedding operator of H_1 into H be compact. Assume that b is a sesquilinear form with domain $\mathcal{D}(b) = H_1$, and that*

$$|b[u,v]| \le c_0 \|u\|_{H_1} \|v\|_{H_1} \qquad (u, v \in H_1), \tag{A.9}$$

$$\operatorname{Re} b[u] \ge c_1 \|u\|_{H_1}^2 - c_2 \|u\|^2 \qquad (u \in H_1), \tag{A.10}$$

where $c_0, c_1 > 0$, $c_2 \ge 0$ are constants.

Then b is a closed sectorial form with a vertex $\gamma = -c_2$. The m-sectorial operator $B = B_b \colon H \to H$ associated with b has a discrete spectrum $\sigma(B_b)$, and $\sigma(B_b) \subset \{\lambda \in \mathbb{C} : \operatorname{Re} \lambda > -c_2\}$. If $\lambda \notin \sigma(B_b)$, then the resolvent $R(\lambda, B_b) \colon H \to H$ is a compact operator. Furthermore, $B_b^* = B_{b^*}$.

For a proof, we note that the inequalities (A.9), (A.10) imply that the form b is closed and sectorial. The remaining statements of Theorem A.14 follow from Theorems A.10, A.8, and A.11.

Theorem A.15. *Let the conditions of Theorem A.14 be fulfilled.*

Then the operator $B_b \colon H \to H$ is Fredholm, and $\operatorname{ind} B_b = 0$.

The proof follows from the compactness of resolvent $R(\lambda, B_b) \colon H \to H$ $(\lambda \notin \sigma(B_b))$, equality $B_b R(\lambda, B_b) = I + \lambda R(\lambda, B_b)$ and Theorem A.1.

Theorem A.16. *Let the conditions of Theorem A.14 be fulfilled, and let $b = b^*$.*

Then the operator $B_b \colon H \to H$ is self-adjoint. The spectrum $\sigma(B_b)$ consists of real isolated eigenvalues $\lambda_s > -c_2$ of finite multiplicity. The eigenfunctions $\{v_s\}$ of the operator B_b form an orthonormal basis in H. Moreover, the functions $\{v_s/\sqrt{\lambda_s + c_2}\}$ form an orthonormal basis in H_1 with inner product

$$(u, v)'_{H_1} = b[u, v] + c_2(u, v). \tag{A.11}$$

Proof. By virtue of Theorem A.14, the operator $B_b \colon H \to H$ is self-adjoint, and the spectrum $\sigma(B_b)$ consists of real isolated eigenvalues $\lambda_s > -c_2$ of finite multiplicity.

The eigenfunction problem

$$B_b v = \lambda v$$

is equivalent to the problem

$$v = (\lambda + c_2)(B_b + c_2 I)^{-1} v.$$

Theorem A.10 (b) and the Hilbert identity imply that a restriction of $(B_b + c_2 I)^{-1}$ to H_1 is a compact operator. Further, by Theorem A.10 (a), we have

$$\begin{aligned}
((B_b + c_2 I)^{-1} u, w)'_{H_1} &= b[(B_b + c_2 I)^{-1} u, w] + c_2((B_b + c_2 I)^{-1} u, w) \\
&= (u, w) = \overline{(w, u)} \\
&= \overline{b[(B_b + c_2 I)^{-1} w, u]} + c_2 \overline{((B_b + c_2 I)^{-1} w, u)} \\
&= (u, (B_b + c_2 I)^{-1} w)'_{H_1}
\end{aligned}$$

for all $u, w \in H_1$. Hence the operator $(B_b + c_2 I)^{-1}: H_1 \to H_1$ is self-adjoint. By the Hilbert–Schmidt theorem, there exists an orthogonal basis in H_1 consisting of eigenfunctions v_s of the operator $(B_b + c_2 I)^{-1}$ corresponding to eigenvalues $(\lambda_s + c_2)^{-1}$. Suppose $\|v_s\| = 1$. By virtue of (A.7), (A.11), we obtain

$$\begin{aligned}(v_s, v_s) = (v_s, v_r)'_{H_1}/(\lambda_s + c_2) &= 0 \qquad (s \neq r),\\ (v_s/\sqrt{\lambda_s + c_2}, v_s/\sqrt{\lambda_s + c_2})'_{H_1} = (v_s, v_s) &= 1 \qquad (s = 1, 2, \ldots).\end{aligned}$$

Hence the functions $v_s/\sqrt{\lambda_s + c_2}$ form an orthonormal basis in H_1. Since H_1 is dense in H, the functions v_s form an orthonormal basis in H. □

The Ritz Method

This approach enables the application of variational methods to the solution of the corresponding linear elliptic boundary value problems.

Let H, H_1 be real Hilbert spaces, and let the imbedding $H_1 \subset H$ be continuous.

We fix a function $f \in H$, and consider a functional

$$E(v) = \|v\|^2_{H_1} - 2(f, v)_H \qquad (v \in H_1). \tag{A.12}$$

There is a constant $c = c(f) > 0$ such that

$$E(v) \geq -c \qquad (v \in H_1). \tag{A.13}$$

Let $d = \inf_{v \in H_1} E(v)$.

The sequence $\{v_n\} \subset H_1$ is called the sequence minimizing the functional E on H_1 if $\lim_{n\to\infty} E(v_n) = d$.

Theorem A.17. *There is a unique function $u \in H_1$ which yields a minimum of the functional E on H_1. If $\{v_n\}$ is a sequence minimizing the functional E on H_1, then $\|v_n - u\|_{H_1} \to 0$ as $n \to \infty$.*

For a proof, see Theorem 4 in S. G. Mikhlin [1], Chapter I, Section 8.

Let $\{\varphi_n\}$ be a sequence of linearly independent vectors in H_1. Suppose that $\{\varphi_n\}$ is complete in H_1. Denote by R_k the linear manifold spanned over the vectors $\varphi_1, \ldots, \varphi_k$.

We find a vector v_k which yields a minimum of the functional E on R_k. By Theorem A.17, such a vector v_k exists. Let $v_k = c_1\varphi_1 + \ldots + c_k\varphi_k$. Denote

$$\begin{aligned}F(c_1, \ldots, c_k) &= E(c_1\varphi_1 + \ldots + c_k\varphi_k)\\ &= \sum_{i,j=1}^{k} c_i c_j (\varphi_i, \varphi_j)_{H_1} - 2\sum_{i=1}^{k} c_i (f, \varphi_i)_H.\end{aligned}$$

If a vector $(c_1, \dots, c_k)$ yields a minimum of the function F, then $\partial F/\partial c_i = 0$ $(i = 1, \dots, k)$. Hence

$$\sum_{j=1}^{k} c_j(\varphi_i, \varphi_j)_{H_1} - (f, \varphi_i)_H = 0. \tag{A.14}$$

Since the vectors $\varphi_1, \dots, \varphi_k$ are linearly independent, the determinant of the system $(A.14)$ is non-zero. Thus the system (A.14) has a unique solution $c_1^k, \dots, c_k^k$. Hence the function $v_k = c_1^k\varphi_1 + \dots + c_k^k\varphi_k$ realizes a minimum of the functional E on R_k. The sequence $\{v_k\}$ is called the *Ritz sequence* for the functional E with respect to the system $\varphi_1, \varphi_2, \dots$.

Theorem A.18. *Let $\{\varphi_k\}$ be an arbitrary linearly independent system of functions, whose linear span is everywhere dense in H_1. Then the Ritz sequence of the functional E, constructed with respect to the system $\{\varphi_k\}$, converges in H_1 to a function which yields a minimum of the functional E on H_1.*

For a proof, see Theorem 3 in S. G. Mikhlin [1], Chapter I, Section 8.

B Functional spaces

L_2-Spaces

An open connected subset Ω of $\mathbb{R}^n$ is called a *domain.* We consider the space of Lebesgue measurable functions f on a domain Ω such that $|f|^p$ is integrable on Ω $(1 \le p < \infty)$. This space is denoted by $L_p(\Omega)$. The space $L_p(\Omega)$ is a Banach space with the norm

$$\|f\|_{L_p(\Omega)} = \left\{ \int_\Omega |f(x)|^p \, dx \right\}^{1/p}.$$

If $p = 2$, then we obtain the Hilbert space $L_2(\Omega)$ with the inner product

$$(f, g)_{L_2(\Omega)} = \int_\Omega f(x)\overline{g(x)} \, dx.$$

Let $L_{p,\text{loc}}(\Omega)$ be the space of Lebesgue measurable functions f on Ω such that $f \in L_p(K)$ for every compact subset K of Ω.

C^k-Spaces

Denote by $C(\Omega)$ (or $C(\overline{\Omega})$) the space of continuous functions on Ω (or on $\overline{\Omega}$).

Let $C^k(\Omega)$ be the space of functions in $C(\Omega)$ having continuous derivatives of order $\le k$ in Ω. Let $C^k(\overline{\Omega})$ be the space of functions in $C^k(\Omega)$ all of whose derivatives of order $\le k$ have continuous extensions to $\overline{\Omega}$.

We denote

$$C^0(\Omega) = C(\Omega), \qquad C^0(\overline{\Omega}) = C(\overline{\Omega}),$$

$$C^\infty(\Omega) = \bigcap_{k=0}^{\infty} C^k(\Omega), \qquad C^\infty(\overline{\Omega}) = \bigcap_{k=0}^{\infty} C^k(\overline{\Omega}).$$

If Ω is bounded and $0 \le k < \infty$, then the space $C^k(\overline{\Omega})$ is a Banach space with the norm

$$\|\varphi\|_{C^k(\overline{\Omega})} = \max_{|\alpha|\le k} \max_{x\in\overline{\Omega}} |\mathcal{D}^\alpha \varphi(x)|.$$

Here $\mathcal{D}^\alpha = \mathcal{D}_1^{\alpha_1} \cdots \mathcal{D}_n^{\alpha_n}$, $\mathcal{D}_j = -i(\partial/\partial x_j)$, $\alpha = (\alpha_1, \dots, \alpha_n)$, $|\alpha| = \alpha_1 + \dots + \alpha_n$, $\alpha_j \ge 0$.

A vector α with non-negative integer coordinates is called a *multi-index.*

Let φ be a continuous function on Ω. A closure of the set $\{x \in \Omega : \varphi(x) \neq 0\}$ in Ω is called a *support* of φ. We denote the support of φ by $\operatorname{supp}\varphi$.

Let $\dot{C}^\infty(\Omega)$ be a set of functions in $C^\infty(\Omega)$ with compact supports in Ω.

Lemma B.1. *Let K be a compact in $\mathbb{R}^n$, and let $\{S_\alpha\}$ be an open covering of K.*

Then there exist a finite subcovering $\{S_{\alpha_j}\}$ and non-negative functions $\varphi_j \in \dot{C}^\infty(\mathbb{R}^n)$ $(j = 1, \dots, q)$ such that:
(a) $\sum_{j=1}^q \varphi_j(x) \le 1$ $(x \in \mathbb{R}^n)$,
(b) $\sum_{j=1}^q \varphi_j(x) = 1$ $(x \in K)$,
(c) $\operatorname{supp}\varphi_j \subset S_{\alpha_j}$ $(j = 1, \dots, q)$.

This family of functions $\{\varphi_j\}$ is called a *partition of unity.*

For a proof, see N. Dunford and J. T. Schwartz [2], Chapter XIV, Section 2, Lemma 4.

Distributions

Let Ω be a domain in $\mathbb{R}^n$. We define a convergence in the space $\dot{C}^\infty(\Omega)$ in the following way:

A sequence $\{\varphi_s\} \subset \dot{C}^\infty(\Omega)$ converges to an element $\varphi \in \dot{C}^\infty(\Omega)$ if there is a compact $K \subset \Omega$ such that $\operatorname{supp}\varphi_s \subset K$ and

$$\lim_{s\to\infty} \|\varphi - \varphi_s\|_{C^k(\overline{\Omega})} = 0$$

for each $k = 0, 1, 2, \dots$.

Denote by $\mathcal{D}(\Omega)$ the linear space $\dot{C}^\infty(\Omega)$ with such convergence.

One can give an equivalent definition of convergence in $\mathcal{D}(\Omega)$, considering $\mathcal{D}(\Omega)$ as a locally convex linear topological space (see K. Yosida [1], Chapter I, Section 1).

A linear functional f on $\mathcal{D}(\Omega)$ is called a *distribution* on Ω if the convergence $\varphi_s \to \varphi$ in $\mathcal{D}(\Omega)$ implies that $\langle f, \varphi_s\rangle \to \langle f, \varphi\rangle$. Denote by $\mathcal{D}'(\Omega)$ the space of distributions with weak convergence.

By virtue of the Schwarz inequality, every function $f \in L_{1,\mathrm{loc}}(\Omega)$ defines a distribution F on Ω by the formula

$$\langle F, \varphi \rangle = \int_\Omega f(x)\varphi(x)\,dx \qquad (\varphi \in \mathcal{D}(\Omega)). \tag{B.1}$$

We say that a distribution defined by (B.1) is a function. In this case we identify the distribution F with the function f.

Let us define some operations on distributions:

1) The derivative $\mathcal{D}^\alpha f$ of a distribution $f \in \mathcal{D}'(\Omega)$ is a distribution on Ω defined by
$$\langle \mathcal{D}^\alpha f, \varphi \rangle = (-1)^{|\alpha|}\langle f, \mathcal{D}^\alpha \varphi \rangle \qquad (\varphi \in \mathcal{D}(\Omega)).$$

2) The product af of a function $a \in C^\infty(\Omega)$ and a distribution $f \in \mathcal{D}'(\Omega)$ is the distribution on Ω defined by
$$\langle af, \varphi \rangle = \langle f, a\varphi \rangle \qquad (\varphi \in \mathcal{D}(\Omega)).$$

3) Let $Q \subset \Omega$ be an open set. The restriction $f|_Q$ of f to Q is the distribution on Ω defined by
$$\langle f|_Q, \varphi \rangle = \langle f, \varphi \rangle \qquad (\varphi \in \mathcal{D}(Q)).$$

A linear mapping A from $\mathcal{D}(\Omega)$ (or $\mathcal{D}'(\Omega)$) into itself is said to be continuous if from the convergence $\varphi_s \to 0$ in $\mathcal{D}(\Omega)$ (or $\mathcal{D}'(\Omega)$) it follows that $A\varphi_s \to 0$ in $\mathcal{D}(\Omega)$ (or $\mathcal{D}'(\Omega)$).

Evidently distributions are infinitely differentiable. The operators of differentiation $\mathcal{D}^\alpha$ and multiplication by a function $a \in C^\infty(\Omega)$ are continuous in the space $\mathcal{D}(\Omega)$ (or $\mathcal{D}'(\Omega)$).

We say that a distribution f is equal to zero in an open set $Q \subset \Omega$, if $\langle f, \varphi \rangle = 0$ for all $\varphi \in \mathcal{D}(Q)$. Two distributions f_1, f_2 on Ω are said to be equal in Q if $f_1 - f_2 = 0$ in Q.

By definition, the *support of a distribution* $f \in \mathcal{D}'(\Omega)$ is the smallest closed subset of Ω outside of which f is zero. We denote the support of f by $\operatorname{supp} f$.

Tempered Distributions

Denote by $S(\mathbb{R}^n)$ the linear space of functions $\varphi \in C^\infty(\mathbb{R}^n)$ such that

$$\sup_{x \in \mathbb{R}^n} |x^\beta \mathcal{D}^\alpha \varphi(x)| = C_{\alpha\beta} < \infty$$

for all multi-indices α, β, where $x^\beta = x_1^{\beta_1} \cdots x_n^{\beta_n}$.

We define a convergence in the space $S(\mathbb{R}^n)$ in the following way:

A sequence $\{\varphi_s\} \subset S(\mathbb{R}^n)$ converges to an element $\varphi \in S(\mathbb{R}^n)$ if $\lim_{s\to\infty} \|x^\beta \mathcal{D}^\alpha \varphi - x^\beta \mathcal{D}^\alpha \varphi_s\|_{C(\mathbb{R}^n)} = 0$ for all α, β.

A linear functional f on $S(\mathbb{R}^n)$ is called a *tempered distribution* on $\mathbb{R}^n$ if the convergence $\varphi_s \to \varphi$ in $S(\mathbb{R}^n)$ implies that $\langle f, \varphi_s \rangle \to \langle f, \varphi \rangle$. Denote by $S'(\mathbb{R}^n)$ the space of tempered distributions with weak convergence.

Similarly to the space $\mathcal{D}'(\Omega)$ we can define the operations of differentiation and restriction.

Suppose that $a \in C^\infty(\mathbb{R}^n)$ and for every α there are an integer m_α and a positive number c_α such that

$$|\mathcal{D}^\alpha a(x)| \le c_\alpha (1+|x|)^{m_\alpha}.$$

Then we define a product of a function a and a distribution $f \in S'(\mathbb{R}^n)$ by the formula

$$\langle af, \varphi\rangle = \langle f, a\varphi\rangle \qquad (\varphi \in S(\mathbb{R}^n)).$$

Clearly $S'(\mathbb{R}^n) \subset \mathcal{D}'(\mathbb{R}^n)$.

We define the Dirac δ-distribution as follows:

$$\langle \delta, \varphi\rangle = \varphi(0) \qquad (\varphi \in S(\mathbb{R}^n)).$$

The Fourier Transform

We define the *Fourier transform* $\widehat{f}$ of a function $f \in L_1(\mathbb{R}^n)$ by the formula

$$\widehat{f}(\xi) = (2\pi)^{-n/2} \int_{\mathbb{R}^n} f(x) \exp(-i(x,\xi))\, dx \qquad (\xi \in \mathbb{R}^n), \tag{B.2}$$

where $(x,\xi) = x_1\xi_1 + \cdots + x_n\xi_n$.

If $g \in L_1(\mathbb{R}^n)$, we define the *inverse Fourier transform* of g by the formula

$$\check{g}(x) = (2\pi)^{-n/2} \int_{\mathbb{R}^n} g(\xi) \exp(i(x,\xi))\, d\xi \qquad (x \in \mathbb{R}^n). \tag{B.3}$$

We denote $\widehat{f}$ by Ff and $\check{g}$ by $F^{-1}g$.

Theorem B.1.

(a) *The transforms F and F^{-1} map $S(\mathbb{R}^n)$ continuously onto itself, and $FF^{-1} = F^{-1}F = I$ on $S(\mathbb{R}^n)$.*

(b) *If $\varphi \in S(\mathbb{R}^n)$, then*

$$(\widehat{\mathcal{D}^\alpha\varphi})(\xi) = \xi^\alpha \widehat{\varphi}(\xi),$$
$$\mathcal{D}^\beta \widehat{\varphi}(\xi) = ((\widehat{-x)^\beta\varphi})(\xi)$$

for all multi-indices α, β.

(c) *If $\varphi, \psi \in S(\mathbb{R}^n)$, then*

$$\int_{\mathbb{R}^n} \varphi(x)\psi(x)\, dx = \int_{\mathbb{R}^n} \widehat{\varphi}(\xi)\widehat{\psi}(\xi)\, d\xi, \tag{B.4}$$

$$\int_{\mathbb{R}^n} \varphi(x)\widehat{\psi}(x)\, dx = \int_{\mathbb{R}^n} \widehat{\varphi}(\xi)\psi(\xi)\, d\xi. \tag{B.5}$$

We define the Fourier transform Fu of $u \in S'(\mathbb{R}^n)$ by the formula

$$\langle Fu, \varphi\rangle = \langle u, F\varphi\rangle \qquad (\varphi \in S(\mathbb{R}^n)). \tag{B.6}$$

By virtue of (B.5), the above definition (B.6) agrees with definition (B.2) if $u \in S(\mathbb{R}^n)$. We also denote Fu by $\widehat{u}$.

Analogously, we define the inverse Fourier transform $F^{-1}v$ of $v \in S'(\mathbb{R}^n)$ by the formula

$$\langle F^{-1}v, \psi\rangle = \langle v, F^{-1}\psi\rangle \qquad (\psi \in S(\mathbb{R}^n)). \tag{B.7}$$

Denote $F^{-1}v$ by $\check{v}$.

Theorem B.2.

(a) *The transforms F and F^{-1} map $S'(\mathbb{R}^n)$ continuously onto itself, and $FF^{-1} = F^{-1}F = I$ on $S'(\mathbb{R}^n)$.*

(b) *If $u \in S'(\mathbb{R}^n)$, then*

$$\begin{aligned} \widehat{D^\alpha u} &= \xi^\alpha \widehat{u}, \\ D^\beta \widehat{u} &= (\widehat{(-x)^\beta u}). \end{aligned}$$

Theorem B.3 (Plancherel). *Let $f \in L_2(\mathbb{R}^n)$.*

Then the Fourier transform Ff in a sense of distributions defines a function $\widehat{f} \in L_2(\mathbb{R}^n)$. This function has the form

$$\widehat{f}(\xi) = \lim_{R\to\infty} (2\pi)^{-n/2} \int_{|x|\le R} f(x)\exp(-i(x,\xi))\,dx \qquad (\textit{in } L_2(\mathbb{R}^n)), \tag{B.8}$$

and

$$\|\widehat{f}\|_{L_2(\mathbb{R}^n)} = \|f\|_{L_2(\mathbb{R}^n)}, \tag{B.9}$$

$$f(x) = \lim_{R\to\infty} (2\pi)^{-n/2} \int_{|\xi|\le R} \widehat{f}(\xi)\exp(i(x,\xi))\,d\xi \qquad (\textit{in } L_2(\mathbb{R}^n)). \tag{B.10}$$

From this theorem follows:

Theorem B.4. *The Fourier transform maps $L_2(\mathbb{R}^n)$ onto itself in a one-to-one manner, and*

$$(f,g)_{L_2(\mathbb{R}^n)} = (\widehat{f},\widehat{g})_{L_2(\mathbb{R}^n)} \tag{B.11}$$

for all $f, g \in L_2(\mathbb{R}^n)$.

For the proofs of Theorems B.1–B.3 the reader is referred to K. Yosida [1], Chapter VI, Sections 1,2.

Sobolev Spaces

The *Sobolev space* $W^s(\mathbb{R}^n)$ $(s \in \mathbb{R})$ is the space of distributions $u \in S'(\mathbb{R}^n)$ such that $(1+|\xi|^2)^{s/2}\widehat{u} \in L_2(\mathbb{R}^n)$ with the norm

$$\|u\|_{W^s(\mathbb{R}^n)} = \left(\int_{\mathbb{R}^n} (1+|\xi|^2)^s |\widehat{u}(\xi)|^2\,d\xi\right)^{1/2}. \tag{B.12}$$

We denote by Ω a bounded domain $Q \subset \mathbb{R}^n$ or a half-space $\mathbb{R}^n_+$, where $\mathbb{R}^n_+ = \{x \in \mathbb{R}^n : x_n > 0\}$, $\mathbb{R}^n_- = \{x \in \mathbb{R}^n : x_n < 0\}$.

Let m be a positive integer. The *Sobolev space* $W^m(\Omega)$ is the space of distributions $u \in \mathcal{D}'(\Omega)$ such that $\mathcal{D}^\alpha u \in L_2(\Omega)$ $(|\alpha| \le m)$ with the norm

$$\|u\|_{W^m(\Omega)} = \left\{ \sum_{|\alpha| \le m} \int_\Omega |\mathcal{D}^\alpha u(x)|^2 \, dx \right\}^{1/2}. \tag{B.13}$$

Let $s = m + \sigma$, where m is a positive integer and $0 < \sigma < 1$. The *Sobolev space* $W^s(\Omega)$ is the space of functions $u \in W^m(\Omega)$ with finite norm

$$\|u\|_{W^s(\Omega)} = \left\{ \|u\|^2_{W^m(\Omega)} + \sum_{|\alpha| = m} \int_\Omega \int_\Omega \frac{|\mathcal{D}^\alpha u(x) - \mathcal{D}^\alpha u(y)|^2}{|x-y|^{n+2\sigma}} \, dx \, dy \right\}^{1/2}. \tag{B.14}$$

The Sobolev spaces $W^s(\Omega)$ and $W^s(\mathbb{R}^n)$ $(s \ge 0)$ are Hilbert spaces. The formulas (B.12) and (B.13), (B.14) with $\Omega = \mathbb{R}^n$ give the equivalent norms in $W^s(\mathbb{R}^n)$ for $s \ge 0$ (see L. N. Slobodetskiĭ [1], Chapter II, Section 2, Lemma 3).

Denote by $W^s_{\text{loc}}(\Omega)$ $(s \ge 0)$ a space of distributions $\{u \in \mathcal{D}'(\Omega) : u \in W^s(K)\}$ for each domain K such that $\overline{K} \subset \Omega$.

We now consider the extension theorem.

Theorem B.5. *Let $s \ge 0$. Suppose that $\Omega \subset \mathbb{R}^n$ is a domain satisfying one of the following conditions*:
(a) $\Omega = R^n_+$;
(b) *Ω is a bounded domain with boundary $\partial\Omega \in C^\infty$*;
(c) *$\Omega = (0, d) \times G$, where $G \subset \mathbb{R}^{n-1}$ is a bounded domain (with boundary $\partial G \in C^\infty$ if $n \ge 3$).*
Then, for every $u \in W^s(\Omega)$, there are $U \in W^s(\Omega)$ such that $u(x) = U(x)$ $(x \in \Omega)$, and

$$\|U\|_{W^s(\mathbb{R}^n)} \le c\|u\|_{W^s(\Omega)}, \tag{B.15}$$

where $c > 0$ does not depend on u.

For a proof, see J. L. Lions and E. Magenes [1], Chapter I, Section 8, Theorem 8.1 and L. N. Slobodetskiĭ [1], Chapter II.

Therefore, if Ω satisfies the conditions of Theorem B.5, then $W^s(\Omega)$ $(s \ge 0)$ is the space of restrictions $U|_\Omega$ $(U \in W^s(\mathbb{R}^n))$ with equivalent norm

$$\|u\|_{W^s(\Omega)} = \inf \|U\|_{W^s(\mathbb{R}^n)} \qquad (U \in W^s(\mathbb{R}^n) : U|_\Omega = u). \tag{B.16}$$

Let m be a positive integer. We denote by $W^{-m}(\Omega)$ the space of distributions $u \in \mathcal{D}'(\Omega)$ such that

$$\|u\|_{W^{-m}(\Omega)} = \sup_{\varphi \in \dot{C}^\infty(\Omega)} \frac{|\langle u, \varphi \rangle|}{\|\varphi\|_{W^m(\Omega)}} < \infty. \tag{B.17}$$

The space $W^{-m}(\Omega)$ is a Hilbert space dual to $\mathring{W}^m(\Omega)$, where $\mathring{W}^m(\Omega)$ is a closure of the set $\dot{C}^\infty(\Omega)$ in $W^m(\Omega)$ (see K. Yosida, Chapter III, Section 10).

Trace Theorems

Let $\Omega \subset \mathbb{R}^n$ be a bounded domain Q with boundary $\partial Q \in C^\infty$ or $\Omega = (0,d) \times G$, where $G \subset \mathbb{R}^{n-1}$ is a bounded domain (with boundary $\partial G \in C^\infty$ if $n \geq 3$). Let M be a boundary $\partial Q \in C^\infty$ of a bounded domain $Q \subset \mathbb{R}^n$ or $(n-1)$-dimensional C^∞ manifold $\Gamma \subset \overline{\Omega}$ with C^∞ boundary $\partial\Gamma$. For each finite open covering $\{U_j\}$ of $\overline{M}$, there exists a partition of unity $\{\varphi_j\}$ such that $\sum_{j=1}^N \varphi_j(x) = 1$ $(x \in \overline{M})$, $\varphi_j \in \dot{C}^\infty(\mathbb{R}^n)$, $\operatorname{supp}\varphi_j \subset U_j$. Let $x \to \omega_j(x) = y$ be a C^∞ diffeomorphism from U_j onto V_j such that

1) $\omega_j(U_j \cap M) \subset \{y : y_n = 0\}$ if $M = \partial Q$ or if $M = \Gamma$, $U_j \cap \partial\Gamma = \emptyset$,
2) $\omega_j(U_j \cap M) \subset \{y : y_{n-1} > 0,\ y_n = 0\}$, $\omega_j(U_j \cap \partial M) \subset \{y : y_{n-1} = 0,\ y_n = 0\}$ if $M = \Gamma$, $U_j \cap \partial\Gamma \neq \emptyset$.

Let $W^s(M)$ be a *Sobolev space* of functions u such that, in local coordinates $y = y^j$, we have $\varphi_j u \in W^s(\omega_s(U_j \cap M))$. The norm in $W^s(M)$ is introduced as follows:

$$\|u\|_{W^s(M)} = \left\{ \sum_{j=1}^{M} \|\varphi_j u\|^2_{W^s(\omega_j(U_j \cap M))} \right\}^{1/2}, \tag{B.18}$$

where the norms $\|\varphi_j u\|_{W^s(\omega_j(U_j \cap M))}$ are calculated in the coordinates $y = y^j$, $s \geq 0$.

Clearly this definition does not depend on the covering U_j or the choice of transformations ω_j.

A space $W^s(M)$ is a Hilbert space.

Denote by $C^\infty(\overline{M})$ a space of functions $u \in C(\overline{M})$ such that in local coordinates $y = y^j$ we have $u \in C^\infty(\omega_j(U_j \cap \overline{M}))$.

We define the trace map

$$\gamma_\mu : C^\infty(\overline{\Omega}) \to C^\infty(\overline{M})$$

by the formula

$$\gamma_\mu u = \mathcal{D}^\mu_\nu u|_{\overline{M}},$$

where μ is a non-negative integer, ν is the vector of the normal to $\overline{M}$ at the point $x \in \overline{M}$.

Theorem B.6. *Let $s > 0$, and let $m < s + 1/2$ be a non-negative integer.*

Then the trace map

$$\gamma = \{\gamma_0, \gamma_1, \dots, \gamma_{m-1}\} : C^\infty(\overline{\Omega}) \to \prod_{\mu=0}^{m-1} C^\infty(\overline{M})$$

extends uniquely to a continuous linear map

$$\gamma : W^s(\Omega) \to \prod_{\mu=0}^{m-1} W^{s-\mu-1/2}(M).$$

Furthermore, this map is surjective, and there exists a linear continuous operator $S: \prod_{\mu=0}^{m-1} W^{s-\mu-1/2}(M) \to W^s(\Omega)$ *such that* $\gamma S g = g$ *for all* $g = (g_0, g_1, \ldots, g_{m-1}) \in \prod_{\mu=0}^{m-1} W^{s-\mu-1/2}(M)$.

A proof is based on a similar result for $\Omega = \mathbb{R}^n_+$, $M = \mathbb{R}^{n-1}$ and partition of unity (see L. N. Slobodetskiĭ [1], Chapter II, Theorem 7 in Section 4 and Theorem 10 in Section 5, J. L. Lions and E. Magenes [1], Chapter I, Section 8, Theorem 8.3).

Denote $\gamma_0 u = u|_M$ for every $u \in W^s(Q)$, where $s > 1/2$. A function $u|_M$ is called a *trace* of u.

Imbedding Theorems

The following theorem states that the elements of the Sobolev space $W^s(Q)$ are differentiable in the classical sense for sufficiently large $s > 0$.

Theorem B.7 (Sobolev). *Let* $\Omega \subset \mathbb{R}^n$ *be a bounded domain* Q *with boundary* $\partial Q \in C^\infty$ *or* $\Omega = (0, d) \times G$, *where* $G \subset \mathbb{R}^{n-1}$ *is a bounded domain (with boundary* $\partial G \in C^\infty$ *if* $n \geq 3$). *Let* $k > n/2 + s$, *where* $s > 0$ *is an integer.*

Then we have the inclusion

$$W^k(Q) \subset C^s(\overline{Q}),$$

and the imbedding operator is continuous.

A proof follows from Theorem 9.8 in J. L. Lions and E. Magenes [1], Chapter I, Section 9 and Theorem B.5.

The next two theorems enable the reduction of the boundary value problems for elliptic equations to Fredholm equations.

Theorem B.8 (Rellich). *Let* $\Omega \subset \mathbb{R}^n$ *be a bounded domain satisfying the conditions of Theorem* B.7, *and let* $k > s \geq 0$.

Then the imbedding operator from $W^k(Q)$ *into* $W^s(Q)$ *is compact.*

A proof follows from Theorem 16.1 in J. L. Lions – E. Magenes [1], Chapter I, Section 16 and Theorem B.5.

Using Theorem B.8, we obtain:

Theorem B.9. *Let* M *be a boundary* $\partial Q \in C^\infty$ *of a bounded domain* $Q \subset \mathbb{R}^n$ *or* $(n-1)$*-dimensional bounded* C^∞ *manifold* Γ *with* C^∞ *boundary* $\partial\Gamma$. *Let* $k > s \geq 0$.

Then the imbedding operator from $W^k(M)$ *into* $W^s(M)$ *is compact.*

The Spaces $\mathring{W}^s(Q)$

The following theorem gives an explicit description of the space $\mathring{W}^m(Q)$.

Theorem B.10. *Let $\Omega \subset \mathbb{R}^n$ be a bounded domain satisfying the conditions of Theorem* B.7, *and let m be a positive integer.*

Then $\mathring{W}^m(\Omega) = \{u \in W^m(\Omega) : \mathcal{D}_\nu^{\mu-1} u|_{\partial Q\setminus K} = 0,\ \mu = 1,\dots,m\}$, where $K = \emptyset$ if $\Omega = Q$ and $K = (\{0\} \times \partial G) \cup (\{d\} \times \partial G)$ if $\Omega = (0,d) \times G$.

For a proof, see J. L. Lions and E. Magenes [1], Chapter I, Section 11, Theorem 11.5.

The next theorem on the equivalent norms in the space $\mathring{W}^m(Q)$ is very useful for the study of the Dirichlet problem.

Theorem B.11. *Let $\Omega \subset \mathbb{R}^n$ be a bounded domain satisfying the conditions of Theorem* B.7, *and let m be a positive integer.*

Then the norm (B.13) *in the space $\mathring{W}^m(\Omega)$ is equivalent to the norm*

$$\left\{ \sum_{j=0}^{n} \int_\Omega |\mathcal{D}_j^m u(x)|^2 \, dx \right\}^{1/2}. \tag{B.19}$$

A proof follows from Theorem 2 in S. L. Sobolev [1], Chapter I, Section 9.

Interpolation

The next two theorems are applied to the investigation of a priori estimates of solutions of elliptic problems.

Theorem B.12. *Let $Q \subset \mathbb{R}^n$ be a bounded domain satisfying the conditions of Theorem* B.7.

Then, for any $u \in W^k(Q)$ and $q \in \mathbb{C}$,

$$|q|^{k-s} \|u\|_{W^s(Q)} \le c(\|u\|_{W^k(Q)} + |q|^k \|u\|_{L_2(Q)}), \tag{B.20}$$

where $0 < s < k$ and $c > 0$ does not depend on u and q.

Theorem B.13. *Let $Q \subset \mathbb{R}^n$ be a bounded domain satisfying the conditions of Theorem* B.7. *If $\partial Q \in C^\infty$, we let $M = \partial Q$. If $Q = (0,d) \times G$, we let $M = \{b\} \times G$, where $0 \le b \le d$.*

Then, for any $u \in W^1(Q)$ and $q \in \mathbb{C}$,

$$|q|^{1/2} \|u|_M\|_{L_2(M)} \le c_1(\|u\|_{W^1(Q)} + |q| \|u\|_{L_2(Q)}), \tag{B.21}$$

where $c_1 > 0$ does not depend on u and q.

Theorems B.12, B.13 are adapted from M. S. Agranovich and M. I. Vishik [1], Chapter I, Section 1.

Having some properties of linear operators in Sobolev spaces, we can extend them to the intermediate spaces. Let B_1, B_2 be Banach spaces. Denote by $B(B_1, B_2)$ a normed linear space of bounded linear operators from B_1 into B_2.

Theorem B.14. *Let* $A \in \bigcap_{i=1,2} B(W^{s_i}(\mathbb{R}^n), W^{s_i}(\mathbb{R}^m))$, *where* $0 \leq s_1 < s_2$.
Then $A \in B(W^{(1-\theta)s_1+\theta s_2}(\mathbb{R}^n), W^{(1-\theta)s_1+\theta s_2}(\mathbb{R}^m))$ *for each* $0 < \theta < 1$.

Theorem B.15. *Let* $Q \subset \mathbb{R}^n$ *be a bounded domain with boundary* $\partial Q \in C^\infty$. *Let* $A \in \bigcap_{i=1,2} B(W^{s_i}(Q), W^{s_i}(\partial Q))$, *where* $0 \leq s_1 < s_2$.
Then $A \in B(W^{(1-\theta)s_1+\theta s_2}(Q), W^{(1-\theta)s_1+\theta s_2}(\partial Q))$ *for each* $0 < \theta < 1$.

Theorems B.14, B.15 follow from Theorems 5.1, 7.7, 9.6 in J. L. Lions and E. Magenes [1], Chapter I.

Differential and Difference Operators

Let $Q \subset \mathbb{R}^n$ be a bounded domain with connected boundary ∂Q such that $0 \in Q$ and for each point $x = (x', x_n) \in Q$ there is a point $(x', -x_n) \in Q$. Let $Q_\delta = \{x \in Q : \rho(x, \partial Q) > \delta\}$ be a domain, where $\delta > 0$ is sufficiently small, $\rho(x, \partial Q) = \inf_{y \in \partial Q} |x - y|$. Denote $Q^+ = Q \cap \{x_n > 0\}$.

Denote

$$\delta_h^k u(x) = (u(x_1, \dots, x_{k-1}, x_k + h, x_{k+1}, \dots, x_n) - u(x))/h, \tag{B.22}$$

where $u \in L_2(Q^+)$, $\operatorname{supp} u \subset \overline{Q^+ \cap Q_\delta}$, $k = 1, \dots, n-1$, $h \neq 0$ is sufficiently small.

Clearly the operators δ_h^k and $-\delta_{-h}^k$ are formally adjoint, i.e.,

$$(\delta_h^k u, v)_{L_2(Q)} = -(u, \delta_{-h}^k v)_{L_2(Q)} \tag{B.23}$$

for all $u, v \in L_2(Q)$ such that $\operatorname{supp} u, \operatorname{supp} v \subset \overline{Q^+ \cap Q_\delta}$.

Theorem B.16. *Let* $u \in L_2(Q^+)$, *and let* $\operatorname{supp} u \subset \overline{Q^+ \cap Q_\delta}$.

(a) *If a distribution* $u_{x_k} \in \mathcal{D}'(Q^+)$ *is such that* $u_{x_k} \in L_2(Q^+)$ $(k < n)$, *then for all sufficiently small* $h \neq 0$

$$\|\delta_h^k u\|_{L_2(Q^+)} \leq \|u_{x_k}\|_{L_2(Q^+)}, \tag{B.24}$$

and

$$\|\delta_h^k u - u_{x_k}\|_{L_2(Q^+)} \to 0 \quad \text{as } h \to 0. \tag{B.25}$$

(b) *If there exists a constant* $c > 0$ *such that for all sufficiently small* $h \neq 0$

$$\|\delta_h^k u\|_{L_2(Q^+)} \leq c \qquad (k < n),$$

then $u_{x_k} \in L_2(Q^+)$, *and we obtain the relation* (B.25) *and the inequality*

$$\|u_{x_k}\|_{L_2(Q^+)} \leq c. \tag{B.26}$$

For a proof, see N. Dunford and J. T. Schwartz [2], Chapter XIV, Section 3, Lemma 50 and V. P. Mikhailov [2], Chapter III, Section 3, Theorem 4.

The One-Dimensional Case

Now we make some remarks on functional spaces in the one-dimensional case.

Denote by $L_p(a,b)$ the space of Lebesgue measurable functions f on (a,b) $(1 \le p < \infty)$. We equip this space with the norm

$$\|f\|_{L_p(a,b)} = \left\{ \int_a^b |f(x)|^p \, dx \right\}^{1/p}.$$

Denote by $C(a,b)$ (or $C[a,b]$) the space of continuous functions on (a,b) (or on $[a,b]$). Let $C^k(a,b)$ (or $C^k[a,b]$) be the space of functions in $C(a,b)$ (or $C[a,b]$) having continuous derivatives of order $\le k$ on (a,b) (or $[a,b]$). We equip the space $C^k[a,b]$ with the norm

$$\|\varphi\|_{C^k[a,b]} = \max_{i \le k} \max_{x \in [a,b]} |\varphi^{(i)}(x)|.$$

As in the multidimensional case we can also introduce the spaces $C^\infty(a,b)$, $C^\infty[a,b]$, $\dot{C}^\infty(a,b)$.

Denote by $\mathcal{D}(a,b)$ the space $\dot{C}^\infty(a,b)$, and by $\mathcal{D}'(a,b)$ the space of distributions on $\mathcal{D}(a,b)$.

For a positive integer k, the space $W^k(a,b)$ is the space of absolutely continuous functions $u(x)$ on $[a,b]$ having absolutely continuous derivatives $u^{(i)}(x)$ on $[a,b]$ for $i \le k-1$ such that $u^{(k)}(x) \in L_2(a,b)$. The norm in $W^k(a,b)$ will have the form

$$\left\{ \sum_{i \le k} \int_a^b |u^{(i)}(x)|^2 \, dx \right\}^{1/2}. \tag{B.27}$$

Thus the trace Theorem B.6 become trivial for $n = 1$.

Let k, s be integers, and let $0 \le s < k$. Then, by virtue of Theorem B.7, for every $u \in W^k(a,b)$

$$\|u\|_{C^s[a,b]} \le c\|u\|_{W^k(a,b)}, \tag{B.28}$$

where $c > 0$. Theorem B.8 implies that the imbedding operator from $W^k(a,b)$ into $W^s(a,b)$ is compact.

We note that

$$\mathring{W}^k(a,b) = \{u \in W^k(a,b) : u^{(i)}(a) = u^{(i)}(b) = 0, \ 0 \le i \le k-1\},$$

where k is a positive integer.

By Theorem B.11, we can introduce the equivalent norm in the space $\mathring{W}^k(a,b)$ by the formula

$$\left\{ \int_a^b |u^{(k)}(x)|^2 \, dx \right\}^{1/2}. \tag{B.29}$$

Theorems B.12, B.13 imply that:

For any $u \in W^k(a,b)$ and $q \in \mathbb{C}$,

$$|q|^{k-s}\|u\|_{W^s(a,b)} \leq c(\|u\|_{W^k(a,b)} + |q|^k\|u\|_{L_2(a,b)}). \tag{B.30}$$

where $0 < s < k$ are integers and $c > 0$ does not depend on u and q;

For any $u \in W^1(a,b)$ and $q \in \mathbb{C}$,

$$|q|^{1/2}|u(d)| \leq c_1(\|u\|_{W^1(a,b)} + |q| \cdot \|u\|_{L_2(a,b)}), \tag{B.31}$$

where $a \leq d \leq b$ and $c_1 > 0$ does not depend on u and q.

C Elliptic Problems

The Dirichlet Problem

We consider the equation

$$\mathcal{A}u = \sum_{|\alpha|,|\beta| \leq m} \mathcal{D}^\alpha a_{\alpha\beta}(x)\mathcal{D}^\beta u(x) = f_0(x) \qquad (x \in Q) \tag{C.1}$$

with the Dirichlet boundary conditions

$$\mathcal{D}_\nu^{\mu-1}u|_{\partial Q} = 0 \qquad (x \in \partial Q,\ \mu = 1, \dots, m), \tag{C.2}$$

where $a_{\alpha\beta} \in C^\infty(\mathbb{R}^n)$, $f_0 \in L_2(Q)$ are complex-valued functions, Q is a bounded domain with boundary $\partial Q \in C^\infty$, and ν is the inner unit normal to ∂Q at the point $x \in \partial Q$.

The equation (C.1) is said to be *strongly elliptic* in $\overline{Q}$ if

$$\operatorname{Re} \sum_{|\alpha|,|\beta| = m} a_{\alpha\beta}(x)\xi^{\alpha+\beta} > 0 \quad \text{for all } x \in \overline{Q},\ 0 \neq \xi \in \mathbb{R}^n. \tag{C.3}$$

We define the unbounded operator $\mathcal{A}: L_2(Q) \to L_2(Q)$ with domain $\mathcal{D}(\mathcal{A}) = \{u \in \mathring{W}^m(Q) : \mathcal{A}u \in L_2(Q)\}$ acting in the space of distributions $\mathcal{D}'(Q)$ by the formula

$$\mathcal{A}u = \sum_{|\alpha|,|\beta| \leq m} \mathcal{D}^\alpha a_{\alpha\beta}(x)\mathcal{D}^\beta u(x). \tag{C.4}$$

In this subsection, we assume that the equation $(C.1)$ is strongly elliptic in $\overline{Q}$. Then we say that the differential operator $\mathcal{A}$ is strongly elliptic.

A function u is called a generalized solution of the boundary value problem (C.1), (C.2) if $u \in \mathcal{D}(A)$ and

$$\mathcal{A}u = f_0. \tag{C.5}$$

We can give the following equivalent definition of generalized solution.

A function u is called a generalized solution of the boundary value problem (C.1), (C.2) if $u \in \mathring{W}^m(Q)$ and for all $v \in \mathring{W}^m(Q)$ we have

$$\sum_{|\alpha|,|\beta|\le m} (a_{\alpha\beta}\mathcal{D}^\beta u, \mathcal{D}^\alpha v)_{L_2(Q)} = (f_0, v)_{L_2(Q)}. \tag{C.6}$$

We introduce the unbounded operator $\mathcal{A}^+ \colon L_2(Q) \to L_2(Q)$ with domain $\mathcal{D}(\mathcal{A}^+) = \{u \in \mathring{W}^m(Q) : \mathcal{A}^+ u \in L_2(Q)\}$ acting in the space of distributions $\mathcal{D}'(Q)$ by the formula

$$\mathcal{A}^+ u = \sum_{|\alpha|,|\beta|\le m} \mathcal{D}^\alpha \overline{a_{\beta\alpha}(x)} \mathcal{D}^\beta u.$$

The operators $\mathcal{A}$ and $\mathcal{A}^+$ are formally adjoint, i.e.,

$$(\mathcal{A}u, v)_{L_2(Q)} = (u, \mathcal{A}^+ v)_{L_2(Q)}$$

for all $u, v \in \dot{C}^\infty(Q)$.

Theorem C.1. *The equation* (C.1) *is strongly elliptic in* $\overline{Q}$ *if and only if there exist constants* $c_1 > 0$, $c_2 \ge 0$ *such that*

$$\mathrm{Re}(\mathcal{A}u, u)_{L_2(Q)} \ge c_1 \|u\|^2_{W^m(Q)} - c_2 \|u\|^2_{L_2(Q)} \tag{C.7}$$

for all $u \in \dot{C}^\infty(Q)$.

Theorem C.1 is due to M. I. Vishik [1] and L. Gårding [1]. For a proof, see also Theorems 4.6, 4.7 in J. Nečas [1], Chapter 3, Section 4.

The inequality (C.7) is called the *Gårding inequality.* The problem (C.1), (C.2) is said to be *coercive* if it satisfies the Gårding inequality. By virtue of Theorem C.1, the problem (C.1), (C.2) is coercive if and only if the equation (C.1) is strongly elliptic in $\overline{Q}$.
Thus we can define strong ellipticity with the help of the Gårding inequality (C.7).

Theorem C.2. *Let the operator* $\mathcal{A} \colon L_2(Q) \to L_2(Q)$ *be strongly elliptic.*

Then the spectrum $\sigma(\mathcal{A})$ *is discrete and* $\sigma(\mathcal{A}) \subset \{\lambda \in \mathbb{C} : \mathrm{Re}\,\lambda > -c_2\}$, *where* $c_2 \ge 0$ *is the constant in inequality* (C.7). *If* $\lambda \notin \sigma(\mathcal{A})$, *then the resolvent* $R(\lambda, \mathcal{A}) \colon L_2(Q) \to L_2(Q)$ *is a compact operator. Furthermore, we have* $\mathcal{A}^* = \mathcal{A}^+$, $(\mathcal{A}^+)^* = \mathcal{A}$.

Theorem C.3. *A strongly elliptic operator* $\mathcal{A} \colon L_2(Q) \to L_2(Q)$ *is Fredholm, and* $\mathrm{ind}\,\mathcal{A} = 0$.

Theorem C.4. *Let the strongly elliptic operator* $\mathcal{A} \colon L_2(Q) \to L_2(Q)$ *be symmetric, i.e.,*

$$(\mathcal{A}u, v)_{L_2(Q)} = (u, \mathcal{A}v)_{L_2(Q)} \tag{C.8}$$

for all $u, v \in \dot{C}^\infty(Q)$.

Then the operator $\mathcal{A}: L_2(Q) \to L_2(Q)$ is self-adjoint, the spectrum $\sigma(\mathcal{A})$ consists of real isolated eigenvalues $\lambda_s > -c_2$ of finite multiplicity. The set of eigenfunctions v_s of the operator $\mathcal{A}$ is an orthonormal basis in the space $L_2(Q)$, while the set of functions $v_s/\sqrt{\lambda_s + c_2}$ is an orthonormal basis in the space $\mathring{W}^m(Q)$ with inner product

$$(u,v)'_{\mathring{W}^m(Q)} = \frac{1}{2} \sum_{|\alpha|,|\beta| \le m} ((a_{\alpha\beta}(x) + \overline{a_{\beta\alpha}(x)})\mathcal{D}^\beta u, \mathcal{D}^\alpha v)_{L_2(Q)} + c_2(u,v)_{L_2(Q)}. \quad \text{(C.9)}$$

Theorems C.2–C.4 are adapted from N. Dunford and J. T. Schwartz [2], Chapter XIV, Section 6, Corollaries 11, 12, 14 and V. P. Mikhailov [2], Chapter IV, Section 1, Theorem 3. We note that the above theorems follow from Theorems A.14–A.16 on sectorial operators and Theorems B.8, C.1.

Now we consider the equation

$$\mathcal{A}u = -\sum_{i,j=1}^{n} \frac{\partial}{\partial x_i} a_{ij}(x) \frac{\partial u(x)}{\partial x_j} + \sum_{i=1}^{n} a_i(x) \frac{\partial u(x)}{\partial x_i} + a_0(x)u(x)$$
$$= f_0(x) \qquad (x \in Q) \quad (C.10)$$

with boundary conditions

$$u|_{\partial Q} = 0, \quad \text{(C.11)}$$

where $Q = (0,d) \times G$ is a cylinder, $G \subset \mathbb{R}^{n-1}$ is a bounded domain, $\partial G \in C^\infty$ if $n \ge 3$; $a_{ij}, a_i, a_0 \in C^\infty(\mathbb{R}^n)$ are real-valued functions and $a_{ij} = a_{ji}$; $f_0 \in L_2(Q)$ is a complex-valued function.

Remark C.1. Let the equation (C.10) be strongly elliptic in $\overline{Q}$, and let $m = 1$. Then Theorems C.1–C.3 are valid for the operator $\mathcal{A}$ corresponding to the boundary value problem (C.10), (C.11). Furthermore, if $a_i = 0$ $(i = 1, \dots, n)$, then Theorem C.4 is also valid.

Smoothness of Solutions

Let us consider the strongly elliptic equation (C.1) with coefficients $a_{\alpha\beta} \in C^\infty(\mathbb{R}^n)$. We suppose that $Q \subset \mathbb{R}^n$ is an arbitrary bounded domain.

One of the most important properties of elliptic differential equations is the smoothness of solutions on compact subsets of the domain Q. This property does not depend on the smoothness of the boundary ∂Q.

Theorem C.5. *Let the equation* (C.1) *be strongly elliptic in $\overline{Q}$. Suppose that $f_0 \in W^k_{\mathrm{loc}}(Q)$, and $u \in W^m_{\mathrm{loc}}(Q)$ is the solution of the equation* (C.1) *in a sense of distributions, where $k \ge 0$ is an integer.*

Then $u \in W^{k+2m}_{\mathrm{loc}}(Q)$.

For a proof, see N. Dunford and J. T. Schwartz [2], Chapter XIV, Section 6, Theorem 2.

Lemma C.1. *Let the equation* (C.1) *be strongly elliptic in* $\overline{Q}$, *and let* $\partial Q \cap S_{2\delta}(x^0) \in C^\infty$ *be the open connected set (in the induced topology). Suppose that* $f_0 \in W^k(Q)$, *and* $u \in W^m(Q)$ *is a solution of* (C.1) *in a sense of distributions satisfying boundary conditions*

$$\mathcal{D}_\nu^{\mu-1} u|_{\partial Q \cap S_{2\delta}(x^0)} = 0 \qquad (\mu = 1, \dots, m),$$

where $x^0 \in \partial Q$, $k \geq 0$ *is an integer,* $S_{2\delta}(x^0) = \{x \in \mathbb{R}^n : |x - x^0| < 2\delta\}$.

Then $u \in W^{k+2m}(Q \cap S_\delta(x^0))$.

Using a partition of unity and Lemma C.1, we obtain:

Theorem C.6. *Let the equation* (C.1) *be strongly elliptic in* $\overline{Q}$, *and let* $\partial Q \in C^\infty$. *Suppose that* $f_0 \in W^k(Q)$, *and* u *is the generalized solution of the boundary value problem* (C.1), (C.2), *where* $k \geq 0$ *is an integer.*

Then $u \in W^{k+2m}(Q)$.

Lemma C.1 and Theorem C.6 are adapted from N. Dunford and J. T. Schwartz [2], Chapter XIV, Section 6, Lemma 19 and Theorem 23.

Remark C.2. Theorem C.6 is also valid for the boundary value problem (C.10), (C.11), where $k = 0$, $m = 1$ (see O. A. Ladyzhenskaya and N. N. Ural'tseva [1], Chapter III, §10, Theorem 10.1).

Elliptic Problems with a Parameter

We shall study the equation

$$Au = \sum_{\beta+|\alpha| \leq 2m} a_{\alpha\beta}(x) q^\beta \mathcal{D}^\alpha u(x) = f_0(x) \qquad (x \in Q) \tag{C.12}$$

with boundary conditions

$$\begin{aligned} B_\mu u &= \left(\sum_{\beta+|\alpha| \leq m_\mu} b_{\mu\alpha\beta}(x) q^\beta \mathcal{D}^\alpha u(x) \right)\Bigg|_{\partial Q} \\ &= f_\mu(x) \qquad (x \in \partial Q,\ \mu = 1, \dots, m). \end{aligned} \tag{C.13}$$

Here $Q \in \mathbb{R}^n$ is a bounded domain (with boundary $\partial Q \in C^\infty$ if $n \geq 2$), $a_{\alpha\beta}, b_{\mu\alpha\beta} \in C^\infty(\mathbb{R}^n)$ are complex-valued functions, $n \geq 1$.

Denote

$$A^0 = A^0(x, \mathcal{D}, q) = \sum_{\beta+|\alpha|=2m} a_{\alpha\beta}(x) q^\beta \mathcal{D}^\alpha,$$

$$B_\mu^0 = B_\mu^0(x, \mathcal{D}, q) = \sum_{\beta+|\alpha|=m_\mu} b_{\mu\alpha\beta}(x) q^\beta \mathcal{D}^\alpha.$$

We consider also the corresponding polynomials

$$A^0(x,\xi,q) = \sum_{\beta+|\alpha|=2m} a_{\alpha\beta}(x) q^\beta \xi^\alpha,$$

$$B^0_\mu(x,\xi,q) = \sum_{\beta+|\alpha|=m_\mu} b_{\mu\alpha\beta}(x) q^\beta \xi^\alpha,$$

where $\xi = (\xi_1, \dots, \xi_n)$, $\xi^\alpha = \xi_1^{\alpha_1} \cdots \xi_n^{\alpha_n}$.

Denote by θ the closed angle in $\mathbb{C}$

$$\theta = \{q : \varphi_1 \le \arg q \le \varphi_2\}.$$

For the operators A^0 and B^0_μ we assume the following conditions:

C.1. *The polynomial $A^0(x, \xi + \tau\nu, q)$ in the variable τ has exactly m roots $\tau_1^+(x,\xi,\nu,q), \dots, \tau_m^+(x,\xi,\nu,q)$ with positive imaginary parts and m roots with negative imaginary parts for any $x \in \overline{Q}$, $q \in \theta$ and ξ such that $|\xi| + |q| \ne 0$, with ξ orthogonal to ν in $\mathbb{R}^n$, $\nu \ne 0$.*

C.2. *The polynomials $\{B^0_\mu(x, \xi + \tau\nu, q)\}$ $(\mu = 1, \dots, m)$ in the variable τ are linearly independent modulo the polynomial $\prod_{j=1}^m (\tau - \tau_j^+(x,\xi,\nu,q))$ for any $x \in \partial Q$, $q \in \theta$ and ξ which is orthogonal to ν such that $|\xi| + |q| \ne 0$, and ν is the inner unit normal vector to ∂Q at x.*

In the Hilbert spaces $W^s(\Omega)$ $(\Omega = Q, \partial Q)$ and $\mathcal{W}^k(Q, \partial Q) = W^k(Q) \times \prod_{\mu=1}^m W^{k+2m-m_\mu-1/2}(\partial Q)$ we introduce the following equivalent norms depending on q:

$$|||u|||_{W^s(\Omega)} = \{\|u\|^2_{W^s(\Omega)} + |q|^{2s}\|u\|^2_{L_2(\Omega)}\}^{1/2}, \tag{C.14}$$

$$|||f|||_{\mathcal{W}^k(Q,\partial Q)} = \left\{ |||f_0|||^2_{W^k(Q)} + \sum_{\mu=1}^m |||f_\mu|||^2_{W^{k+2m-m_\mu-1/2}(\partial Q)} \right\}^{1/2}, \tag{C.15}$$

where $s > 0$, $k \ge \max\{0, m_\mu - 2m + 1\}$.

We define a bounded operator $\mathcal{L} = \mathcal{L}(q) \colon W^{k+2m}(Q) \to \mathcal{W}^k(Q, \partial Q)$ by the formula $\mathcal{L}u - (Au, B_1 u, \dots, B_m u)$.

Theorem C.7. *Let conditions* C.1, C.2 *be fulfilled.*

Then there exists $q_0 > 0$ such that for $q \in \{q \in \theta : |q| \ge q_0\}$ and $u \in W^{k+2m}(Q)$

$$c_1 |||\mathcal{L}u|||_{\mathcal{W}^k(Q,\partial Q)} \le |||u|||_{W^{k+2m}(Q)} \le c_2 |||\mathcal{L}u|||_{\mathcal{W}^k(Q,\partial Q)}, \tag{C.16}$$

where $c_1, c_2 > 0$ do not depend on q and u.

Theorem C.8. *Let conditions* C.1, C.2 *hold.*

Then $\mathcal{L}(q) \colon W^{k+2m}(Q) \to \mathcal{W}^k(Q, \partial Q)$ is a Fredholm operator, and we have $\operatorname{ind} \mathcal{L}(q) = 0$ for all $q \in \mathbb{C}$. Moreover, there exists $q_0 > 0$ such that for all $q \in \{q \in \theta : |q| \ge q_0\}$ the operator $\mathcal{L}(q)$ has a bounded inverse $\mathcal{L}^{-1}(q) \colon \mathcal{W}^k(Q, \partial Q) \to W^{k+2m}(Q)$.

Now we shall study the Dirichlet problem. The following assumptions are imposed on A^0 and B^0_μ:

C.3. $A^0 u = \sum_{|\alpha|=2m} a_\alpha(x)\mathcal{D}^\alpha u(x) + q^{2m}u(x)$ *is a differential operator with real coefficients* $a_\alpha \in C^\infty(\mathbb{R}^n)$ *such that* $\sum_{|\alpha|=2m} a_\alpha(x)\xi^\alpha > 0$ *for* $0 \neq \xi \in \mathbb{R}^n$, $x \in \overline{Q}$.

C.4. $B^0_\mu u = (-i\partial/\partial\nu)^{\mu-1}u$ $(\mu = 1,\dots,m)$ *are boundary operators of the Dirichlet problem.*

Obviously, under conditions C.3, C.4 for any $\varepsilon > 0$ and $q \in \{\lambda \in \mathbb{C} : |\arg q| \leq (\pi-\varepsilon)/2m\} \cup \{\lambda \in \mathbb{C} : |\arg q - \pi| \leq (\pi - \varepsilon)/2m\}$ the operators A^0, B^0_μ satisfy conditions C.1, C.2. Hence the following theorem holds.

Theorem C.9. *Let conditions* C.3, C.4 *be fulfilled.*

Then we have:

(a) $\mathcal{L}(q)\colon W^{k+2m}(Q) \to \mathcal{W}^k(Q,\partial Q)$ *is a Fredholm operator, and* $\operatorname{ind}\mathcal{L}(q) = 0$ *for all* $q \in \mathbb{C}$.

(b) *For every* $\varepsilon > 0$, *there exists* $q_\varepsilon > 0$ *such that for* $|q| \geq q_\varepsilon$, $q \in \{\lambda \in \mathbb{C} : |\arg q| \leq (\pi-\varepsilon)/2m\} \cup \{\lambda \in \mathbb{C} : |\arg q - \pi| \leq (\pi-\varepsilon)/2m\}$ *the operator* $\mathcal{L}(q)$ *admits a bounded inverse* $\mathcal{L}^{-1}(q)\colon \mathcal{W}^k(Q,\partial Q) \to W^{k+2m}(Q)$.

(c) *A priori estimate* (C.16) *is valid for all* $u \in W^{k+2m}(Q)$.

Theorems C.7–C.9 are adapted from M. S. Agranovich and M. I. Vishik [1], Chapter I, Theorems 4.1, 5.1, 5.2.

We introduce the unbounded operator $\mathcal{A}\colon W^k(Q) \to W^k(Q)$ defined by

$$\mathcal{A}u = \sum_{|\alpha|\leq 2m} a_\alpha(x)\mathcal{D}^\alpha u(x),$$

$$\mathcal{D}(\mathcal{A}) = \{u \in W^{k+2m}(Q) : (B^0_\mu u)|_{\partial Q} = 0,\ \mu = 1,\dots,m\},$$

where $a_\alpha \in C^\infty(\mathbb{R}^n)$ are real and the operators A^0, B^0_μ satisfy the conditions C.3, C.4.

From Theorems C.9, A.8, and B.8 we obtain:

Theorem C.10. *Let conditions* C.3, C.4 *hold.*

Then we have:

(a) $\mathcal{A}\colon W^k(Q) \to W^k(Q)$ *is a Fredholm operator, and* $\operatorname{ind}\mathcal{A} = 0$.

(b) *The spectrum* $\sigma(\mathcal{A})$ *is discrete.*

(c) *For* $\lambda \notin \sigma(\mathcal{A})$, *the resolvent* $R(\lambda,\mathcal{A})\colon W^k(Q) \to W^k(Q)$ *is a compact operator.*

(d) *For any* $\varepsilon > 0$, *all except perhaps a finite number of points of* $\sigma(\mathcal{A})$ *belong to the angle* $|\arg\lambda| < \varepsilon$.

Theorem C.10 is adapted from N. Dunford and J. T. Schwartz [2], Chapter XIV, Section 6, Theorems 23, 27.

One-Dimensional Case

We introduce the bounded linear operator $\mathcal{L} = \mathcal{L}(\lambda): W^2(0,d) \to \mathcal{W}(0,d) = L_2(0,d) \times \mathbb{C} \times \mathbb{C}$ by the formulas $\mathcal{L}u = (Au - \lambda u, u(0), u(d))$, where $Au = -a_0(x)u''(x) + a_1(x)u'(x) + a_2(x)u(x)$, $a_i \in C[0,d]$ are real-valued functions, $a_0(x) \geq k > 0$ $(x \in [0,d])$, $\lambda \in \mathbb{C}$.

We use the norms

$$|||u|||_{W^2(0,d)} = \{\|u\|^2_{W^2(0,d)} + |\lambda|^2\|u\|^2_{L_2(0,d)}\}^{1/2},$$
$$|||f|||_{\mathcal{W}(0,d)} = \{\|f_0\|^2_{L_2(0,d)} + |\lambda|^{3/2}(|f_1|^2 + |f_2|^2)\}^{1/2}$$

depending on a parameter λ, where $f = (f_0, f_1, f_2)$, $|\lambda| > 0$.

From Theorem C.9 it follows:

Theorem C.11. *The operator $\mathcal{L}(\lambda): W^2(0,d) \to \mathcal{W}(0,d)$ is Fredholm, and $\operatorname{ind}\mathcal{L}(\lambda) = 0$ for all $\lambda \in \mathbb{C}$. For every $\varepsilon > 0$, there exists $q_\varepsilon > 0$ such that for $\lambda \in \Omega_{\varepsilon,q_\varepsilon} = \{\lambda \in \mathbb{C} : |\arg\lambda| \geq \varepsilon,\ |\lambda| \geq q_\varepsilon\}$ the operator $\mathcal{L}(\lambda)$ has a bounded inverse $\mathcal{L}^{-1}(\lambda): \mathcal{W}(0,d) \to W^2(0,d)$, and each $u \in W^2(0,d)$ satisfies inequality*

$$c_1|||\mathcal{L}u|||_{\mathcal{W}(0,d)} \leq |||u|||_{W^2(0,d)} \leq c_2|||\mathcal{L}u|||_{\mathcal{W}(0,d)}, \tag{C.17}$$

where $c_1, c_2 > 0$ do not depend on λ and u.

We now introduce the unbounded operator $\mathcal{A}: L_2(0,d) \to L_2(0,d)$ defined by

$$\mathcal{A}u = Au, \qquad \mathcal{D}(\mathcal{A}) = \{u \in W^2(0,d) : u(0) = u(d) = 0\}.$$

From Theorem C.10 we obtain:

Theorem C.12. *The operator $\mathcal{A}: L_2(0,d) \to L_2(0,d)$ is Fredholm, and $\operatorname{ind}\mathcal{A} = 0$. The spectrum $\sigma(\mathcal{A})$ is discrete. For $\lambda \notin \sigma(\mathcal{A})$, the resolvent $R(\lambda, \mathcal{A}): L_2(0,d) \to L_2(0,d)$ is a compact operator. For any $\varepsilon > 0$, all except perhaps a finite number of points of $\sigma(\mathcal{A})$ belong to the angle $|\arg\lambda| < \varepsilon$.*

Bibliography

Agmon, S.

[1] "The coerciveness problem for integro-differential forms", *J. Analyse Math.* **6** (1958), 183–223.

Agranovich, M. S. and M. I. Vishik

[1] "Elliptic problems with a parameter and parabolic problems of general type", *Uspekhi Mat. Nauk* **19** (1964), 53–161; English transl. in *Russian Math. Surveys* **19** (1964).

Akhiezer, N. I. and I. M. Glazman

[1] *Theory of Linear Operators in Hilbert Space.* Vol. 2, Vishcha Shkola, Kharkov, 1978 (Russian); English translation: Pitman Press, Boston-London-Melbourne, 1981.

Antonevich, A. B.

[1] "The index and the normal solvability of a general elliptic boundary value problem with a finite group of translations on the boundary", *Differentsial'nye Uravneniya* **8** (1972), 309–317; English transl. in *Differential Equations* **8** (1974).

Arnol'd, V. I.

[1] *Geometrical Methods in the Theory of Ordinary Differential Equations*, Nauka, Moscow, 1978 (Russian); English translation: Springer-Verlag, New York-Heidelberg-Berlin, 1983.

Banks, H. T. and J. Q. Jacobs

[1] "An attainable sets approach to optimal control of functional differential equations with function space terminal conditions", *J. of Differential Equations* **13** (1973), 127–149.

Banks, H. T. and G. A. Kent

[1] "Control of functional differential equations of retarded and neutral type to target sets in function space", *SIAM J. Control* **10** (1972), 567–593.

Baumstein, A. and A. L. Skubachevskiĭ

[1] "On smooth solutions of the boundary value problems for the systems of differential-difference equations", *J. of Nonlinear Analysis: Theory, Methods and Applications* **25** (1995), 655–668.

Bellman, R. and K. Cooke

[1] *Differential-Difference Equations*, Academic Press, New York-London, 1963.

Bellman, R. and J. M. Danskin

[1] *A Survey of the Mathematical Theory of Time Lag, Retarded Control, and Hereditary Processes*, The RAND Corporation, R-256, 1954.

Bitsadze, A. V.

[1] "On the theory of nonlocal boundary value problems", *Dokl. Akad. Nauk SSSR* **277** (1984), 17–19; English transl. in *Soviet Math. Dokl.* **30** (1984).

[2] "On some class of conditionally solvable nonlocal boundary value problems for harmonic functions", *Dokl. Akad. Nauk SSSR* **280** (1985), 521–524; English transl. in *Soviet Math. Dokl.* **31** (1985).

Bitsadze, A. V. and A. A. Samarskiĭ

[1] "On some simple generalizations of linear elliptic boundary value problems", *Dokl. Akad. Nauk SSSR* **185** (1969), 739–740; English transl. in *Soviet Math. Dokl.* **10** (1969).

Blekher, P. M.

[1] "Operators depending meromorphically on a parameter", *Vestnik Moskov. Univ., Ser.* I *Math. Mekh.*, no. 5 (1969), 30–36; English transl. in *Moscow Univ. Math. Bull.* **24** (1969).

Bony, J. M., P. Courrege et P. Priouret

[1] "Semi-groups de Feller sur une variété à bord compacte et problèmes aux limites intégro-différentiels du second ordre donnant lieu au principe du maximum", *Ann. Inst. Fourier* (*Grenoble*) **18** (1968), 369–521.

Browder, F.

[1] "Non-local elliptic boundary value problems", *Amer. J. Math.* **86** (1964), 735–750.

Cancelier, C.

[1] "Problèmes aux limites pseudo-différentiels donnant lieu au principe du maximum", *Comm. P. D. E.* **11** (1986), 1677–1726.

Carleman, T.

[1] "Sur la théorie des équations intégrales et ses applications", *Verhandlungen des Internat. Math. Kongr.*, Zürich, Bd. 1 (1932), 132–151.

Colonius, F.

[1] *Optimal Periodic Control.* Lecture Notes in Mathematics, Springer-Verlag, Berlin, 1988.

Cooke, K. and J. Wiener

[1] "Distributional and analytic solutions of functional differential equations", *J. of Mathematical Analysis and Applications* **98** (1984), 111–129.

Dunford, N. and J. T. Schwartz

[1] *Linear Operators. Part 1: General Theory*, Interscience Publishers, New York-London, 1958.

[2] *Linear Operators. Part 2: Spectral Theory*, Interscience Publishers, New York-London, 1963.

Eĭdel′man, S. D. and N. V. Zhitarashu

[1] "Nonlocal boundary value problems for elliptic equations", *Mat. Issled.* **6** (1971), no. 2 (20), 63–73 (Russian).

Feller, W.

[1] "The parabolic differential equations and the associated semi-groups of transformations", *Ann. of Math.* **55** (1952), 468–519.

[2] "Diffusion processes in one dimension", *Trans. Amer. Math. Soc.* **77** (1954), 1–30.

Feynman, R. P. and J. A. Wheeler

[1] "Classical electrodynamics in terms of direct interparticle actions", *Rev. of Modern Phys.* **21** (1949), 425–433.

Figueiredo, D. G.

[1] "The coerciveness problem for forms over vector-valued functions", Comm. Pure Appl. Math. **16** (1963), 63–94.

Gabasov, R. and S. V. Churakova

[1] "On the controllability theory of linear systems with delay", *Izv. Akad. Nauk SSSR, Tekhnicheskaya Kibernetika* No. 4 (1969), 17–28 (Russian).

Gabasov, R. and F. M. Kirillova

[1] *Qualitative Theory of Optimal Processes*, Nauka, Moscow, 1971 (Russian).

Gårding, L.

[1] "Dirichlet's problem for linear elliptic partial differential equations", Math. Scand. **1** (1953), 55–72.

Gohberg, I. C. and M. G. Kreĭn

[1] "The basic propositions on defect numbers, root numbers and indices of linear operators", *Uspekhi Mat. Nauk* **12:2** (74) (1957), 43–118; English transl. in *Amer. Math. Soc. Transl. Ser.* 2, **13** (1960), 185–264.

Guseva, O. V.

[1] "On the boundary value problems for strongly elliptic systems", *Dokl. Akad. Nauk SSSR* **102** (1955), 1069–1072 (in Russian).

Halanay, A.

[1] "Optimal controls for systems with time lag", *SIAM J. Control* **6** (1968), 213–234.

Hale, J.

[1] *Theory of Functional Differential Equations*, Springer-Verlag, New York-Heidelberg-Berlin, 1977.

Il'in, V. A. and E. I. Moiseev

[1] "Nonlocal boundary-value problem of the second kind for a Sturm-Liouville operator", *Differentsial'nye Uravneniya* **23** (1987), 1422–1431; English transl. in *Differential Equations* **23** (1988).

[2] "An a priori bound for a solution of the problem conjugate to a nonlocal boundary-value problem of the first kind", *Differentsial'nye Uravneniya* **24** (1988), 795–804; English transl. in *Differential Equations* **24** (1988).

Ishikawa, Y.

[1] "A remark on the existence of a diffusion process with non-local boundary conditions", *J. Math. Soc. Japan* **42** (1990), 171–184.

Kamenskiĭ, A. G.

[1] "Boundary value problems for equations with formally symmetric differential-difference operators", *Differentsial'nye Uravneniya* **12** (1976), 815–824; English transl. in *Differential Equations* **12** (1977).

Kamenskiĭ, G. A.

[1] "Variational and boundary value problems with deviating argument", *Differentsial'nye Uravneniya* **6** (1970), 1349–1358; English transl. in *Differential Equations* **13** (1971).

[2] "A variational method of solving boundary-value problems for certain linear differential equations with deviating arguments", *Differentsial'nye Uravneniya* **13** (1977), 1185–1191; English transl. in *Differential Equations* **13** (1978).

Kamenskiĭ, G. A. and A. D. Myshkis

[1] "Formulation of boundary-value problems for differential equations with deviating arguments containing highest-order terms", *Differentsial'nye Uravneniya* **10** (1974), 409–418; English transl. in *Differential Equations* **10** (1975).

Kamenskiĭ, G. A., Myshkis, A. D., and A. L. Skubachevskiĭ

[1] "Minimum value of a quadratic functional and linear elliptic boundary value problems with deviating arguments", *Differentsial'nye Uravneniya* **16** (1980), 1469–1473; English transl. in *Differential Equations* **16** (1981).

[2] "Smooth solutions of a boundary value problem for differential-difference equation of neutral type", *Ukrainskiĭ Matematicheskiĭ Zhurnal* **37** (1985), 581–589; English transl. in *Ukrain. Math. Journal* **37** (1986).

[3] "Generalized and smooth solutions of boundary value problems for functional differential equations with many senior members", *Časopis pro pěstovani matematiky* **111** (1986), 254–266.

Karlovich, Yu. I., Kravchenko V. G., and G. S. Litvinchuk

[1] "Noether's theory of singular integral operators with shift", *Izv. Vyssh. Uchebn. Zaved. Mat.* **27** (1983), 3–27; English transl. in *Soviet Math. (Iz. VUZ)* **27** (1983).

Kato, T.

[1] *Perturbation Theory for Linear Operators*, Springer-Verlag, Berlin-Heidelberg-New York, 1966.

Kent, G. A.

[1] "A maximum principle for optimal control problems with neutral functional differential systems", *Bull. Amer. Math. Soc.* **77** (1971), 565–570.

Kharatishvili, G. L. and T. A. Tadumadze

[1] "Nonlinear optimal control systems with variable delays", *Mat. Sb.* **107 (149)** (1978), 613–628; English transl. in *Mathem. USSR Sb.* **35** (1979).

Kirillova, F. M. and S. V. Churakova

[1] "On the controllability problem of linear systems with delay", *Differentsial'nye Uravneniya* **3** (1967), 436–445; English transl. in *Differential Equations* **3** (1968).

Kondrat′ev, V. A.

[1] "Boundary value problems for elliptic equations in domains with conical or angular points", *Trudy Moskov. Mat. Obshch.* **16** (1967), 209–292; English transl. in *Trans. Moscow Math. Soc.* **16** (1967).

Krall, A. M.

[1] "The development of general differential and general differential-boundary systems", *Rocky Mountain J. of Math.* **5** (1975), 493–542.

[2] "Stieltjes differential-boundary operators, III. Multivalued operators – linear relations", *Pacific J. Math.* **59** (1975), 125–134.

Krasovskiĭ, N. N.

[1] *Control Theory of Motion*, Nauka, Moscow, 1968 (Russian).

Kreĭn, S. G.

[1] *Linear Equations in Banach Spaces*, Nauka, Moscow, 1971 (Russian); English translation: Birkhäuser, Boston, 1982.

Ladyzhenskaya, O. A. and N. N. Ural′tseva

[1] *Linear and Quasilinear Elliptic Equations*, Nauka, Moscow, 1964 (Russian); English translation: Academic Press, New York-London, 1968.

Lions, J. L. and E. Magenes

[1] *Non-Homogeneous Boundary Value Problems and Applications*, Vol. 1, Springer-Verlag, New York-Heidelberg-Berlin, 1972.

Mikhlin, S. G.

[1] *The Problem of the Minimum of a Quadratic Functional*, Gostekhizdat, Moscow-Leningrad, 1952 (Russian); English transl.: Holden-Day, San Francisco, 1965.

Mikhailov, V. P.

[1] "Riesz bases in $L_2(0,1)$", *Dokl. Akad. Nauk SSSR* **144** (1962), 981–984; English transl. in *Soviet Math. Dokl.* **3** (1962).

[2] *Partial Differential Equations*, Nauka, Moscow, 1983 (Russian).

Morrey, C. B.

[1] *Multiple Integrals in the Calculus of Variations*, Springer-Verlag, Berlin-Heidelberg-New York, 1966.

Myshkis, A. D.

[1] *Linear Differential Equations with Retarded Argument*, Nauka, Moscow, 1972 (Russian); Deutscher transl.: Lineare Differentialgleichungen mit nacheilenden Argumentom, Deutscher Verlag Wiss., Berlin, 1955.

Nečas, J.

[1] *Les Méthodes Directes en Théorie des Equations Elliptiques*, Academia Tchecoslovaque des Sciences, Prague, 1967.

Oleĭnik, O. A. and E. V. Radkevič

[1] *Second Order Equations with Nonnegative Characteristic Form*, Itogi Nauki, Moscow, 1971 (Russian); English translation: Amer. Math. Soc., Providence, Rhode Island and Plenum Press, New York, 1973.

Onanov, G. G. and A. L. Skubachevskiĭ

[1] "Differential equations with displaced arguments in stationary problems in the mechanics of a deformed body", *Prikladnaya Mekhanika* **15** (1979), 39–47; English transl. in *Soviet Applied Mech.* **15** (1979).

Osipov, Yu. S.

[1] "Stabilization of control systems with delays", *Differentsial'nye Uravneniya* **1** (1965), 605–618; English transl. in *Differential Equations* **1** (1966).

Paneyakh, B. P.

[1] "Certain nonlocal boundary-value problems for linear differential operators", *Mat. Zametki* **35** (1984), 425–434; English transl. in *Math. Notes* **35** (1984).

Picone, M.

[1] "I teoremi d'esistenza per gl'integrale di una equazione differenziale lineare ordinaria soddisfacenti ad una nuova classe di condizioni", *Rend. Acc. Lincei* **17** (1908), 340–347.

[2] "Equazione integrale traducente il più generale problema lineare per le equazioni differenziali lineari ordinarie di qualsivoglia ordine", *Academia nazionale dei Lincei. Atti dei convegni* **15** (1932), 942–948.

Pontryagin, L. S., Boltyanskiĭ, V. G., Gamkrelidze R. V., and E. F. Mishchenko

[1] *The Mathematical Theory of Optimal Processes*, Interscience, 1962.

Przeworska-Rolewicz, D.

[1] *Equations with Transformed Argument*, PWN, Warszawa, 1973.

Rabinovič, V. S.

[1] "On the solvability of differential-difference equations on $\mathbb{R}^n$ and in a half-space", *Dokl. Akad. Nauk SSSR* **243** (1978), 1498–1502; English transl. in *Soviet Math. Dokl.* **19** (1978).

Roĭtberg, Ya. A. and Z. G. Sheftel′

[1] "On a class of general nonlocal elliptic problems", *Dokl. Akad. Nauk SSSR* **192** (1970), 511-513; English transl. in *Soviet Math. Dokl.* **11** (1970).

[2] "Nonlocal problems for elliptic equations and systems", *Sib. Mat. Zh.* **13** (1972), 165–181; English transl. in *Siberian Math. J.* **13** (1972).

Samarskii, A.A.

[1] "Some problems of the theory of differential equations", *Differentsial′nye Uravneniya* **16** (1980), 1925–1935; English transl. in *Differential Equations* **16** (1980).

Sato, K. and T. Ueno

[1] "Multi-dimensional diffusion and the Markov process on the boundary", *J. Math. Kyoto Univ.* **4** (1965), 529–605.

Schul′man, L. S.

[1] "Some difference-differential equations containing both advance and retardation", *J. Math. Phys.* **15** (1974), 295–298.

Shteingol′d, S. D.

[1] "On the smoothness of generalized solutions of the boundary value problems for functional differential equations", *Uspekhi Mat. Nauk* **46** (1991), 203–204 (Russian).

Skubachevskiĭ, A. L.

[1] "On the spectrum of some nonlocal elliptic boundary value problems", *Mat. Sb.* **117** (**159**) (1982), 548–558; English transl. in *Math. USSR Sb.* **45** (1983).

[2] "Some nonlocal elliptic boundary-value problems", *Differentsial′nye Uravneniya* **18** (1982), 1590–1599; English transl. in *Differential Equations* **18** (1983).

[3] "Nonlocal elliptic boundary-value problems with degeneration", *Differentsial′nye Uravneniya* **19** (1983), 457–470; English transl. in *Differential Equations* **19** (1983).

[4] "Nonlocal elliptic problems with a parameter", *Mat. Sb.* **121** (**163**) (1983), 201–210; English transl. in *Math. USSR Sb.* **49** (1984).

[5] "Smoothness of generalized solutions of the first boundary-value problem for an elliptic differential-difference equation", *Mat. Zametki* **34** (1983), 105–112; English transl. in *Math. Notes* **34** (1984).

[6] "The elliptic problems of A. V. Bitsadze and A. A. Samarskiĭ", *Dokl. Akad. Nauk SSSR* **278** (1984), 813–816; English transl. in *Soviet Math. Dokl.* **30** (1984).

[7] "Solvability of elliptic problems with Bitsadze-Samarskiĭ boundary conditions", *Differentsial′nye Uravneniya* **21** (1985), 701–706; English transl. in *Differential Equations* **21** (1985).

[8] "Nonlocal boundary-value problems with a shift", *Math. Zametki* **38** (1985), 587–598; English transl. in *Math. Notes* **38** (1986).

[9] "Elliptic problems with nonlocal conditions near the boundary", *Mat. Sb.* **129** (**171**) (1986), 279–302; English transl. in *Math. USSR Sb.* **57** (1987).

[10] "The first boundary value problem for strongly elliptic differential-difference equations", *J. of Differential Equations* **63** (1986), 332–361.

[11] "Solvability of elliptic problems with nonlocal boundary conditions", *Dokl. Akad. Nauk SSSR* **291** (1986), 551–555; English transl. in *Soviet Math. Dokl.* **34** (1987).

[12] "Eigenvalues and eigenfunctions of some nonlocal boundary-value problems", *Differentsial'nye Uravneniya* **25** (1989), 127–136; English transl. in *Differential Equations* **25** (1989).

[13] "On some problems for multidimensional diffusion processes", *Dokl. Akad. Nauk SSSR* **307** (1989), 287–292; English transl. in *Soviet Math. Dokl.* **40** (1990).

[14] "Model nonlocal problems for elliptic equations in dihedral angles", *Differentsial'nye Uravneniya* **26** (1990), 119–131; English transl. in *Differential Equations* **26** (1990).

[15] "Truncation-function method in the theory of nonlocal problems", *Differentsial'nye Uravneniya* **27** (1991), 128–139; English transl. in *Differential Equations* **27** (1991).

[16] "On the stability of index of nonlocal elliptic problems", *J. of Mathematical Analysis and Applications*, **160** (1991), 323–341.

[17] "Boundary value problems for differential-difference equations with incommensurable shifts", *Dokl. Akad. Nauk Ross.*, **324** (1992), 1155–1158; English transl. in *Russian Acad. Sci. Dokl. Math.* **45** (1992).

[18] "Generalized and classical solutions of boundary value problems for differential-difference equations", *Dokl. Akad. Nauk Ross.* **334** (1994), 433–436 (Russian).

[19] "On the damping problem of control system with delay", *Dokl. Akad. Nauk Ross.* **335** (1994), 157–160 (Russian).

[20] "Nonlocal elliptic problems and multidimensional diffusion processes", *Russian J. of Math. Physics* **3** (1995), 327–360.

[21] "On some properties of elliptic and parabolic functional differential equations", *Uspekhi Mat. Nauk* **51** (1996), 169–170; English transl. in *Russian Math. Surveys* **51** (1996).

Skubachevskiĭ, A. L. and G. M. Steblov

[1] "On the spectrum of differential operators with domain nondense in $L_2(0,1)$", *Dokl. Akad. Nauk SSSR* **321** (1991), 1158–1163; English transl. in *Soviet Math. Dokl.* **44** (1992).

Skubachevskiĭ, A. L. and E. L. Tsvetkov

[1] "Second boundary-value problem for elliptic differential-difference equations", *Differentsial'nye Uravneniya* **25** (1989), 1766–1776; English transl. in *Differential Equations* **25** (1990).

Slobodetskiĭ, L. N.

[1] "Generalized Sobolev spaces and their application to boundary problems for partial differential equations", *Leningrad. Gos. Ped. Inst. Uchen. Zap.* **197** (1958), 54–112; English transl. in *Amer. Math. Soc. Transl.* (2) **57** (1966).

Sobolev, S. L.

[1] *Applications of Functional Analysis in Mathematical Physics*, Leningrad St. Univ. Press, Leningrad, 1950 (Russian); English translation: AMS, Providence, 1963.

Sommerfeld, A.

[1] "Ein Beitrag zur hydrodynamischen Erklärung der turbulenten Flüssigkeitsbewegungen", *Proc. Intern. Congr. Math.* (Rome, 1908), Vol. III, Reale Accad. Lincei, Roma, 1909, 116–124.

Stein, E. M.

[1] *Singular Integrals and Differentiability Properties of Functions*, Princeton Univ. Press, Princeton, 1970.

Taira, K.

[1] *Diffusion Processes and Partial Differential Equations*, Academic Press, New York-London, 1988.

[2] "On the existence of Feller semigroups with boundary conditions", *Mem. Amer. Math. Soc.* **99** (1992), 1–65.

Tamarkin, J. D.

[1] *Some General Problems of the Theory of Ordinary Linear Differential Equations and Expansion of an Arbitrary Function in Series of Fundamental Functions*, Petrograd, 1917; adridged English transl. in *Math. Z.* **27** (1928), 1–54.

[2] "The notion of the Green's function in the theory of integro-differential equations", *Trans. Amer. Math. Soc.* **29** (1927), 755–800.

Tsvetkov, E. L.

[1] "Solvability and spectrum of the third boundary value problem for elliptic differential-difference equation", *Mat. Zametki* **51** (1992), 107–114 (Russian).

Ueno, T.

[1] "Multi-dimensional wave equation and an associated group of operators on the boundary", *J. Fac. of Science Univ. Tokyo* **26** (1979), 413–442.

[2] "Representation of the generator and the boundary condition for semigroups of operators of kernel type", *Proc. Japan Acad.* **59** (1983), 414–417.

Ventsel′, A. D.

[1] "On boundary conditions for multidimensional diffusion processes", *Teoriya Veroyatn. i ee Primen.* **4** (1959), 172–185; English transl. in *Theory Prob. and its Appl.* **4** (1959).

Vishik, M. I.

[1] "On strongly elliptic systems of differential equations", *Mat. Sb.* No. 5, **29** (71) (1951), 615–676 (Russian).

Volterra, V.

[1] "Sur les équations intégro-différentielles et leurs applications", *Acta Math.* **35** (1912), 295–356.

[2] "The general equations of biological strife in the case of historical actions", *Proc. Edinburgh Math. Soc.* **6** (1939), 4–10.

Vorontsov, M. A., Ricklin, J. C., and G. W. Carhart

[1] "Optical simulation of phase-distorted imaging systems: nonlinear and adaptive optics approach", *Optical Engineering* **34** (1995), 3229–3238.

Watanabe, S.

[1] "Construction of diffusion processes with Wentzell's boundary conditions by means of Poisson point processes of Brownian excursions." In *Probability Theory*, 255–271. Banach Center Publications, Vol. 5, PWN – Polish Scientific Publishers, Warsaw, 1979.

Weiss, L.

[1] "On the controllability of delay-differential systems", *SIAM J. Control* **5** (1967), 575–587.

Wiener, J.

[1] "Generalized-function solutions of differential and functional differential equations", *J. of Mathematical Analysis and Applications* **88** (1982), 170–182.

Yakubov, S. Ya. and E. Ch. Ibragimov

[1] "Bounds for solutions of an ordinary differential equation with a parameter and functional conditions", *Differentsial'nye Uravneniya* **26** (1990), 1551–1562; English transl. in *Differential Equations* **26** (1991).

Yosida, K.

[1] *Functional Analysis*, Springer-Verlag, Berlin-Heidelberg-New York, 1971.

List of Symbols

The number opposite each symbol is the page on which it is defined or explained.

Index

Titles previously published in the series

OPERATOR THEORY: ADVANCES AND APPLICATIONS

BIRKHÄUSER VERLAG

Edited by **I. Gohberg,**

School of Mathematical Sciences, Tel-Aviv University, Ramat Aviv, Israel

80. **I. Gohberg, H. Langer** (Eds): Operator Theory and Boundary Eigenvalue Problems International Workshop in Vienna, July 27-30, 1993, 1995 (ISBN 3-7643-5259-0)

81. **H. Upmeier**: Toeplitz Operators and Index Theory in Several Complex Variables, 1995 (ISBN 3-7643-5280-5)

82. **T. Constantinescu**: Schur Parameters, Factorization and Dilation Problems, 1996 (ISBN 3-7643-5285-X)

83. **A.B. Antonevich**: Linear Functional Equations. Operator Approach, 1995 (ISBN 3-7643-2931-9)

84. **L.A. Sakhnovich**: Integral Equations with Difference Kernels on Finite Intervals, 1996 (ISBN 3-7643-5267-1)

85. **Y.M. Berezansky, G.F. Us, Z.G. Sheftel**: Functional Analysis, **Vol. I**, 1996 (ISBN 3-7643-5344-9)

86. **Y.M. Berezansky, G.F. Us, Z.G. Sheftel**: Functional Analysis, **Vol. II**, 1996 (ISBN 4-7643-5345-7)

87. **I. Gohberg / P. Lancaster / P.N. Shivakumar** (Eds): Recent Developments in Operator Theory and Its Applications, International Conference in Winnipeg, October 2–6, 1994, 1996 (ISBN 3-7643-5413-5)

88. **J. van Neerven** (Ed.): The Asymptotic Behaviour of Semigroups of Linear Operators 1996 (ISBN 3-7643-5455-0)

89. **Y. Egorov / V. Kondratiev**: On Spectral Theory of Elliptic Operators, 1996 (ISBN 3-7643-5390-2)

90. **A. Böttcher / I. Gohberg** (Eds): Singular Integral Operators and Related Topics. Joint German-Israeli Workshop, Tel Aviv, March 1-10, 1995, 1996 (ISBN 3-7643-5466-6)